62
5300508

63
5301518

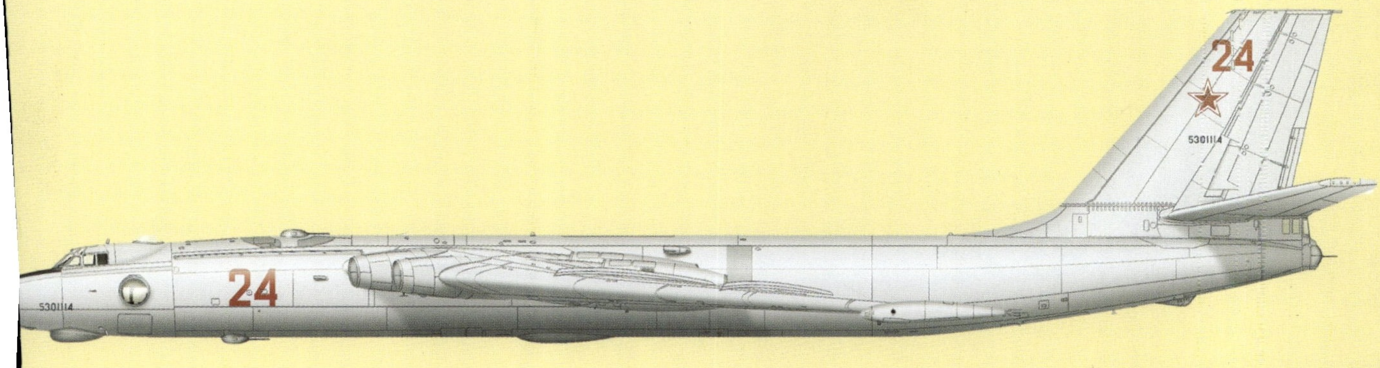
24
5301114
24
5301114

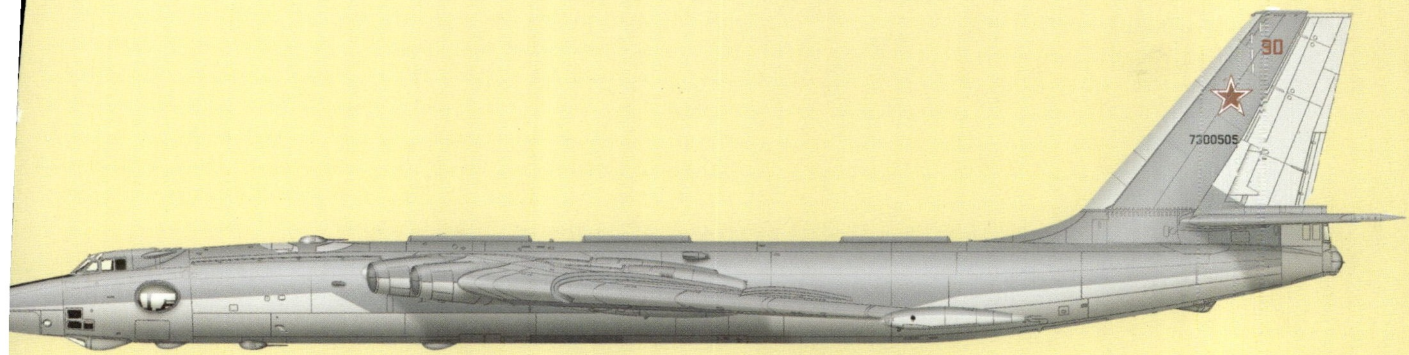
90
7300505

MYASISHCHEV

M-4 and 3M

  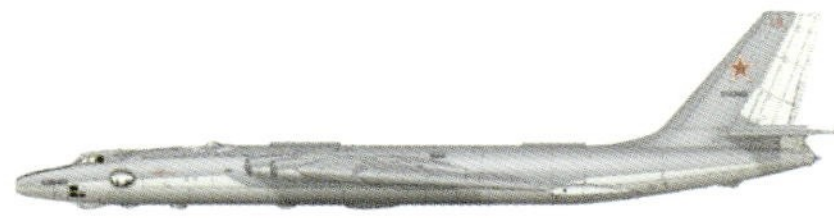

MYASISHCHEV

M-4 and 3M

THE FIRST SOVIET STRATEGIC JET BOMBER

YEFIM GORDON
DIMITRIY KOMISSAROV

SCHIFFER MILITARY
4880 Lower Valley Road Atglen, PA 19310

Library of Congress Control Number: 2020943649

Designed by Polygon Press Ltd., Moscow
Cover Designed by Jack Chappell
Type set in Times New Roman/Titan Text/Tasse/Optima Lt Std/Square 721 BT

ISBN: 978-0-7643-6182-1
Printed in India

Published by Schiffer Publishing, Ltd.
4880 Lower Valley Road
Atglen, PA 19310
Phone: (610) 593-1777; Fax: (610) 593-2002
E-mail: Info@schifferbooks.com
www.schifferbooks.com

Contents

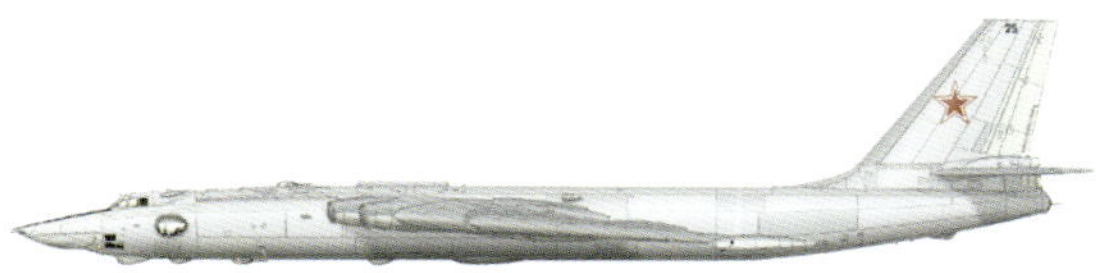

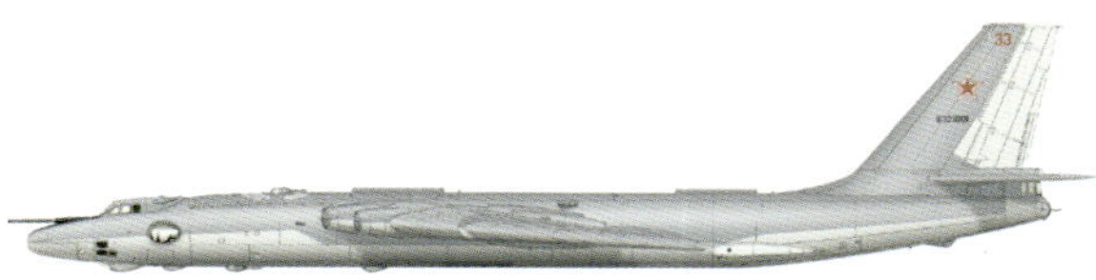

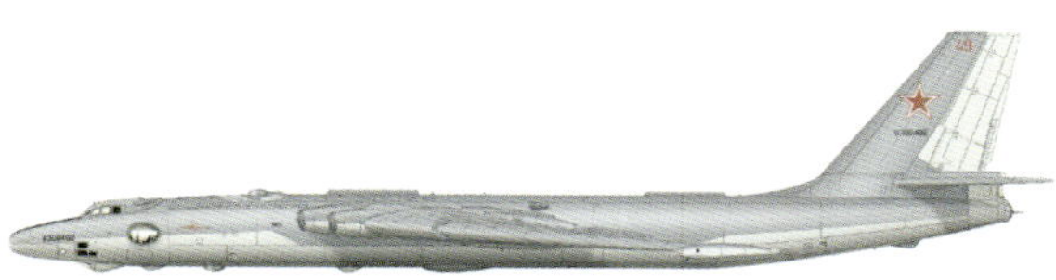

Acknowledgments

This book is illustrated with photos by Yefim Gordon, Sergey Aleksandrov, Anatoliy Andreyev, Mikhail Gribovskiy, Igor' Kolokolov, Vasiliy Kunyayev, Sergey Popsuyevich, Sergey Skrynnikov, Yevgeniy Sologoobov, Helmut Walther, and the US Office of Naval Research / Special Devices Center, as well as from the archives of the Myasishchev EMZ, the M. M. Gromov Flight Research Institute (LII), the TASS News Agency, the Russian Aviation Research Trust (RART), and the personal archive of Yefim Gordon.

Special thanks are due to Avico-Press / Konstantin Udalov, who provided a lot of pictures and drawings, and to Sergey Komissarov, who provided valuable materials from Russian archives.

The authors have also used the following web sources: www.airforce.ru, www.russianplanes.net, www.testpilot.ru, www.airwar.ru, www.repo.ssau.ru, www.npomash.ru, www.npo-saturn.ru, www.buran.ru, www.allmines.net, www.rbase.new-factoria.ru, www.kraeved.rounb.ru, and www.forumavia.ru.

Line drawings by the Myasishchev EMZ, Avico-Press, and Ivnamin Sultanov.

Colour profiles by Andrey Yurgenson.

Head-on view of a taxiing 3MS-2 tanker, the last version in air force service. The landing lights create a curious 'triple X' effect caused by the camera lens.

Introduction

During the immediate prewar period and in the years of the Great Patriotic War (1941–45), the attitude of the Soviet leader Iosif V. Stalin toward long-range bombers was inconsistent, to say the least. Production of the Petlyakov Pe-8 four-engined bomber, an innovative design by the standards of the late 1930s, was by turns terminated and relaunched several times. The Pe-8 was previously known as the TB-7 (*tyazholyy bombardirovshchik* – heavy bomber) and, before that, as the ANT-42 because Vladimir Mikhaïlovich Petlyakov, the bomber's project chief, then worked as an aide to Andrey Nikolayevich Tupolev. Failing to find government support, another Soviet aircraft designer, Vladimir Mikhaïlovich Myasishchev, proceeded at his own risk with the development of the DVB-102 (***dahl'****niy* ***vysotnyy bombardirovshchik*** – long-range high-altitude bomber), which first flew in February 1942. This aircraft incorporated an amazing combination of advanced features, including a tricycle landing gear, two pressurised cabins for the crew, three remote-controlled powered gun barbettes, and integral fuel tanks in the wing torsion box instead of the usual bladder tanks. The design effort, which proceeded throughout the war, culminated in flight tests of a prototype, earning Myasishchev and his team the personal gratitude of Stalin and various government awards.

Nonetheless, in April 1946 the DVB-102 programme was cancelled – officially for want of a suitable power plant. (Development problems with the bomber's intended 1,800 hp Klimov M-120TK 18-cylinder Y-block in-line engines meant that initial flight testing had to be undertaken with Shvetsov M-71 radials, which were not only less powerful (1,600 hp) but also troublesome, and the aircraft could not meet the design specifications.) At the same time, the Moscow-based OKB-482 design bureau, headed by Myasishchev (***opytno-konstrooktorskoye byuro*** – experimental design bureau; the number is a code allocated for security reasons), was dissolved. Its premises and manufacturing facilities (plant No. 482) were transferred to Sergey Vladimirovich Ilyushin's OKB-240, as were most of the defunct Myasishchev OKB's staff and the plant's workforce; the rest were distributed between the Tupolev design bureau (OKB-156), the Yakovlev design bureau (OKB-115), and other aviation industry enterprises across the Soviet Union.

In passing, such closures of Soviet aircraft design bureaus in the immediate postwar years were usually not the result of a perceived 'reduced need for military hardware in peacetime conditions' but rather the result of personal rivalry between the OKB chiefs and fighting over resources within the Ministry of Aircraft Industry (MAP – *Ministerstvo aviatsionnoy promyshlennosti*), which controlled the OKBs and the aircraft factories. The winner took over the assets of the defunct competitor, and *vae victis*!

A few years later, Ilyushin, who had several bomber programmes of his own, found himself in a similar situation: his OKB was not closed down, of course, but he received *an offer that he could not refuse* (from the government, that is) not to develop any more strategic bombers. The totalitarian state had its own methods of 'fair competition.' Whereas in the notorious 1930s, massive repressions and purges were used to clear the way for whichever design team was in favour at the time, in the immediate postwar years (the late 1940s), the Soviet government displayed a little more finesse, reducing the number of defence industry enterprises on the pretext that in peacetime there was no longer a need for so many. (Somehow this brings to mind a line from a song by Sting called 'Murder by Numbers': *There isn't really any need for bloodshed, // Just do it with a little more finesse; // If you can slip a tablet into someone's coffee, // Then it avoids an awful lot of mess.*) In so doing, the capabilities of the Soviet aircraft industry were severely restricted by funding cutbacks; as a result, experimental aircraft construction was limited to a handful of major design bureaus. OKB-156, led by Andrey N. Tupolev, a man who enjoyed considerable authority in the Kremlin, became the 'official purveyor to the court' insofar as strategic bomber development was concerned.

As for peacetime conditions, few – if any – in the Soviet political establishment (and the MAP top brass) had any illusions on that score. The Second World War had hardly ended when the world found itself in the throes of a new conflict – the confrontation between the East and the West known as the Cold War. This term, coined in 1947, denoted political, economic, ideological, and military confrontation, stopping short of outright armed conflict. Some hotheads, however, were not content to stop short. Since a new and fearsome weapon – the nuclear bomb – was now the decisive argument in the dialogue of the two superpowers, this was the time when the USA started drawing plans of preemptive nuclear strikes against the Soviet Union and its allies. Operation *Dropshot* alone (approved at the end of 1949) envisaged dropping 300 nuclear bombs and 29,000 conventional high-explosive (HE) bombs on 200 targets in 100 cities and towns to wipe out 85% of the Soviet Union's industrial potential in a single blow. Between 75 and 100 of these 300 nukes were intended for destroying Soviet combat aircraft on the ground. Quite apart from the intercontinental bombers of the US Air Force's Strategic Air Command, there was

General designer Andrey N. Tupolev (1888–1972), head of OKB-156 and Myasishchev's chief competitor

Below: Early-production Tu-4s built by the Kazan' aircraft factory No. 22 on the flight line of a Soviet heavy bomber regiments. The aircraft, serialled '39 Black' (c/n 220804), has canvas covers on the wings and the horizontal tail.

the US Navy with its shipboard strategic bombers – the mixed-power North American AJ Savage and the twin-turbojet Douglas A3D Skywarrior, which could be quickly brought within striking distance of the Soviet navy bases in Murmansk, Tallinn, Kaliningrad, Sevastopol', and Odessa.

What could the USSR oppose to these bloodthirsty plans and the armadas of NATO bombers? Only a similar force, of course, which would enable a retaliatory strike against the USA. As the call, so the echo.

In the early 1950s the Soviet Union's sole means of delivering nuclear munitions was the Tupolev Tu-4 strategic bomber (NATO reporting name *Bull* in the 'B for Bomber' series). This was a reverse-engineered version of the Boeing B-29A Stratofortress, four of which (including a crashed example) had fallen into Soviet hands in 1944 after raids on Japan and had been interned because the Soviet Union still had a pact of neutrality with Japan at the time. However, born in a remarkably short time, the piston-engined Tu-4 making up the backbone of the Soviet strategic bomber force was growing obsolete just as rapidly. The Soviet aircraft industry responded immediately by launching development of new bomber types; however, the legacy of the Tu-4 – or perhaps the inertia of thought – was so strong that in designing new bombers capable of striking across the North Pole the Tupolev OKB still relied on piston engines for a while. The result was predictable: the Tu-80 – a considerably reworked Tu-4 powered by 2,400 hp Shvetsov ASh-73TKFN radials – showed disappointing performance, and only a single prototype was built. The OKB made a last throw, evolving the Tu-80 design into the Tu-85 *Barge* long-range bomber, powered by four 2,500 hp Dobrynin VD-4K 24-cylinder X-block in-line engines. Even though this aircraft, which first flew in January 1951, did achieve the specified performance (indeed, preparations for series production were in hand for a brief period), it was clear that the days of the 'corncob engines' were over. To top it all, the Tu-85 used an outdated configuration with inefficient tractor propellers. In the USA the piston age in strategic aviation had ended by then; consider also that the ultimate American piston-engined bombers – the Northrop XB-35 'flying wing' and the famous 'six-turning, four-burning' Convair B-36 Peacemaker – utilised a more complex but also more efficient arrangement with pusher propellers. Also, the outbreak of the Korean War in 1950, a conflict in which propeller-driven aircraft were pitted against jets (and the latter almost invariably emerged victorious), made the military on both sides of the Iron Curtain change their attitude toward piston-engined aircraft.

On 28 February 1951, Col.-Gen. Pavel F. Zhigarev, the then commander in chief of the Soviet air force (VVS – ***Voyenno**-voz-**doosh**nyye **seely***), reported to Stalin as follows:

'Combat experience gained in Korea when using [Mikoyan/Gurevich] MiG-15 jet fighters against American B-29 aircraft shows that bombers of this kind are relatively defenceless [sic] *when confronted by modern high-speed jet fighters.*

Captured American airmen from a [downed] B-29 state that the crew of an American bomber is unable to track attacking MiG-15s with their guns and fire on them accurately.

This circumstance makes it much easier for MiG-15 pilots to attack B-29 bombers due to the latter's relatively low speed. As a result, the MiG-15 usually comes out as the winner in air-to-air combat between the two types. . . . In five aerial engagements with a numerically superior adversary, MiG-15 fighters have shot down 10 American B-29s and one [Lockheed] F-80 [Shooting Star]. No MiG-15s have been lost in these engagements.

In view of the facts stated above, it is alarming that our Tu-4 bomber possessing closely comparable performance to that of the B-29 [and well it might – *Auth.*], *including a top speed of 560 km/h [350 mph], may prove to be relatively defenceless in wartime when attacking enemy targets that are defended by modern jet fighters.*

It is equally alarming that the new four-engined bomber developed by Comrade Tupolev, which is currently undergoing trials [the Tu-85 – *Auth.*] *and has shown a maximum range of 12,000 km [7,450 miles] and a top speed of 600 km/h [370 mph], will have much the same deficiencies as the Tu-4.*

Having a considerable difference in speed with jet fighters [in plain language, this means 'being much slower than' – *Auth.*], *during deep penetration missions the Tu-4 and the new bomber will be repeatedly attacked by enemy fighters, to which, as Korean combat experience shows, they will be unable to put up effective resistance and thus reach their target without suffering heavy losses.'*

Meanwhile, the Western world was steadily building up its combat jet fleets. Soviet intelligence reports indicated that development of jet-powered heavy bombers outmatching current and even future fighters had started in the USA as early as 1944. The Boeing Model 450 (B-47 Stratojet) first flew in December 1947, entering service in 1950, by which time the Boeing Company was already working on the much more capable Model 464 – the future B-52 Stratofortress. In Great Britain the famous 'V-Bombers' (the Avro 698 Vulcan, Handley Page HP.80 Victor, and Vickers Type 660/667/674/706/710/733/758 Valiant) were developed and put into service with the Royal Air Force; so was the smaller English Electric Canberra bomber.

True enough, in mid-1950 the Tupolev OKB had started design work on a twin-turbojet medium bomber known in house as 'aircraft 88'; the first prototype was completed some 18 months later and first flew on 27 April 1952. Yet, this aircraft, which entered production and service in the mid-1950s as the Tu-16 *Badger*, did not have intercontinental range; what the VVS needed was an aircraft capable of delivering a strike against the continental US and, it was hoped, coming back. Again, the Soviet Union was facing the dire necessity not only to keep up with the Joneses but to surpass them in this field if at all possible. According to the deadline set by the Soviet government (some sources say by Stalin himself), such an aircraft was to be created within two years.

The Soviet political and military leaders were extremely worried about the national aircraft industry being late in developing heavy swept-wing jets. Much later, in the 1980s, using the Soviet press as an instrument, certain people tried to create the impression that Andrey N. Tupolev, the patriarch and effective monopolist in heavy aircraft development in the Soviet Union, flatly refused to design swept-wing jet aircraft. Tupolev was even quoted as say-

General designer Vladimir M. Myasishchev (1902–1978), head of OKB-23 (here in his major general's uniform)

Below: The sole prototype of the Tu-85 ('aircraft 85'), Tupolev's last piston-engined bomber. Even though it met the design target, the Tu-85 was not put into production because piston engines were considered obsolete by the early 1950s.

Top and above: Artist's impressions of two early designs proposed by Myasishchev in 1949 – the DVB-202 (*top*) and DVB-302. They look like improved versions of the Tu-4, differing from each other only in engine type; in fact, this was not the case.

Below and bottom: Two different display models of the DVB-302. Note the raised flight deck on the second model.

ing at a Kremlin conference chaired by Stalin himself that *'I will never ever design such an aircraft because the flutter phenomenon in large swept wings is totally unexplored and cannot be overcome at transonic speeds.'* Tupolev allegedly backed this statement with well-founded arguments based on research conducted in the USSR. Another publication claimed in earnest that Tupolev derided early reports on the B-52 as *'a lot of bluff from across the ocean'* and even remarked in a high-handed manner at the end of the abovesaid conference, *'Come on, you already have me developing the long-range Tu-16 jet, and the piston-engined Tu-85 will suit our needs for years to come!'* The author alleged that Stalin was angered by this remark and made it patently clear that Tupolev would have to take on the job, whether he wanted it or not: *'If you have trouble, we will help you; if you refuse, we will force you!'*

To be sure, Andrey N. Tupolev was often conservative in his views; it is equally true that his character was difficult to deal with. This, understandably enough, brought him his share of detractors; still, it is sad that after Tupolev's death in 1972, such people would take pleasure in kicking the dead lion.

In reality, things were a bit different, to say the least. The source that the newspaper authors alluded to was a letter from Tupolev to Stalin written in 1951, in which the chief designer of OKB-156 stated his views on the subject. Suffice it to quote an abstract from this letter – the way it was really written. Tupolev wrote:

'The biggest problem to be overcome in the creation of a long-range fast heavy bomber is the need to reconcile high speed with long range. Therefore the first stage of our work was to determine the maximum speeds a swept-wing heavy aircraft can currently achieve, regardless of the power of the engines installed. It turned out that, judging by our experience and extensive research conducted by TsAGI, the following maximum speeds can be attained with assurance, given the current state of aero- and hydrodynamics development:

- *950–960 km/h [590–596 mph] at 8,000 m [26,250 ft.]*
- *885–900 km/h [550–560 mph] at 12,000 m [39,370 ft.]*

Note: Even-higher speeds are possible at lower altitudes, but these are not much use from a tactical standpoint. [TsAGI (*Tsen**trahl'**nyy **a**ero- i **ghid**rodina**mich**eskiy insti**toot*** – the Central Aero- & Hydrodynamics Institute, named after Nikolay Ye. Zhukovskiy) was the leading Soviet authority on aerodynamics research and development and continues to hold this position in modern Russia – *Auth.*]

We have considered the possibility of achieving our goal by using jet engines. This research allowed us to determine the effect of the aircraft's size, weight, and engine power on speed, altitude, and range. Two different calculation methods were used in all cases to preclude incidental errors.

The conclusion we reached is this: it is possible to create a fast bomber if jet engines are used. However, it would be extremely difficult to obtain a range in excess of 11,000–12,000 km [6,830–7,450 miles] from such a bomber, since this would require development of a unique aircraft of extraordinary weight and size.

Thus, we are unable to find a reasonable solution to the task if jet engines are to be used.

The appearance in the Soviet Union of indigenous turboprop engines developed by [Nikolay Dmitriyevich] Kuznetsov, which have

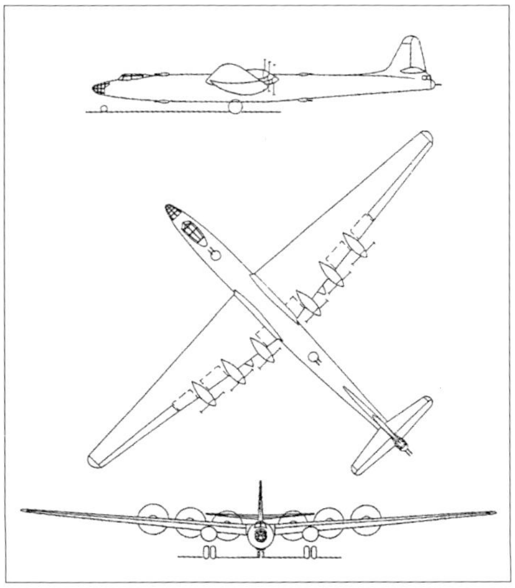

A late 1940s PD project of a strategic bomber developed by OKB-482 with six turboprops in pusher configuration

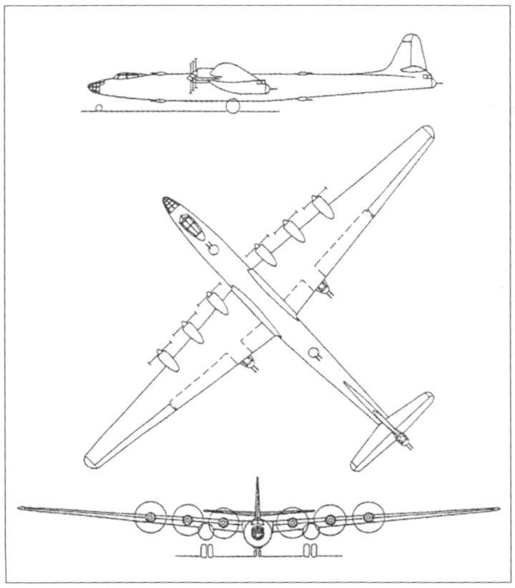

A version of the same project, with six piston engines and wing-mounted gun turrets at the tips of the main gear fairings

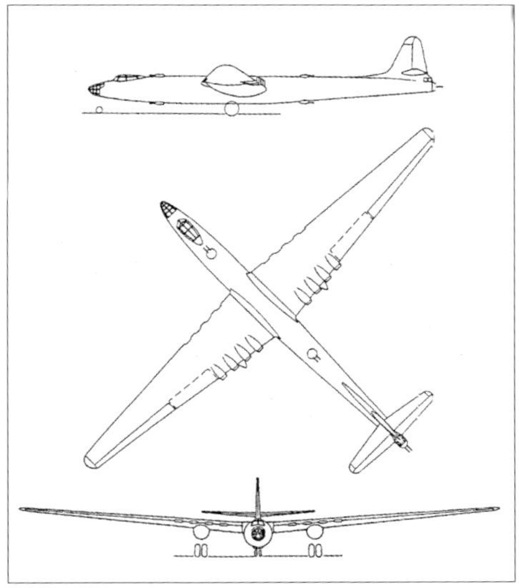

A further variation on the same project, with eight turbojets buried in the wing roots

passed state acceptance trials on a test bench, has created the prerequisites for developing viable long-range bombers powered by turboprops. Calculations show that a reasonably sized bomber (that is, with a weight of not more than 130–160 tons [286,600–352,730 lbs.]) powered by turboprop engines can attain much-longer range than a jet aircraft; the range could be as long as 14,000–15,000 km [8,700–9,320 miles] or even 18,000 km [11,180 miles].'

See the difference?

That said, you will surely agree it would be unfair to say Tupolev was refusing to design a jet-powered strategic bomber on the grounds that 'the flutter phenomenon in large swept wings is totally unexplored and cannot be overcome at transonic speeds'! What really made the patriarch back off was the lack of conviction (no pun intended) that the then-current state of the art in jet engines would allow the bomber to achieve the required range. This is why he placed his bets on the 12,000 ehp Kuznetsov TV-12 (NK-12) turboprop, designing his 'aircraft 95' (Tu-95 *Bear*) bomber around these engines. On the other hand, Tupolev was undoubtedly aware that developing this engine would take time.

Indeed, at that time the experts of MAP and the Soviet air force were divided over the feasibility of creating a long-range swept-wing jet bomber. Not only Tupolev but other prominent Soviet aviation scientists had their misgivings, and the Soviet government was influenced to no small degree by these sceptics. In 1949–50 the Tupolev OKB, TsAGI, and the Central Aero Engine Research Institute (TsIAM – *Tsentrahl'nyy institoot aviatsionnovo motorostroyeniya*) had undertaken a joint study on the feasibility of a jet-powered strategic bomber, which showed that the maximum range such an aircraft could achieve would be no more than 6,000 km (3,730 miles). The study reinforced the Tupolev OKB's belief in propeller-driven aircraft.

As an aside, we may mention that the Boeing Company, too, toyed with the idea of a turboprop-powered strategic bomber for a while. The Model 474 (XB-55) project, developed in 1948, envisaged an aircraft basically similar to the B-47 but having less wing sweep and four 4,100 ehp Allison T40-A-2 coupled turboprops installed under the wings in pylon-mounted nacelles (!) to drive six-blade tractor contraprops. However, getting the required reliability level from the engine would take time; it soon became clear that the XB-52 would both outperform the XB-55 and be flying at least a year before the T40 could be brought up to scratch. In January 1949, the XB-55 was cancelled.

Conversely, Vladimir M. Myasishchev was one of the supporters of jets. However, to prove the sceptics were wrong, he had to see for himself if he was right and prove his point with rock-solid evidence. The first research and preliminary design (PD) studies of a strategic jet bomber began at his initiative as early as 1948.

Here we have to go back a little. After the closure of OKB-482 and his ouster from MAP, Myasishchev went on to work as dean of the aircraft design faculty at the Moscow Aviation Institute (MAI), and it was there that he made his first design studies on the subject. Among other things, he calculated the design parameters of a jet bomber giving it a chance of penetrating enemy air defences if its route lay 200–300 km (124–186 miles) from a base where Shooting Stars were deployed. However, he soon found himself at odds with the institute's Communist Party bosses (in Soviet times, each enterprise or institution invariably had a, so to say, 'chapter' of the Communist Party, and Myasishchev was not a member of the party). As a result, he was demoted to head of the fixed-wing aircraft design chair within the faculty. In this capacity he made an acquaintance of Gheorgiy N. Nazarov, a highly skilled aviation engineer who would later become his closest aide.

Myasishchev's proposal to draw up a joint MAI/TsAGI research and development (R&D) work plan on advanced aspects of aircraft design was an important step toward his return to the aircraft industry together with Nazarov. A major section of the plan was devoted to determining the parameters of future aircraft. This work was aimed at obtaining objective data that – now we come to the main point – would prove the feasibility of developing a long-range bomber with jet engines and high-aspect-ratio swept wings. The R&D effort involved a lot of calculations, graphs, and PD layouts. In the course of 18 months (some sources say two years), the chair's engineering staff, assisted by ordinary and post-graduate students, managed to amass the necessary data from which to proceed with the strategic jet.

Then Vladimir M. Myasishchev delivered a report on the results of the parametric research programme at a session of TsAGI's Scientific and Technical Council. The session was chaired by academician Aleksandr I. Makarevskiy and was attended by such notable aviation scientists as Sergey A. Khristianovich, Vladimir V. Strominskiy, Gheorgiy S. Büschgens (all three later became academicians as well), and others. After listening carefully to the report (followed by Gheorgiy N. Nazarov's own report) and going over the details, the Soviet aircraft industry's most authoritative institute gave the go-ahead, stating it was possible to create a strategic jet bomber with the stated performance characteristics obtained in the course of lengthy research. This was, in effect, the official launch of the aircraft initially known simply as the SDB (*strategicheskiy dahl'niy bombardirovshchik* – strategic long-range bomber).

The first proposals concerning the SDB were submitted by Myasishchev in February 1951, when he was still heading the fixed-wing aircraft design chair at MAI. The Russian popular press quoted varying descriptions and performance estimates for the projected bomber, which was developed in several configurations. One source describes the SDB preliminary design project as having shoulder-mounted wings, a T tail (!), a bicycle undercarriage, and four Mikulin AM-TKRD-03 turbojets in underwing nacelles; AM was the initials of OKB-300's chief designer Aleksandr A. Mikulin, while TKRD stood for *toorbokompressornyy re'aktivnyy dvigatel'* ('turbocompressor'; that is, air-breathing jet engine). The inboard engines were pylon mounted, while the outer nacelles adhered directly to the wing underside at the wingtips and accommodated the landing gear's outrigger struts in addition to the engines, B-47 style. The fuselage was of basically circular cross-section, changing to oval at the extremities. The wings, equipped with slotted flaps, were swept back 35° at quarter chord (this sweep angle was typical of many early Soviet jets) and had a cranked leading edge, with 45° leading-edge sweep inboard and 40° outboard, plus 3° negative camber on the outer portions.

Another PD project configuration envisaged a power plant of six 3,100 kgp (6,830 lbst) Klimov VK-5 centrifugal-flow turbojets. With a maximum takeoff weight of 90 tons (198,400 lbs.), the aircraft was to deliver a 3-ton (6,610 lb.) bombload over a maximum distance of 11,000–12,000 km (6,840–7,450 miles) at a cruising speed of 750–800 km/h (465–496 mph). The proposal concerning the SDB bomber envisaged an even bigger and heavier aircraft; according to PD documents, it would have a wingspan of 50 m (164 ft.), a 300 m² (3,225 sq. ft.) wing area, an overall length of up to 44 m (144 ft., 4¼ in.), a normal takeoff weight of 110 tons (242,500 lbs.), and an MTOW of 140 tons (308,640 lbs.). The design range was again 12,000 km but with a bombload increased to 5 tons (11,020 lbs.). The maximum ordnance load was 20 tons (44,090 lbs.), and the aircraft could carry bombs of any calibre between 500 kg (1,102 lbs.) and 9,000 kg (19,840 lbs.). The cruising speed was set at 800-850 km/h (496-530 mph).

According to yet another description, the SDB was a large aircraft utilising shoulder-mounted swept wings, a conventional swept tail unit, and a power plant consisting of four Mikulin AMRD-03 axial-flow turbojets then under development. With an estimated takeoff thrust of 8,200 kgp (18,080 lbst), the AMRD-03 (later redesignated AM-3 in production form) was then the world's most powerful jet engine. The only decidedly unconventional feature at first glance was the bicycle landing gear with outrigger struts retracting into fairings at the wingtips.

The wings had a cranked leading edge, with 45° leading-edge sweep on the inboard portions and 40° outboard. The trailing edge featured a double kink due to different airfoil sections being used along the span and to structural features intended to ensure adequate stiffness; the entire trailing edge was occupied by flaps and ailerons. The fuselage was of basically elliptical cross-section, providing a bomb bay that was large enough to accommodate two bombs of the largest calibre (FAB-9000 HE bombs) above one another. The rest of the internal volume was occupied by the two pressurised crew cabins, fuel tanks, wheel wells, and equipment bays. Structurally the fuselage was built in five sections that were joined by bolts and flanges. The crew of seven comprised two pilots, a navigator / bomb aimer, two gunners using beam blisters ahead of the wings for sighting, a radio operator, and a tail gunner.

The SDB was the largest and heaviest aircraft that had been designed in the Soviet Union until then. It is certainly not common for a company developing the aircraft to take operational requirements to a new and higher level, but with the SDB this was exactly the case. The Soviet government and the air force's top brass had little choice but to agree to these new standards and officially endorse them by issuing a directive ordering the development and construction of the SDB strategic bomber.

As soon as Mikhail V. Khroonichev, the then minister of Aircraft Industry – the one who had ordered the liquidation of OKB-482 in 1946 and effectively drummed Myasishchev out of the industry – got wind of the project, he did his best to stop it. Trying to forestall Myasishchev, he approached Tupolev, suggesting that the latter develop a jet-powered strategic bomber. Tupolev, who was then having health problems, answered that there were many questions associated with such a project, and he would look into the possibility when his physician allowed him to start work.

After obtaining TsAGI approval for his concept, Myasishchev submitted the project for review to the Soviet Union's Council of Ministers (that is, government) on 1 March 1951. He then approached Khroonichev, proposing to build such an aircraft. Since Tupolev had not yet presented his ideas on such an aircraft, Khroonichev apparently had no choice but to report the idea upstairs to Stalin. So did Vyacheslav A. Malyshev, vice chairman of the Council of Ministers.

Later, at a Kremlin conference that outlined the direction in which the Soviet Union's strategic aviation should develop, Stalin spoke a famous phrase: *'Let's believe Comrade Myasishchev and entrust him with developing such an aircraft.'* Thus Myasishchev received carte blanche in recruiting specialists for his reborn OKB – and earned himself a bitter rival in the form of Tupolev, who not only did not take kindly to competition but also knew Myasishchev's engineering potential better than anyone else.

Chapter 1

Shaping the Bomber

By early 1951 the choice of jet engines and swept wings for Myasishchev's new bomber had been virtually agreed. On 24 March 1951 the Council of Ministers and the Communist Party Central Committee issued joint directive No. 949-469, ordering the creation of GSOKB-23 (*Gosu**dar**stvennoye* ***soyooz**noye* ***op**ytno-kon-**strook**torskoye byu**ro*** – State All-Union Experimental Design Bureau) and appointing Vladimir M. Myasishchev as its head. Two days later it was followed up by an MAP order to the same effect. These documents assigned MAP plant No. 23 – an aircraft production facility in Fili, then a western suburb of Moscow (now long since absorbed by the city) – to the newly established OKB for prototype and production bomber construction. This plant was then manufacturing Tu-4s, and production of the Bull was discontinued in favour of the future jet bomber. Appropriate funding was assigned, and other aircraft industry enterprises acting as subcontractors were specified. The 'all-Union' bit (meaning that the OKB had nationwide importance) spoke volumes for how seriously the government took the new bomber project.

However, Mikhail V. Khroonichev did not change his malicious attitude and would not give up. On 29 March 1951 he wrote thus to Nikolay A. Bulganin, first vice chairman of the Council of Ministers: *'I believe that chief designer Comrade Tupolev's offer to develop a new long-range jet bomber* [sic – *Auth.*] *with a speed of 850–950 km/h [527–590 mph] and a range of 12,000–14,000 km [7,450–8,695 miles] merits attention, but since it came late . . . , launching a parallel bomber project* [to Myasishchev's

A still from a unique documentary showing Vladimir M. Myasishchev in his study at a time when his first strategic bombers were being developed

project – *Auth.*] *with the flight performance Comrade Tupolev is proposing is inexpedient, considering that the top speed of the bomber being developed by Comrade Myasishchev is 900 km/h [559 mph] for the first prototype and 1,000 km/h [621 mph] for the second prototype. Therefore, if we are to accept Tupolev's offer . . . , we should by all means demand higher performance – the top speed should be at least 1,000 km/h and the range . . . at least 13,000 km [8,075 miles] with a bombload of at least 5 tons [11,020 lbs.].'* In other words, get Myasishchev's project rejected in favour of a more capable competing design.

Thus, Tupolev was not out of the race. Just under three months later, on 11 July 1951, the Council of Ministers and the Communist Party Central Committee issued joint directive No. 2396-1137, which tasked OKB-156 with designing and building a long-range high-speed bomber – the Tu-95. Since the TV-12 turboprop was suffering development delays, the first prototype was to enter flight test by September 1952 with a provisional power plant consisting of four 12,500 ehp Kuznetsov 2TV-2F–coupled turboprops; the second prototype with the definitive power plant comprising four TV-12s was to follow a year later. Thus began an unofficial competition for the best strategic bomber.

The abovesaid directive No. 949-469 stated that the reborn Myasishchev OKB was to specialise in long-range heavy bombers and what was referred to vaguely as special-mission aircraft (the latter kind later turned out to be high-flying strategic reconnaissance aircraft, among other things). The directive further tasked GSOKB-23 with developing a strategic bomber powered by either six 5,000 kgp (11,020 lbst) Lyul'ka AL-5 axial-flow turbojets or four 8,000 kgp (17,640 lbst) Mikulin AM-3 turbojets.

Pursuant to the directive, the aircraft was to be capable of covering a distance of 11,000–12,000 km (6,830–7,450 miles) in maximum-range cruise at 13,000–16,000 m (42,650–52,490 ft.), with a speed of 800 km/h (496 mph) while carrying 5 tons (11,020 lbs.) of bombs and have a maximum ordnance load of 24 tons (52,910 lbs.). The top speed was stipulated as 900 km/h (590 mph), and the takeoff run as 2,000 m (6,560 ft.) without rocket boosters and 1,600 m (5,250 ft.) with rocket boosters. Apart from free-fall bombs of 500–9,000 kg (1,102–19,840 lb.) calibre, the envisaged weapons options included UB-2000F ***Chai**ka* (Seagull) and UB-5000F Kondor radio-controlled gliding bombs intended for destroying small but high-value targets (UB = *oopravlyayemaya **bom**ba* – guided bomb), naval mines and torpedoes. The guided bombs had large wings spanning 2.1 m (6 ft., 10$^{43}/_{64}$ in.) for the Chaika and 2.67 m (8 ft., 9$^{7}/_{64}$ in.) for the Kondor; hence they could be carried only externally under the bomber's wings because the bomb bay was only 1.8 m (5 ft., 10$^{55}/_{64}$ in.) wide. This meant a lot of extra drag, and the idea was soon dropped.

The first prototype was to enter flight test in December 1952, followed a year later by a second prototype featuring aerodynamic improvements. This sure was a tough schedule, especially for a 'clean sheet of paper' project, but the Soviet government was aware that the American plans of a nuclear first strike against the USSR were still on the agenda.

With this historic directive in place, work soon started on a plot of unused land between the Moskva River and the premises of plant No. 23 (nowadays the State Scientific & Production Centre named after Mikhail V. Khroonichev, an enterprise concerned with ballistic missiles and space launch vehicles) as the Myasishchev OKB re-created its research, development, and experimental production facilities. An MAP order requiring other aviation OKBs to transfer part of their engineers and other workers to the new OKB (subject to approval by chief designer Vladimir M. Myasishchev) was issued – and, it should be noted, the order was obeyed without delay. Designers, aerodynamicists, and structural-strength engineers were transferred from the Yakovlev OKB, Vladimir N. Chelomey's OKB-52 missile design bureau, etc., literally together with their desks, drawing boards, and pencils! Additionally, the Myasishchev OKB actively recruited new cadre fresh from higher- and technical-education establishments; in particular, nearly all graduates of the Moscow Aviation Institute, class of 1951, and some graduates of the Khar'kov Aviation Institute were sent to OKB-23.

Not all in the ministry were happy with this rapid buildup. In October 1951, Myasishchev's bête noire Mikhail V. Khroonichev wrote a report to Nikolay A. Bulganin, stating: *'The number of [the Myasishchev OKB's] engineering staff is currently 146 persons more than required by the CoM directive dated 18 May 1951. Deputy chief designer Comrade [Nikolay M.] Glovatskiy's request to assign an additional 75–100 engineers to Comrade Myasishchev's OKB and increase the total number of employees to 400 is thus unfounded.'*

Apparently Myasishchev was well aware of Stalin's famous maxim 'The personnel is the decisive factor,' and he tried to get hold of as many skilled engineers and workers as he possibly could. The people whom he, in the words of his close aide Leonid Leonidovich Selyakov, had set his sights on and whom he wanted to work for OKB-23 included aerodynamicist Vsevolod N. Matveyev of MAP's Flight Research Institute (LII – ***Lyot**no-is**sled**ovatel'skiy insti**toot***), engineers Yevgeniy S. Fel'sner and Vladimir M. Baryshev of the Ilyushin OKB, engineer Leonid L. Kerber of the Tupolev OKB, and the future academician Mikhail K. Yangel' from the Ministry of Armaments' NII-88 (*naoochno-is**sled**ovatel'skiy insti**toot*** – research institute). The chief designer had worked with all of them before and knew their worth, but not all of these people would eventually join the Myasishchev OKB.

By the end of April 1952 the Myasishchev OKB had virtually regained its former self – especially since in accordance with the abovementioned MAP order, the designers and production engineers who had worked for the defunct OKB-482 were to be transferred to Myasishchev's new OKB by all means. These included Ya. B. Nodel'man and G. G. Matveyev working in the wing design team, K. I. Popov in the fuselage design team, G. I. Arkhangel'skiy in the landing-gear design team, and so on. Due to this and the 'infusion of fresh blood,' the staff of the OKB and plant No. 23 totalled more than 4,000 by the time the design bureau's new premises and the factory were ready for action.

The new 200-ton (440,900 lb.) jet bomber bore the in-house designation *izdeliye* M (sometimes referred to as 1M) and was also known as '***te**ma* (subject, or topic) 25' (***izdel**iye* (product) such and such is a term often used for coding Soviet/Russian military hardware items).

Initially a handful of designers had to shoulder the main burden of the task. Leonid L. Selyakov, who drew the first sketches of the future bomber, was forced to be a jack-of-all-trades, being responsible for general arrangement design, structural-strength calculations, aerodynamic design, and strategy development. Vladimir M. Myasishchev was busy with major organisational tasks at the time, forming the sections, departments, and teams of the new enterprise. These are Selyakov's recollections of those days:

'My temporary "office" – the place I had to work in – was a tiny cubicle near the office of the plant's former vice director, which was (also temporarily) occupied by V. M. Myasishchev. All the furniture there consisted of a small table, a chair, and a drawing board. I was all by myself. I had to perform the duties of a designer, aerodynamicist, structural-weight-calculations engineer, structural-strength engineer, and strategist. The task was to portray the outlook of the new aircraft on paper – including its basic dimensions, weights, and performance figures. I had to create the first general arrangement of the new aircraft. Afterward the huge design team would come in, making such alterations as would be necessary. Huge battles would rage as the engineers would discuss the strengths and weaknesses of this or that design feature. Meanwhile, however, I had to decide everything myself, choosing the road that we would take.'

Another excerpt from Selyakov's memoirs:

'Vladimir Mikhaïlovich spent days and nights (yes, nights!) solving immense organisational tasks. He was hugely assisted in this complicated and painstaking job by Gheorgiy Nikolayevich Nazarov (his first deputy) and Nikolay Matveyevich Glovatskiy. People kept coming and applying for jobs; they had to be hired and organised into sections, departments, and teams. We needed rooms to put them all up in, and there were none. As a temporary measure we put them up in the storage rooms of the assembly shop. Time kept slipping away; the team needed some work to do, and it was up to me to prepare the work for them.

Make no mistake: the job was an extremely complex one. All we had was questions but no answers. What engine type should we use? How refined was the aircraft from an aerodynamic and structural-weight standpoint? Within an unbelievably short time frame we created the general arrangement – or, to be precise, the general outlines from which to proceed. Concurrently we devised the project's technological philosophy, so to say. This was because the aircraft was huge by the standards of the day and required new standard sizes and thicknesses of metal stampings, sheet metal, forgings, etc. This crucial aspect was the responsibility of a remarkable person, Z. Nepomnyashchiy, the chief technologist of our new organisation. S[ergey] M. Leshchenko, the then director of plant No. 23, played a major part in this and other matters.

As the organisation grew, the scope of the work widened. For the first six months, however, the entire plant was mostly concerned with manufacturing a tremendous number of wind tunnel and scaled-strength models, which were necessary for project development.'

As the work on the *izdeliye* M bomber progressed, the team headed by Vladimir M. Myasishchev evolved from the former technical draft section of plant No. 23 into a fully fledged enterprise numbering 700 employees and capable of developing virtually any kind of aircraft. Incidentally, this put it in a unique position by Soviet standards – the Myasishchev OKB was based at

Above: An early PD configuration of the 'aircraft M' featuring a butterfly tail and unswept wings

Above: This display model shows another early project configuration of the 'aircraft M' bomber, likewise featuring a butterfly tail but having swept wings. Note the flush air intakes, aft-retracting nose gear unit, and the apparently 'solid' nose lacking a glazed navigator's station.

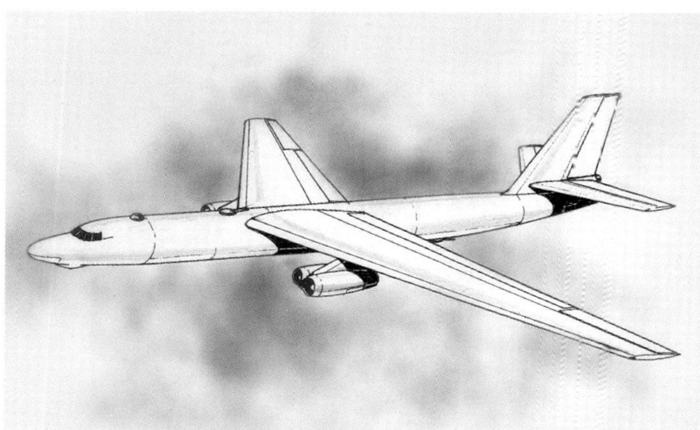

An artist's impression of yet another PD project configuration of the 'M.' The aircraft has the definitive conventional tail, but the engines are carried in side-by-side nacelles under the shoulder-mounted wings.

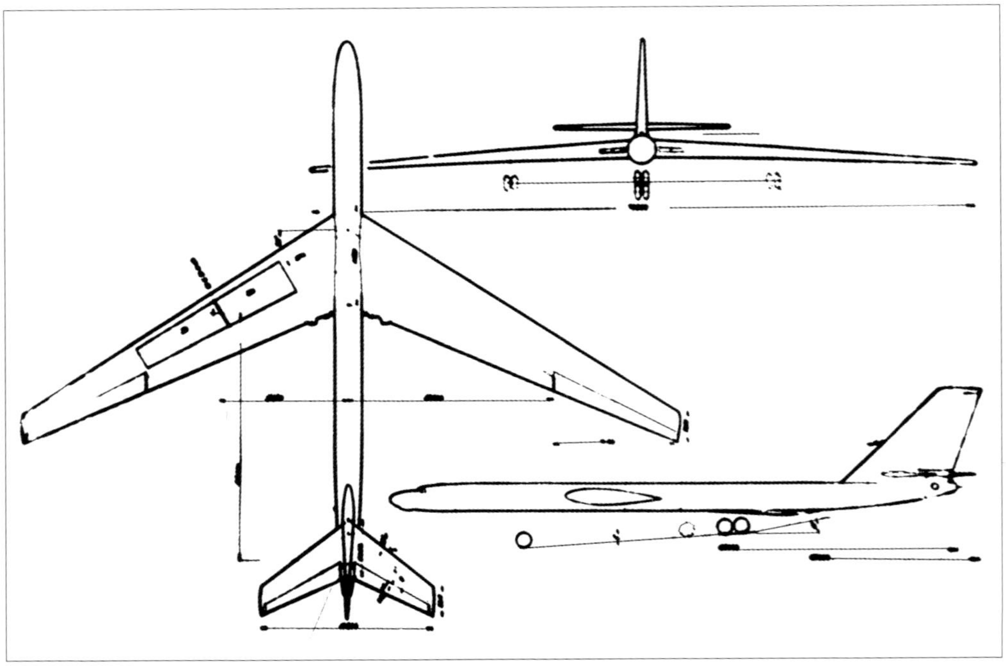

Left: An OKB-23 diagram showing one of the *izdeliye* M bomber's early project configurations with buried engines and 'V-Bomber'-style common intakes. The nose gear unit has twin wheels; the outrigger struts are placed well inboard and equipped with large-diameter twin wheels. Note also the extremely high fuselage fineness ratio and the wing fuel cell bay placement.

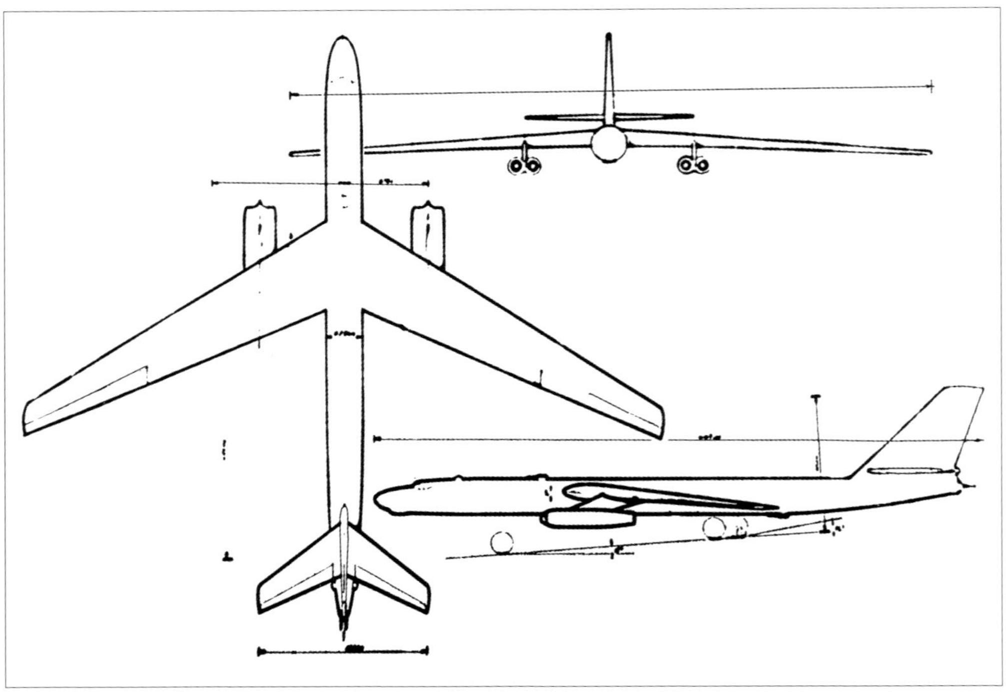

A three-view drawing of the project configuration of *izdeliye* M shown at the bottom of the previous page, featuring B-47-style paired nacelles on sharply swept pylons. The large-span ailerons are noteworthy.

a major production factory that was earmarked for producing the aircraft it designed (usually Soviet aircraft design bureaus had only a small prototype-manufacturing facility unsuitable for mass production). OKB-23 consisted of five sections: the Aerodynamics, Stability, and Control Department (including the wind tunnel model section), headed by Isaak Ye. Baslavskiy; the Static Strength, Dynamic Strength, and Aeroelasticity Department (including the scaled-strength model section), headed by Lev I. Balabookh; the Design Department (including the Advanced Development Projects Section), headed by S. A. Sidorov; the Information Department and technical library, headed by N. I. Biryukov; and, finally, the aerodynamics laboratory. The mockup construction section, the scale model section, and the static-test laboratory were the domain of the Design Department.

According to Leonid L. Selyakov, the information department played a major part in making the Myasishchev OKB what it was. Correctly processing and arranging information and conveying it in timely fashion to every designer was an issue taken very seriously at OKB-23. All organisational issues were handled by Selyakov's aide A. D. Murav'yov.

Many other research-and-development establishments in the Soviet aircraft industry were actively involved in the creation of the new bomber. These included TsAGI, LII, TsIAM, and the All-Union Research Institute of Aviation Materials (VIAM – *Vsesoy-*

***ooznyy institoot** aviatsionnykh materiahlov*), to name but a few. In particular, a whole team of VIAM experts was seconded to plant No. 23 to help with the issue of whether magnesium alloys should be used in the bomber's construction along with conventional aluminium alloys.

According to the specific operational requirement (SOR) endorsed by the Soviet air force's commander in chief Pavel F. Zhigarev on 23 June 1951, the new bomber was to have a 12,000 km (7,450-mile) range with a 5-ton (11,020 lb.) minimum bombload. The latter figure had been chosen because this was the weight of the first Soviet nuclear bomb, the RDS-1 having a yield of 22 kilotons. The aircraft was required to fly over the target at 11,000 m (36,090 ft.) with a speed of Mach 0.75. This altitude, which was some 4,000 m (13,120 ft.) lower than that of the Boeing B-52 and only a little higher than that of the B-29, was determined by the Soviet bomber's smaller wing area – 350 m² (3,760 sq. ft.) versus 371.6 m² (4,000 sq. ft.) for the B-52 – and power plant performance. The maximum ordnance load was specified as 24 tons (52,910 lbs.).

The success achieved by MiG-15*bis Fagot-B* fighters in combat against slow-flying and ill-armed B-29s in the course of the Korean War was given due regard by the Soviet military, who demanded that Myasishchev's bomber should have potent defensive armament. This was to comprise four twin-cannon barbettes (two dorsal barbettes, a ventral barbette, and a tail barbette) but was eventually reduced to three barbettes. Due attention was also given to operational aspects: the new bomber was to operate from newly constructed bases with concrete runways 3,000–5,000 m (9,840–16,400 ft) long and 50–100 m (164–330 ft) wide.

In the course of PD studies, the OKB considered numerous layouts for the new bomber. These differed in a number of important respects, but what they had in common was the bomber's large size, high-aspect-ratio wings, and bicycle landing-gear arrangement (more about this later). Here are but a few of these preliminary configurations (the enumeration given is for the sake of convenience only and does not reflect the order in which they actually appeared).

Version 1 had unswept wings and a swept butterfly tail, looking like a scaled-up competition sailplane. The wing leading edge was kinked, featuring about 30° sweep at the root and very slight sweep over the rest of the span; the trailing edge was at right angles to the fuselage. The engines were buried in the wing roots, the engine cowlings protruding above the wing upper surface at the trailing edge, just like on the aircraft that was built eventually (the future M-4). The wingtips mounted the outrigger gear fairings, which protruded halfway beyond the wing leading edge.

Version 2 likewise had a butterfly tail but featured swept wings. Again the four engines were buried in the wing roots. The landing gear included a twin-wheel nose unit and a four-wheel main bogie, both retracting aft; the outrigger struts and their fairings were moved slightly inboard from the wingtips. The defensive armament was reduced to three turrets, the upper turret being placed ahead of the wings and the lower one aft of the main gear unit. The aircraft was 50.5 m (165 ft., 8³⁄₁₆ in.) long and 12.1 m (39 ft., 8⅜ in.) high, with a wingspan of 51.54 m (169 ft., 1⁹⁄₆₄ in.), a wing area of 304 m² (3,272.23 sq. ft.), and a maximum takeoff weight of 177 tons (390,210 lbs.).

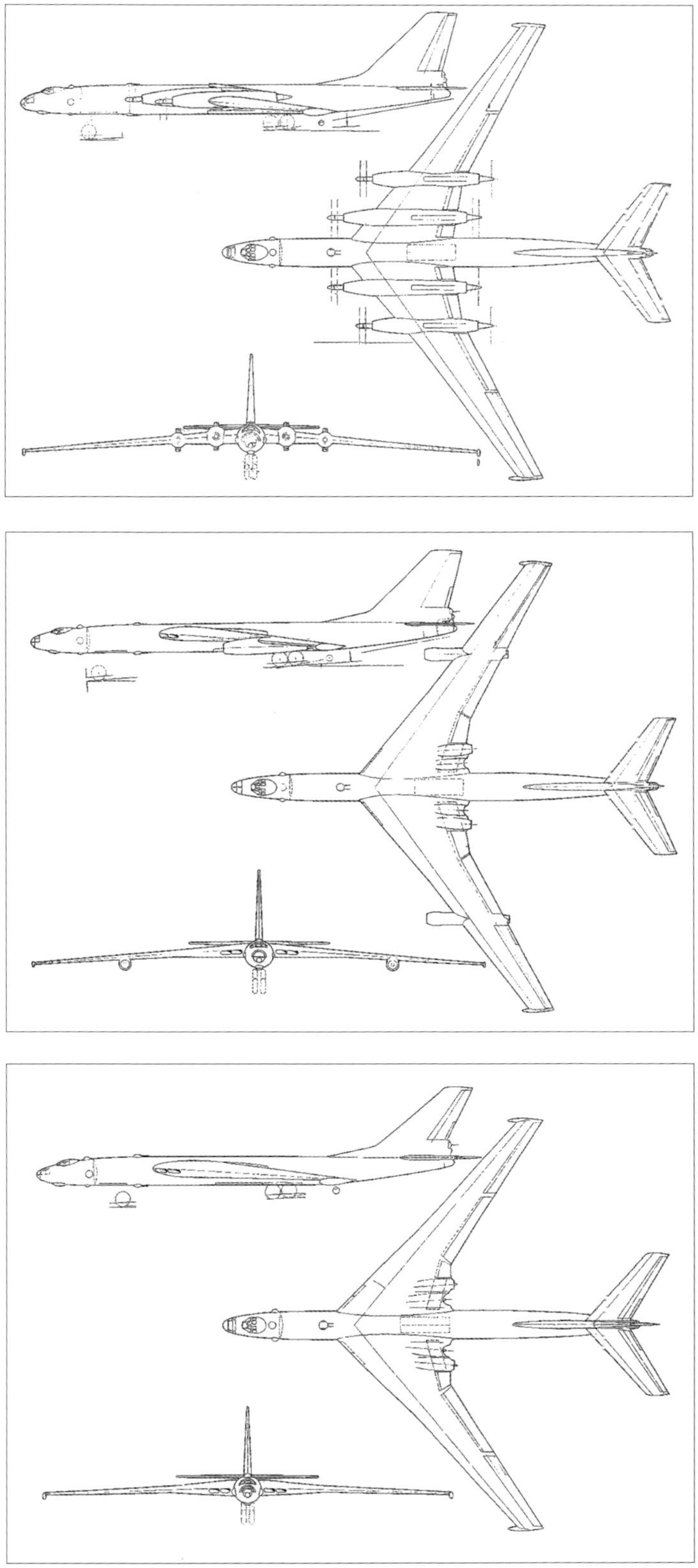

Three provisional general arrangements of the *izdeliye* M bomber combining the same basic airframe with different power plants. The one at the top has eight TV-2F turboprops in tractor/pusher pairs, the centre one has six AL-5 turbojets (four inboard and two outboard), and the lower one has four AM-03 turbojets (and a forward-retracting nose gear unit).

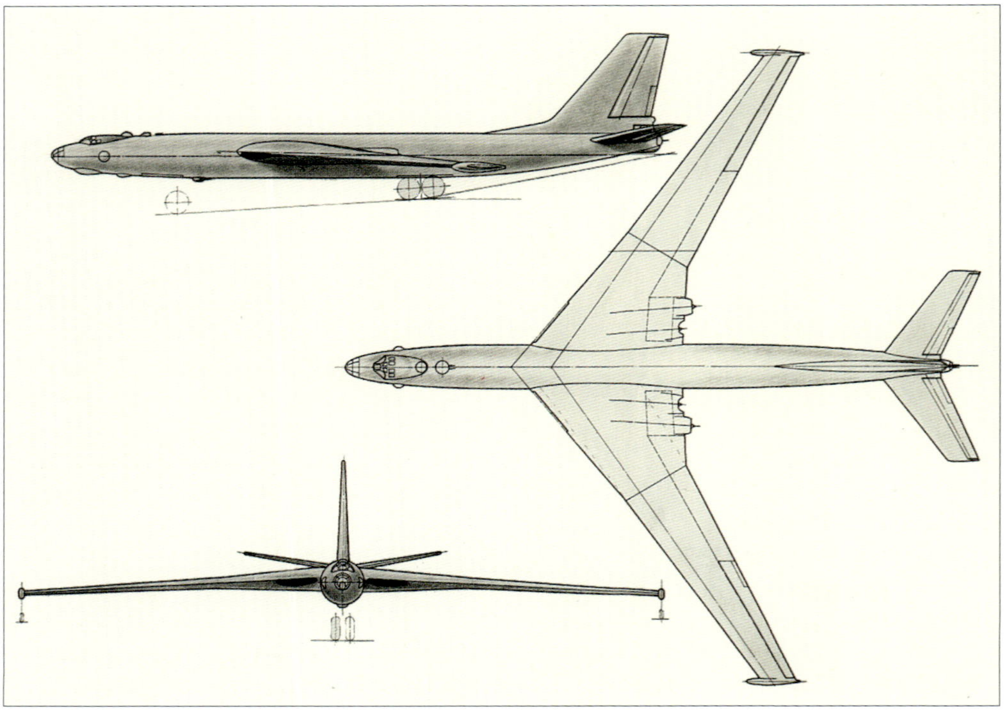

Almost there: This three-view drawing from the PD project shows the aircraft with four AM-03s breathing through flush intakes, a dorsal cannon barbette repositioned farther forward, and a redesigned tail gunner's station (the tail turret is aft of the pressurised cabin rather than above it – compare this with the drawing on the preceding page). Note the cranked-wing trailing edge with no sweepback on the inboard flaps, and the addition of dihedral to the horizontal tail, which was still fuselage-mounted at this stage.

Version 3 featured swept wings and conventional swept tail surfaces with no dihedral on the fuselage-mounted tailplanes. The wing thickness at the roots amounted to nearly 70% of the fuselage diameter. As in Version 2, the aircraft had a twin-wheel nose unit and a four-wheel main bogie, but this time the outrigger struts were mounted at the wingtips and augmented by two auxiliary struts at 40% of the wingspan, which were jettisoned after takeoff (a feature that was soon rejected); the outrigger struts had no fairings. The defensive armament was identical to Version 2. The aircraft was 50.6 m (166 ft., 0⅛ in.) long and 11.7 m (38 ft., 4⅝ in.) high, with a wingspan of 46.3 m (151 ft., 10 53/64 in.), a wing area of 280 m² (3,013.89 sq. ft.), and a maximum takeoff weight of 186 tons (410,050 lbs.).

Version 4 again had a conventional tail unit, but the wings were nearly shoulder-mounted and the engines were carried in Boeing B-47-style paired pods on swept pylons located at approximately 25% span. The wing planform and fuselage design were very similar to that of the future M-4, except for the twin-wheel nose gear unit and the defensive armament arrangement, which was identical to versions 2 and 3.

Version 5 differed radically in that it was a turboprop aircraft; apparently the Myasishchev OKB had decided not to dismiss this option altogether. Eight Kuznetsov TV-2F engines, each rated at 6,250 ehp and driving eight-bladed contrarotating propellers, were installed back to back in nacelles going clean through the wing torsion box (that is, the forward engine in each pair was in tractor con-

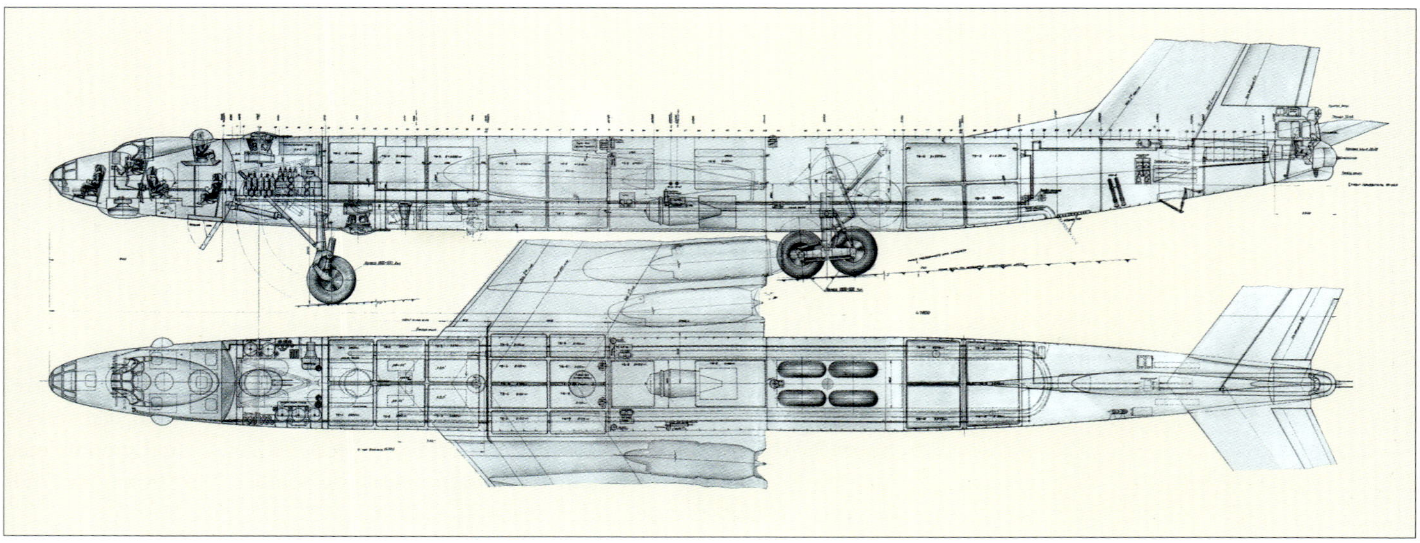

A cutaway drawing of the same project version, showing the fuselage fuel tanks, the aft-retracting main gear unit, and the retractable tail bumper

figuration, with an annular air intake around the propeller spinner, while the aft engine was in pusher configuration, breathing through a dorsal air scoop). The aircraft was 50.5 m (165 ft., $8\frac{3}{16}$ in.) long and 12.1 m (39 ft., $8\frac{3}{8}$ in.) high, with a wingspan of 51.54 m (169 ft., $1\frac{9}{64}$ in.); wing area was 304 m^2 (3,270 sq. ft.) and the MTOW was 175 tons (385,800 lbs.). The aircraft was dimensionally identical to version 2 but had a slightly lower MTOW of 175 tons (385,800 lbs.).

Version 6 was a swept-wing aircraft powered by six Lyul'ka AL-5 turbojets; four engines were buried in the wing roots, breathing through elliptical intakes (as on the future M-4), while the other two were mounted in conformal underwing nacelles at about 60% of the wingspan. The external dimensions and wing area were identical to the previous version, and the maximum takeoff weight was increased to 178 tons (392,420 lbs.).

Version 7 was almost identical to the previous one but was powered by four Mikulin AMRD-03 turbojets buried in the wing roots; the nose gear unit was redesigned to retract forward. The MTOW was 177 tons (390,210 lbs.).

Version 8 was very similar but with slightly different dimensions. The aircraft was 50.6 m (166 ft., $\frac{1}{8}$ in.) long and 11.7 m (38 ft., $4\frac{5}{8}$ in.) high, with a wingspan of 46.3 m (151 ft., $10\frac{53}{64}$ in.); wing area was 280 m^2 (3,010 sq. ft.), and the MTOW was increased to 186 tons (410,050 lbs.).

Other versions were also considered. One was a much-larger aircraft with a 65 m (213 ft., $3\frac{3}{64}$ in.) overall length, a 76.2 m (250 ft.) wingspan, a 19.85 m (65 ft., $1\frac{1}{2}$ in.) stabiliser span, and a 3.2 m (10 ft., $5\frac{63}{64}$ in.) fuselage diameter; the wings and horizontal tail were swept back 30° at quarter chord, while vertical tail sweep was 40°. Yet another version was 40.5 m (132 ft., $10\frac{31}{64}$ in.) long, with a 51 m (167 ft., $3\frac{7}{8}$ in.) wingspan, a 13.3 m (43 ft., $7\frac{5}{8}$ in.) stabiliser span, and a 2.9 m (9 ft., $6\frac{11}{64}$ in.) fuselage diameter; the engines were carried on pylons positioned 6.76 m (22 ft., $2\frac{9}{64}$ in.) from the fuselage sides.

Power plants under consideration included the Mikulin AM-3 turbojet, the AM-13 coupled engine consisting of two 4,100 kgp (9,040 lbst) AM-11 axial-flow turbojets, the 6,830 kgp (15,060 lbst) Lyul'ka AL-7 axial-flow turbojet, and the mighty TV-12 (NK-12) turboprop. In the latter case, four of these engines were housed in spaced nacelles on the wing leading edge, driving AV-60 eight-bladed contrarotating propellers.

Upon careful consideration, the engineers chose the optimum layout – a swept-wing aircraft with moderately swept conventional tail surfaces (the horizontal tail was now dihedral and was set above the rear fuselage at the base of the fin) and four AM-3s housed in the wing roots. Using the smaller AL-5 engines did not allow the aircraft to meet the government specifications, especially as regards field performance – the takeoff run was estimated as an absolutely unacceptable 3,500 m (11,490 ft.). The wings were swept back 35° at quarter chord, with an aspect ratio of 8.65; by comparison, it was 8.44 for the B-52 and 8.83 for the Tu-95. The ventral cannon barbette was repositioned ahead of the nosewheel well.

However, now we have to go back a little. The advanced development project (ADP) of the *izdeliye* M took four months to complete. The definitive version featuring four AM-3s was presented in August 1951; the Soviet air force's commander in charge, Col.-Gen. Pavel F. Zhigarev, endorsed the ADP documents on 19 September. The conclusive report on the ADP of the *izdeliye* M long-range fast bomber stated that while the aircraft's speed, armament, and equipment met the specifications set forth in Council of Ministers directive No. 949-469, the range and service ceiling were insufficient and the takeoff run was excessively long. The same report approved the choice of the AM-3 turbojet as the bomber's power plant and recommended that the engine be further improved.

A full-size wooden engineering mockup was built by plant No. 23 for finalising the bomber's internal layout and the positioning of specific equipment items; this inherited the provisional designation SDB of the bomber's very first project studies. The layout basically followed the one set forth in the ADP documents. Outwardly the mockup differed from the subsequent real aircraft in having 'V-Bomber'-style flush engine air intakes. Separate mockups of the forward crew cabin and the tail gunner's station were also built. On 22 October 1951 the SDB and the project materials were presented to the so-called mockup review commission (*maketnaya komissiya*), an expert panel composed of industry and MoD representatives. This stage (similar to the so-called 'gates' of today) ensured that any obvious errors were detected and eliminated before the first metal was cut, thereby avoiding waste of time and resources. On 30 November the commission gave its general approval, signing a protocol that said the mockup could serve as a basis for the prototype's manufacturing drawings.

Concurrently the OKB built a number of ground test rigs for the *izdeliye* M. The SM propulsion test rig (*stend motornyy*) represented a wing section with two engines, a section of the fuel system and lubrication system, engine controls, etc. Among other things, this rig was used to check the intakes of the inboard and outboard engines for harmful mutual influence in various engine operating modes, including start-up and shutdown. Another installation called BV (*[stend] bombovovo vo'oruzheniya* – bomb armament rig) was used to check the bomb-loading options; it represented a centre fuselage section with the bomb bay, complete with bomb racks. Another ground test article built by plant No. 23 was the *izdeliye* IM, a chunk of fuselage complete with the wing centre section and wing root portions incorporating the inlet ducts, anticipating the usual static tests of a full-size airframe. Known colloquially as the *krest* (cross), this structural test article, nearly 25 m (82 ft.), long passed its tests at TsAGI successfully, proving that the wing/fuselage joint of the *izdeliye* M would stand up to the loads. This, in turn, allowed the OKB to manufacture the aforementioned scaled-strength models and check the future bomber's wings for flutter resistance.

Special calculation methods were also devised to determine the loads that the high-aspect-ratio swept wings would be subjected to by gusts of wind in takeoff/landing and cruise modes. They showed that the dynamic loads from single gusts would be compensated by the aircraft's vertical speed imparted by the gust even as the wings flexed; generally the wings of the Myasishchev bomber had adequate – perhaps even excessive – strength reserves.

Thus, by the end of 1951, OKB-23 had virtually completed development of its first jet strategic bomber. Next thing, the OKB started detail design of the *izdeliye* M.

The aircraft was literally packed with new features – the huge swept wings of high aspect ratio, the swept tail surfaces, and, of course, the landing gear, the likes of which had not been seen on

A partial mockup of the wing root section, featuring the original flush engine air intakes with internal splitter

heavy bombers before. The designers were facing the daunting task of obtaining a high lift/drag ratio coupled with a high speed, a low specific fuel consumption (SFC), and low structural weight. The correlation and mutual effect of the future aircraft's various properties had still to be determined. For instance, for the first time it was established that the influence of wing aspect ratio on structural weight and lift/drag ratio depended on wing loading (that is, it transpired that it was possible to create a lightweight high-aspect-ratio structure only if the wing loading was fairly high). A sufficiently high fuel load / structural-weight ratio could also be achieved only with a high wing loading.

To alleviate the bending loads on high-aspect-ratio wings, the Boeing Military Aircraft Company had opted for underwing engine nacelles spread out spanwise when developing the B-47 and B-52. This arrangement gave a bonus, facilitating engine maintenance and change; however, it also created major problems associated with dynamic wing stability as the wing/nacelle assembly flexed in flight and with the aerodynamic interference between the wings and the nacelles. (Later, the underwing placement of jet engines became the most common on heavy aircraft as progress in aerodynamics and structural-design theory eliminated these problems.)

Conversely, the Old World preferred the buried engine placement. Avro, Handley Page, and Vickers chose it for their 'V-Bombers.' TsAGI also favoured this arrangement, as evidenced by a whole series of PD projects of the aforementioned 'aircraft 88' twin-jet medium bomber leading up to the definitive Tu-16. Wing aerodynamics were the factor that tipped the scales in favour of this power plant arrangement in the case of the Myasishchev bomber. However, as distinct from the 'V-Bombers' and the Tu-16, on the *izdeliye* M the Myasishchev OKB decided to house the engines completely inside the thick wing root section. By thickening this part of the wings, the designers managed to make it stiffer, which confined aeroelasticity and dynamic wing stability problems to the

A large model of the *izdeliye* M with functional flaps and control surfaces, built by OKB-23 for testing in TsAGI's T-101 wind tunnel

The nose of the full-size mockup of *izdeliye* M. The poster reads 'Labourers of the USSR! Boldly implement the achievements of science and progressive experience!'

The tail gunner's station of the mockup, with the tail turret partly uncowled to show the ammunition sleeves. Note the lateral escape hatch.

outer portions of the wings. This made it possible to design a relatively lightweight wing structure with a high resistance to flutter and turbulence. (The same design philosophy of rigid inner-wing portions and flexible outer wings was later used by the Boeing Commercial Aircraft Group for the Boeing 737 short-haul airliner.)

The thick inner wings also made it possible to achieve the optimum placement of the engines with respect to each other. As on the Tu-16, the engines were placed aft of the wing torsion box for easy maintenance access, their inlet ducts of necessity passing straight through the wing spars, whose root portions were designed as hollow frames. Importantly, the inlet ducts were designed in such a way as to enable potential installation of more-powerful engines with a greater mass flow in the future. Unlike the Tu-16, however, the engines were not 'squeezed' into the sides of the area-ruled

An overall view of the wooden mockup of *izdeliye* M; note the slightly different air intake configurations to port and starboard

Right: An early wind tunnel model of the M; oddly, it has only two engines with protruding air intakes.

Below: A display model of the M in the final configuration with individual air intakes, four-wheel nose gear bogie, and three pairs of wing fences

Bottom: A photo of a scaled-strength model of the M from a TsAGI bulletin distributed on a need-to-know basis

Bottom right: A wind tunnel model of the M with protruding air intakes

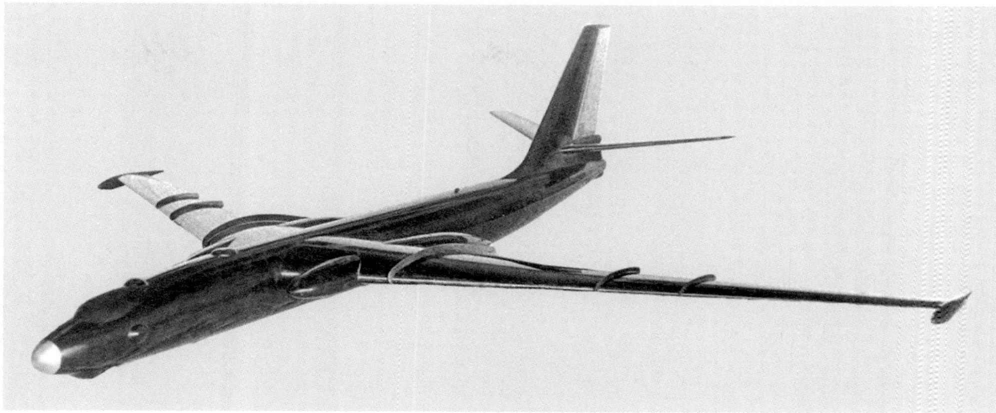

fuselage but placed well away from it. Each engine had its individual elliptical air intake protruding beyond the wing leading edge (the Vulcan/Victor/Valiant-style paired flush intakes with splitters, considered at an early stage, were ultimately rejected). The intakes were stepped to minimise pressure losses in the inlet ducts and preclude harmful interference in all engine operation modes. (By comparison, the Mikoyan/Gurevich MiG-19 *Farmer* tactical fighter, which had a nose air intake divided by a splitter into air ducts for the two engines, sometimes ran into problems during engine starting. In a crosswind the downwind engine had to be started first; otherwise the upwind engine would literally rob it of breath, making start-up impossible!) The engines were installed with a perceptible 'toe-in' angle and horizontal stagger; the jet pipes pointed 4° outward and 4° downward to avoid the exhaust impinging on the rear fuselage sides and horizontal tail, which might cause fatigue problems. Also, the nozzle shrouds of the inboard engines were notched at the top and bottom (their trailing edge was concave in plan view); this caused the jet plume to spread out vertically, reducing the effect of the hot gases on the rear fuselage skin.

The wings were of two-spar construction. The entire trailing edge was occupied by high-lift devices (hinged flaps inboard and TsAGI area-increasing flaps outboard) and ailerons incorporating geared tabs (Flettner tabs); there were no leading-edge devices. To delay airflow departure at high angles of attack, the wings featured geometrical camber, the incidence being much less at the tips; this delayed tip stall, allowing the ailerons to work normally and ensuring roll control. Also, three pairs of wing fences were fitted to reduce spanwise airflow. To alleviate the bending loads on the wings, the outer wing torsion boxes housed fuel tanks along most of the span (14 bladder tanks in each wing); a further 20 bladder tanks were accommodated in the inner wing portions and the fuselage. In addition to theoretical wing design work, OKB-23

The photos on the opposite page show the interior of the *izdeliye* M mockup.

Top left: The captain's workstation and centre instrument panel; note the throttles on the port sidewall
Top right: The copilot's workstation and folding centre pedestal

Second row, left: The interior of the tail gunner's station, showing the gunsight
Second row, right: The radio operator's workstation and port beam blister with movable gunsight

Third row, left: The starboard side of the navigator's station
Third row, right: The portside control panel and OPB-1 bombsight in the navigator's station

Bottom left: The navigator/operator's station with PR-1 ECM/ELINT equipment
Bottom right: The RSIU-4 radio at the radio operator's workstation

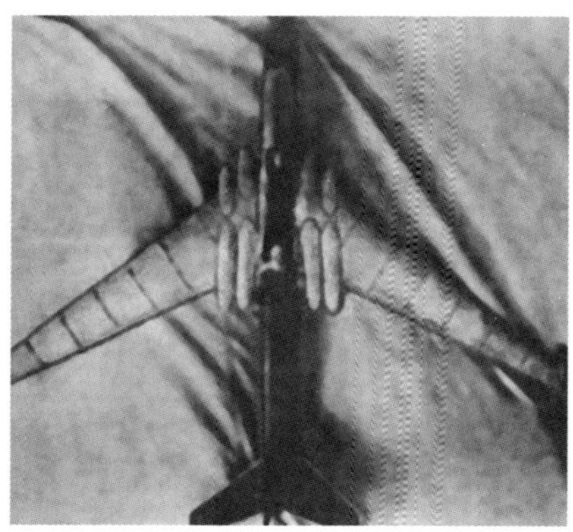

engineers undertook a lot of research on wing dynamics and flutter, using scaled-strength models. It turned out that the deep and voluminous root sections of the high-aspect-ratio wings sort of reduced the computed wingspan, increasing the wings' flexural and torsional stiffness.

It may be said that while the takeoff weight of the projected *izdeliye* M was similar to that of the B-52, its wing area was about 150 m² (1,610 sq. ft.) less, resulting in an exceptionally high wing loading – about 600 kg/m² (123 lbs./sq. ft.). By comparison, the B-52's larger span and wing area gave a wing loading of about 425 kg/m² (87 lbs./sq. ft.). This efficient wing structure, together with an optimum fuselage diameter, ensured that the Myasishchev bomber a relatively low structural weight.

The large fuselage was of basically circular cross-section, with a maximum diameter of 3.5 m (11 ft., 5$^{51}/_{64}$ in.) and a slightly

Top: Another partial mockup of the wing root section featuring the definitive version of the air intakes

Above: A large model of the bomber in TsAGI's T-101 wind tunnel. Note the three pairs of wing fences; the innermost pair would eventually be deleted and the other two pairs moved outward on the actual aircraft.

Above right: The SM power plant test rig built at plant No. 23, featuring a section of the fuselage side, the nos. 1 and 2 engines complete with inlet ducts and stepped air intakes, and a section of the inner-wing fuel tanks.

Right: A three-view drawing from OKB-23 documents showing the final project form of the *izdeliye* M

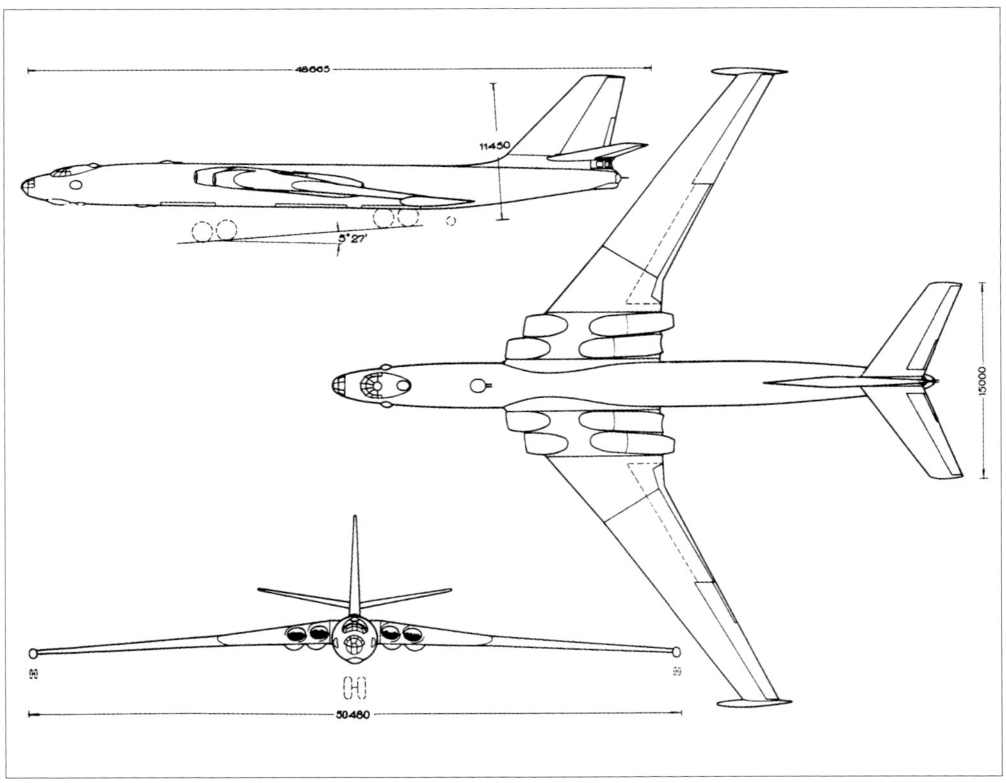

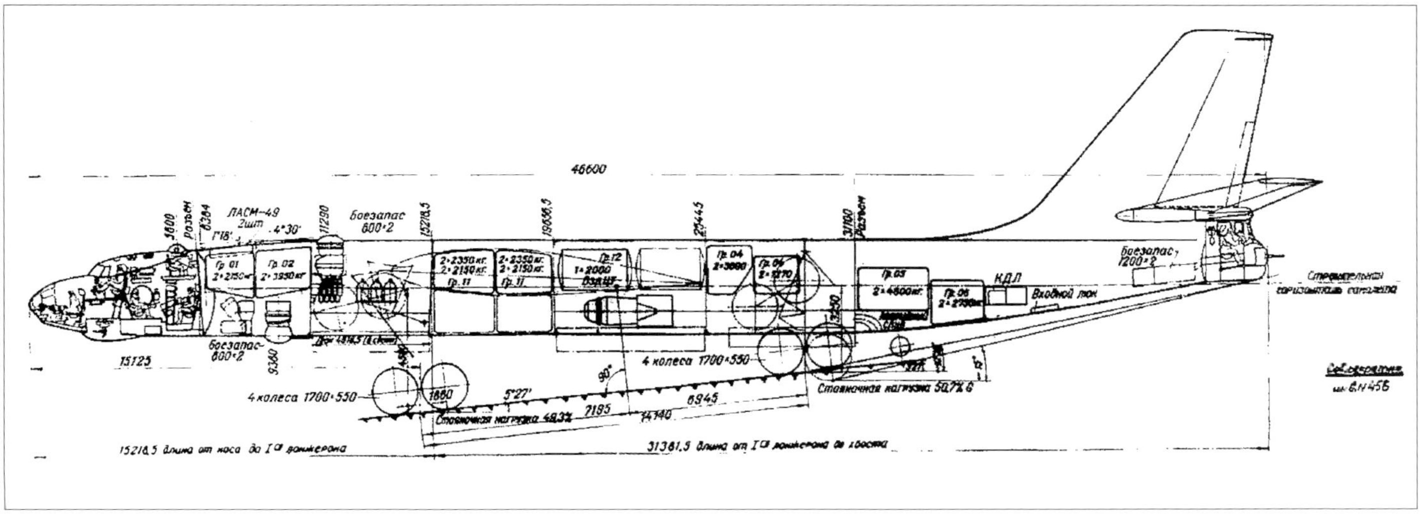

A cutaway side view of the *izdeliye* M as actually built from the ADP documents, showing the redesigned landing gear

upswept rear section. It incorporated a capacious bomb bay aft of the wing centre section torsion box, fuel tank bays, and avionics/ equipment bays. The crew of eight was accommodated in two pressurised cabins with ventral entry hatches. The front cabin commenced with an extensively glazed navigator / bomb aimer's station, with a Rubidiy-MM2 (Rubidium) ground-mapping radar in a chin radome having a flattened teardrop shape aft of it. In clear weather the bomb aimer could use an OPB-11R vector-synchronised bombsight (*opticheskiy pritsel bombardirovochnyy* – optical bombsight; R = *[svyazanyy s] radahrom* – linked to the radar). Developed by OKB-283 in Leningrad under the guidance of V. S. Dekhtyaryov, the radar could detect large targets, such as

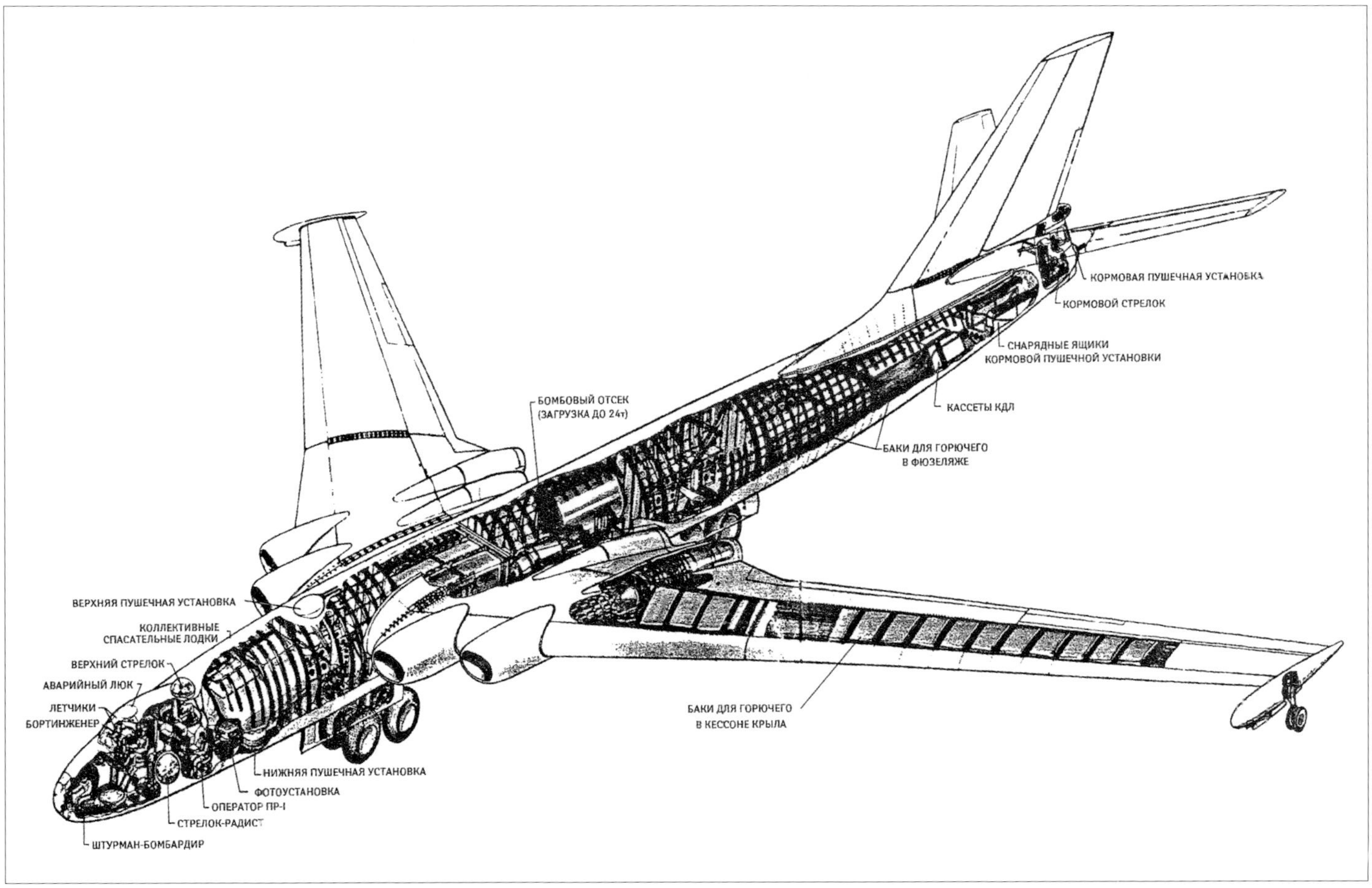

A different cutaway drawing from the ADP documents. The wing fences are not shown.

OKB-23 Leonid L. Selyakov, who played a decisive role in shaping the M bomber's general arrangement

major industrial centres, at 150–180 km (93–111 mile) range and had a maximum targeting range of some 70 km (43.5 miles); it permitted bomb delivery from altitudes of 2,000–15,000 m (6,560–49,210 ft.) at flight speeds of 300–1,250 km/h (186–776 mph). For better electronic countermeasures (ECM) resistance, the radar worked in two selectable frequencies.

The two pilots were seated side by side, with the flight engineer, radio operator (both facing aft), dorsal gunner, and navigator/operator behind them. The Nav/Op worked the PR-1 combined electronic intelligence (ELINT) / ECM equipment (*pomekhovo-razvedyvatel'naya [apparatoora]*). The tail gunner had the other pressure cabin located at the aft extremity of the fuselage all to himself. Both cabins incorporated aluminium armour to protect the crew. All crew members were provided with downward-firing ejection seats; the dorsal and tail gunner ejected through the entry hatches, while the other crew members used separate hatches, the pilots ejecting in turn though the same hatch.

Designing the bomb bay proved to be a challenge, not least because the designers had a vague idea of the future nuclear bombs' size and weight. Also, it was important to avoid a shift in the bomber's CG when the big bomb was released. The bomb bay door design was a separate bucket of worms. At first the designers wanted to incorporate inward-sliding doors that would not increase drag when open, but this proved to be too much of a technical challenge, and conventional outward-opening doors were used instead; they were bulged to accommodate the biggest bombs.

The SPV-25 defensive armament suite (*sistema pushechnovo vo'oruzheniya* – cannon armament system for project 25) consisted of six 23 mm (.90-calibre) Afanas'yev/Makarov TKB-495A fast-firing cannons in dorsal-, ventral-, and rear-powered barbettes, with an overall ammunition supply of 4,400 rounds. The cannon was developed by Nikolay M. Afanas'yev and Nikolay F. Makarov at the firearms design bureau in the city of Tula (hence TKB for *Tool'skoye konstrooktorskoye byuro* – Tula Design Bureau), and the version with a lengthened barrel (indicated by the *A* suffix) later entered production as the AM-23. The dorsal and ventral barbettes located aft of the front pressure cabin were remote-controlled, using an astrodome, hemispherical lateral sighting blisters in the front pressure cabin (the flight engineer and radio operator could use these), and a periscopic sight for the lower hemisphere. Additionally, a PRS-1 Argon gun-ranging radar (*pritsel rahdiolokatsionnyy strelkovyy* – radio gunsight) was mounted above the tail gunner's station, detecting targets within a 70° cone. The carefully chosen placement of the cannons afforded a blind spot that was 25% smaller than that of the Tu-95.

To improve the huge bomber's field performance, the designers incorporated a three-canopy brake parachute system and intended to equip the bomber with jet-assisted takeoff (JATO) boosters. The latter were then considered a viable option for heavy bombers in the days when adequately powerful afterburning turbojets or turbofans had not yet been developed.

Earlier, we mentioned VIAM. The *izdeliye* M was the first heavy bomber to make use of several new structural materials developed by this institute, including the V65 and V95 aluminium alloys (V stood for *vysokoprochnyy* – high-strength). They were stronger than the customary D16T series duralumin while having the same specific weight. On the minus side, it turned out that the V65 and V95 had poor fatigue resistance. This proved fatal for some other Soviet aircraft utilising these materials in their airframe structure – structural failure caused by fatigue cracks ended in crashes. Still, the Myasishchev OKB succeeded in designing an airframe with an acceptable fatigue life, controversial structural materials notwithstanding.

The new bomber's landing-gear design was of special interest. A major contribution to its development was made by the Myasishchev OKB's landing-gear section chief G. I. Arkhangel'skiy, structural-strength department chief Lev I. Balabookh, and two talented young engineers, V. K. Karrask and G. D. Dermichev. In their quest for the optimum landing-gear arrangement for a 200-ton aircraft, the engineers considered the relative merits of the classic tricycle layout, a quad layout patterned on the B-52 and the bicycle layout. A special theoretical research group was set up to investigate the 'dynamics of a heavy aircraft's travel on a runway.' Eventually the team selected a bicycle landing gear in which the simple twin nosewheels were replaced with a four-wheel bogie; at the same time the landing gear was redesigned so that all four units retracted forward. The front bogie had a special tilting design; this feature served the same purpose as the telescopic 'jump strut' of carrier-based aircraft, increasing the angle of attack during the takeoff run to achieve extra lift. The twin-wheel outrigger struts were placed at the wingtips, retracting into teardrop fairings. The chosen layout required wide taxiways and runways but provided maximum stability during the takeoff/landing run and enabled automatic takeoff of a heavy aircraft within the specified runway length.

The bicycle landing gear so boldly proposed by Myasishchev was not exactly terra incognita to Soviet aircraft designers by then. Aside from Robert L. Bartini's prewar fighter prototypes – the Stal'-6 ('Steel-6', a.k.a. EI, *eksperimentahl'nyy istrebitel'* – experimental fighter) of 1933 and the unfinished Stal'-8 (I-240) of 1934, both of which featured a retractable monowheel supplemented by a tail skid and outrigger skids, the first Soviet aircraft to have a bicycle undercarriage dated back to 1941, when Igor' V. Chetverikov proposed his I-1AM-37 fighter.

The next attempt was undertaken in 1949, when TsAGI wanted to convert a Tupolev UTB-2 bomber trainer (*oochebno-treniro-*

Right: The Myasishchev OKB-23 armament design team in 1952, with team chief A. A. Zavgorodniy in the middle of the front row

Below right: Another design team within OKB-23 ('team 714') in 1951, with team leader Ye. V. Voznesenskiy third from left in the front row

Bottom right: Team 321 in February 1952. This design team at the Myasishchev OKB was responsible for the pressure cabins and rendered assistance to the wing design team.

*vochnyy bombardi**rov**shchik*) into a bicycle landing-gear test bed for accumulating research data to be used in the development of other aircraft. (The UTB-2 was a dual-control derivative of the Great Patriotic War–vintage Tu-2 *Bat* medium bomber, powered by 575 hp Shvetsov ASh-21 seven-cylinder radials driving two-blade propellers instead of 1,850 hp ASh-82FN 14-cylinder radials with three- or four-blade props.) The work proceeded under the direction of Igor' P. Tolstykh. The modified UTB-2 was to have a special truss mounted under the fuselage, permitting installation of experimental landing-gear struts at various distances from the CG, thus allowing the optimum position and incline of the main gear units of future aircraft to be determined. The work was commissioned by OKB-1, a design bureau based at plant No. 256 in Doobna, near Moscow, where captive German engineers had been at work since early 1947. OKB-1 was then developing the 'aircraft 150' twin-jet medium bomber, an unsuccessful competitor to the Tu-16 featuring a bicycle undercarriage; the work was supervised by Semyon M. Alekseyev (the former aide of fighter designer Semyon A. Lavochkin), who had been appointed head of the 'captive German' OKBs in 1948 and OKB-1's chief designer for jet bombers in 1949.

Above: The ShR-1 test bed jacked up in the hangar of plant No. 23. The massive nosewheels, the four-wheel main bogie, and the crude outrigger struts with stock Tu-4 nosewheels are clearly visible.

Above right: Close-up of the ShR-1's main gear unit, showing the tandem oleos and the mounting truss allowing installation in three different positions

Left: The crew of the ShR test bed. *Left to right*: Captain Fyodor F. Opadchiy, navigator Aleksandr I. Pomazunov, and copilot Aleksandr S. Rozanov.

Below: The SUR takeoff booster test bed sits under wraps at Zhukovskiy, showing the nose gear bogie with small wheels (identifying it as the former ShR-2 gear test bed), the cutaway rear fuselage housing the JATO booster, and the photo calibration crosses.

However, the landing-gear test bed based on the UTB-2 was never built, even though Vladimir M. Myasishchev and Gheorgiy N. Nazarov (both MAI employees at the time) were asked to participate in the development of the test bed at TsAGI's request. To speed up verification of the bicycle landing gear on a real-life combat aircraft, Alekseyev offered to convert his I-215 twin-jet multirole fighter, whose takeoff weight was similar to that of the UTB-2. This obviated the need to design and build a specialised aircraft for this mission. The I-215 existed and was flyable; all the engineers needed to do was install a twin-wheel main gear strut (with larger wheels than hitherto) under the centre fuselage and small outrigger struts beneath the engine nacelles, eliminating the existing mainwheel wells in the wings.

It was decided to complete the unfinished second prototype I-215, which was languishing at MAP's aircraft factory No. 21 in Gor'kiy (now renamed back to Nizhniy Novgorod) after the liquidation of Alekseyev's OKB-21 in 1948, with a bicycle landing gear. Bearing the designation I-215D (*doob**lyor*** – lit. 'understudy' or 'backup,' the Soviet term for second prototypes used until the late 1960s), it differed considerably from the first prototype. The new twin-wheel main gear unit was placed aft of the aircraft's centre of gravity (CG) and incorporated a 'kneeling' feature to increase the angle of attack (AoA) on takeoff by 3°. The wings had an incidence of 3°30'; this, together with the 'kneeling' feature, provided an adequate AoA and sufficient lift for takeoff. Hence, unlike aircraft with a tricycle landing gear, the I-215D could become airborne without rotation, not requiring the pilot to haul back on the control stick. Thus, takeoff was virtually automatic; all the pilot had to do was keep the aircraft from side-slipping and monitor engine operation at maximum power.

OKB-1 test pilot Col. Ivan Ye. Fyodorov, Hero of the Soviet Union, flew the I-215D to the OKB-1 flight test facility in Doobna (Moscow Region); the flight went perfectly, the aircraft behaving as if it had been designed as a 'bicycle' from the outset. The tests were so successful that several other design bureaus were able to develop viable combat aircraft with a bicycle landing gear. As for the original 'customer,' OKB-1 scaled up the I-215D's landing-gear design and used it on the 'aircraft 150,' which was 10 times heavier. This both proved the feasibility of using a bicycle under-carriage on a large aircraft and verified the means of reducing the takeoff run and simplifying the takeoff procedure as compared to aircraft with a conventional tricycle landing gear.

Obtaining MAP permission, Myasishchev and Nazarov carefully studied the know-how OKB-1 had accumulated; to this end, Nazarov was seconded to plant No. 256 in Doobna, participating in the trials of 'aircraft 150' as a backup for the engineer in charge. This experience came in handy when designing the *izdeliye* M.

Careful analysis of the results obtained with the I-215D and 'aircraft 150' showed that because of runway loading restrictions, the 'kneeling' feature was unusable for an aircraft having thrice the takeoff weight of the 'aircraft 150.' Yet, there is a saying: 'If the mountain will not come to the Prophet Mohammed, then Mohammed will go to the mountain.' Using the experience gained by OKB-1, the Myasishchev OKB 'reversed' the idea: since the main gear could not 'kneel,' the nose gear would 'stand on tiptoe' to achieve the desired result. Thus the massive 12-wheel landing gear of the Myasishchev bomber created the same effect as the delicate gear of the I-215D single-seat fighter.

In addition to the advantages described above, Vladimir M. Myasishchev pointed out as early as the SDB project's PD stage that the bicycle undercarriage offered a substantial weight saving and thus was the best choice for the *izdeliye* M bomber. Such a landing gear accounted for no more than 3.0%–3.5% of the aircraft's weight versus 4.0%–6.5% for a conventional tricycle landing gear. This was because a bicycle undercarriage retracted into the fuselage, not into the high-set wings (which are the best choice for a bomber, since they maximise usable fuselage volume); thus the main gear unit could be made shorter and consequently lighter. Besides, with a bicycle landing gear the wings would be unencumbered by main gear fairings creating extra drag. (On the *izdeliye* M, main bogies stowing in the thick wing roots were ruled out because of the chosen engine placement, and any other tricycle-gear option would either drastically reduce the volume available for internal fuel [and hence, range] or require large fairings, as on the Tu-16. To be sure, the wing-mounted outrigger struts of a bicycle landing gear also required fairings, but these would be much smaller.) Most importantly, however, a bicycle landing gear saved the wing structure from shock loads during taxiing, takeoff, and especially landing.

Compared to other landing-gear arrangements, the bicycle landing gear was better suited for operation from unpaved airstrips, which might be a necessity even for heavy bombers in wartime conditions. Finally, the bicycle landing gear made it possible to place the bomb bay close to the CG, distributing the fuel tanks, equipment, etc. almost 50/50 within and outside the wheelbase.

In order to test a bicycle landing gear that was similar in size and weight to that of the real thing, OKB-23 developed a special version of the Tu-4 bomber. In late 1951 a brand-new *Bull* manufactured by plant No. 23 (no serial number; construction number 230322 – that is, plant No. 23, third aircraft out of five in production batch 22) was delivered to the OKB for conversion. Completed in January 1952, this test bed bore the in-house designation *izdeliye* ShR, the *Sh* denoting *shassee* (landing gear) and the *R* referring to the Tu-4's product code '*izdeliye* R.' The standard nose gear unit with 950 × 350 mm (37.40 × 13.77 in.) wheels gave way to a new levered suspension strut with twin 1,450 × 520 mm (57 × 20.4 in.) wheels – that is, stock Tu-4 mainwheels. The scratch-built main gear unit was a pair of Tu-4 main gear oleos mounted in tandem and rigidly connected; to these was hinged a bogie with four 1,450 × 520 mm wheels. This assembly was attached to a hefty tubular truss installed near the aft bomb bay and could be fitted in three different ways to absorb 72%, 85%, or 90% of the total weight. This unusual feature was meant to show how the changing weight distribution and wheelbase affected field performance and ground manoeuvrability (this was because different PD project versions of the jet bomber had an aft-retracting or forward-retracting nose gear unit, with differing wheelbase and weight distribution). The nosewheels and mainwheels were equipped with hydraulic brakes.

The outrigger struts installed a short way outboard of the nos. 1 and 4 engines consisted of stock Tu-4 nose gear units mated with special truss-type mountings, which allowed the length of the outriggers to be adjusted. In this configuration (with the twin-wheel nose gear unit), the test bed was also referred to as the ShR-1. Interestingly, the nonretractable experimental landing gear could be removed and the normal gear reinstalled for positioning flights.

The conversion job was completed in January 1952; the test programme was held by LII at the institute's airfield near the town of Zhukovskiy (Moscow Region). Stage 1 lasted from April to June; the ShR-1 made 50 high-speed runs and 34 flights to check stability and handling, the operation of the nose gear steering mechanism, and the optimum position of the main and outrigger struts. Fyodor F. Opadchiy was the project test pilot.

Meanwhile, the original twin-wheel nose unit had been rejected in favour of a four-wheel bogie with electrically steerable front wheels. Hence the test bed was modified accordingly; the front bogie had four 950 × 350 mm stock Tu-4 nosewheels. In this guise the aircraft was redesignated ShR-2 and underwent further tests in 1953, making 24 high-speed runs and 17 flights; the nose gear bogie absorbed 20% of the weight versus 40% on the real thing. The tests fully confirmed the anticipated behaviour and allowed the takeoff and landing technique to be mastered well in advance of the *izdeliye* M's maiden flight. The aircraft required no rotation to become airborne, taking off almost of its own will. LII test pilots who flew the test bed rated the bicycle landing gear very highly because it offered excellent ground handling.

Even before the ShR test bed was built, the Myasishchev OKB manufactured an electrically controlled scale model of the four-wheel nose gear unit that served to test the bogie tilting system. This was followed by a special simulator in which pilots perfected the jet bomber's automatic takeoff procedure; as the aircraft accelerated down the runway, the growing wing lift offloaded the nose gear bogie, and the pressure in the nose gear oleo pushed the nose up, causing the bogie's front wheels to leave the ground and tilting the aircraft to the correct AoA. The simulator helped many pilots overcome the psychological barrier associated with flying bicycle-gear aircraft – it was hard not to haul back on the control yoke in order to initiate rotation on takeoff, which is standard operational procedure on aircraft with a conventional tricycle undercarriage.

The nose gear deserves special attention for yet another reason – its unusual steering system. For all practical purposes the *izdeliye* M bomber behaved like a tractor-trailer combination (a 'big rig') while taxiing – the forward pair of wheels on the nose gear bogie was steerable, the bogie itself turning on the axis of the shock absorber strut; this drastically reduced the power required for steering. Thus when the wheels were turned, the bogie behaved like an automobile, the rest of the aircraft tagging along like a semi-trailer. As the bogie tilted to induce rotation on takeoff, the steering mechanism was disabled automatically. The result was a simple but effective system that gave the bomber high manoeuvrability when taxiing and directional stability on takeoff/landing.

Several other test beds based on the Tu-4 were involved in the development of *izdeliye* M. For the first time in Soviet aircraft design practice, the aircraft incorporated fully powered controls with reversible and irreversible hydraulic actuators in the aileron, rudder, and elevator control circuits, plus an artificial-feel mechanism. Testing these features on a ground rig ('iron bird,' to use Boeing terminology) was not enough, and the actuators had to be verified in flight. Hence the Myasishchev OKB converted another Tu-4 (identity unknown) into the *izdeliye* UR control system test bed; *U* stood for *[sistema] oopravleniya* (control system) and the *R* was again a reference to the Tu-4. The first version designated UR-1 featured reversible actuators everywhere; tests in this configuration were completed in March 1952, involving 12 flights totalling about 20 hours. The tests showed that the powered controls worked acceptably, even though a few failures did occur. Later the aircraft was refitted with irreversible actuators as the UR-2 and tested in April 1952. Further tests with a pneumatically operated automatic artificial-feel system were conducted on the UR-2 in September 1952.

Project performance of the Myasishchev jet bomber

	SDB	1M
Power plant	4 × AM-TKRD-03	4 × AM-03
Maximum thrust, kgp (lbst)	4 × 8,250 (18,190)	4 × 8,500 (18,740)
Maximum speed, km/h (mph):		
at sea level	700 (435)	700 (435)
at high altitude	900 (559)	930 (577)
Service ceiling, m (ft.)	11,000 (36,090)	11,000 (36,090)
Range, km (miles)	12,000 (7,745)	12,000 (7,745)
Takeoff run, m (ft.)	2,000 (6,560)	2,200 (7,220)
Landing run, m (ft.)	1,200 (3,940)	1,300 (4,265)
Wing area, m^2 (sq. ft.)	300 (3,229)	309.8 (3,334)
Takeoff weight, kg (lbs.):		
normal	110,000 (242,510)	118,000 (260,145)
maximum	140,000 (308,650)	155,000 (341,720)
Wing loading, kg/m^2 (lbs./sq. ft.):		
normal	366.7 (75.18)	380.9 (78.0)
maximum	473.9 (97.16)	503.9 (103.3)
Thrust/weight ratio	0.236	0.22
Crew	7	7

Later, the UR-2 was further modified for testing the crew escape system of *izdeliye* M and redesignated *izdeliye* KR, the *K* denoting *katapool'**teeruyemoye** **kreslo*** (ejection seat). This was necessitated by the bomber's unusual escape system with sliding ejection seats designed in-house; as mentioned earlier, the captain and copilot were to eject consecutively through a common hatch, the navigator / bomb aimer, flight engineer, radio operator, and navigator/operator using their own hatches.

The KR test bed featured a reduced set of ejection seats for the pilots, navigator, flight engineer, and dorsal gunner. Between October 1952 and January 1953, it performed 12 test flights involving 20 ejections (13 with dummies and seven with live parachutists). As a result, some changes were introduced into the seat's design.

Finally, the Myasishchev OKB, together with LII, converted another Moscow-built Tu-4 (no serial, c/n 230113) into one of the institute's seven Tu-4LL engine test beds. *LL* stood for *letayushchaya laboratoriya* – lit. 'flying laboratory'; this Russian term is used indiscriminately to denote any kind of flying test bed or research/survey aircraft.

The first configuration of this particular aircraft, known at the Myasishchev OKB as *izdeliye* DR-1 (*D* stood for ***dvigatel'*** – engine), served to test the Lyul'ka AL-5 turbojet, which was then considered as a possible power plant for the new bomber. In January 1952, when the much more powerful Mikulin AM-3 was selected, this replaced the AL-5 development engine on Tu-4LL c/n 230113, and the in-house designation at OKB-23 was changed to DR-2.

The turbojet was installed in the forward bomb bay in a special nacelle that could move up and down on a system of levers. It was semirecessed in the fuselage for takeoff and landing to give adequate ground clearance, extending into the slipstream before start-up; it was also possible to extend the engine for ground runs when the aircraft was parked over a special trench with concrete-lined walls and a jet blast deflector. When the nacelle was retracted, the air intake was closed by a movable shutter to prevent windmilling and foreign-object damage. In an emergency (for instance, if the hydraulic retraction mechanism failed), the development engine could be jettisoned to permit a safe landing, special pyrotechnical guillotines cutting the fuel and electric lines. The test engineer and his assistant sat in the No. 2 pressurised cabin just aft of the wings; the cabin floor incorporated a window for inspecting the nozzle of the development engine. In 1952 the AM-3 completed its state acceptance (= certification) trials, showing a maximum thrust of 8,700 kgp (19,180 lbst), and entered production at aero-engine plant No. 300 in Moscow.

Chapter 2

Prototypes and Testing

Major changes in the aircraft's dimensions took place before the design was frozen. This was a result of the AM-3A engine's actual specific fuel consumption (SFC) of 1.0 kg/kgp·hr. (lbs./lbst·hr.) at full power, which was higher than anticipated, and the available manufacturing technology, which caused the bomber to grow steadily heavier. The wing area was increased to 326.35 m² (3,512.8 sq. ft.); to achieve the required flexural and torsional stiffness, the designers were forced to reduce the wing taper and reinforce the entire wing structure. Similar changes were made to the tail surfaces, which had already been altered in comparison with the original project as a result of tests in MAI's T-1 wind tunnel: the vertical tail was moved slightly forward to where the fuselage was thicker and stiffer, and its area was increased 7% by extending the fin leading edge forward, while the horizontal tail was moved aft. One thing leads to another: to achieve the required range with the redesigned wings and tail unit, the wing tanks were enlarged and the fuselage was stretched by more than 4 m (13 ft., 1³¹⁄₆₄ in.) to make room for extra fuel. The landing-gear wheelbase was increased to obtain equal weight distribution. To ensure directional stability during the takeoff and landing run, the outrigger struts were placed as far as possible from the main gear bogie and aligned with the aircraft's CG; their fairings acted as wing end plates, cutting induced drag. The innermost pair of wing fences just outboard of the engines was deleted, and the remaining two were moved outward.

The SOR for the *izdeliye* M bomber was endorsed by the mockup review commission on 30 November 1951. That same

With landing-gear retraction almost complete, the first prototype M-4 ('aircraft M') is shown here wearing the tactical code '25 Red.' The code is symbolic, referring to the programme's designation 'project 25' (VM-25).

month, when preparation of the manufacturing drawings was still in progress, the Myasishchev OKB's experimental shop cut the first metal for two prototypes and a static test airframe. The complete set of manufacturing drawings was delivered by 1 April 1952, by which time the work was well advanced; final assembly of the first prototype commenced on 15 May that year.

At factory No. 23, the aircraft received the product code *izdeliye* 25, whereas some Myasishchev OKB documents referred to the bomber as the VM-25 (that is, Vladimir Myasishchev, Project 25). The air force, however, allocated the service designation M-4, under which the bomber would be best known. Incidentally, ***tema*** (subject) 25 was also the general designation of the strategic jet bomber programme, which was designated as a top-priority programme for the other MAP and defence industry enterprises involved.

New production technologies were introduced as the M-4 programme gathered pace. For instance, to cut structural weight by reducing the number of skin panel joints, the engineers decided to use duralumin sheets 2 m (6 ft., $6^{47}/_{64}$ in.) wide instead of the stock 1.5 m (4 ft., 11 in.) items offered by the Soviet metal industry; hence, rolling mills had to be reconstructed and new thermal treatment kilns built. There were dozens – perhaps even hundreds! – of such cases when Myasishchev's new bomber programme brought about a technology upgrade. Nevertheless, by the time the first prototype had been completed and passed ground checks, its empty weight had risen from the intended 70 tons (154,320 lbs.) to 79.7 tons (175,705 lbs.), despite the fact that the armament and some of the equipment was missing (in particular, the nose gear bogie-tilting mechanism and the brake parachutes were not installed). The maximum takeoff weight rose from the 155 tons (341,710 lbs.) envisaged by the 1M project to more than 180 tons (396,825 lbs.).

Test crew training and preparation for the M-4's first flight proceeded in parallel with prototype construction. The first prototype (c/n 0001 – that is, batch 00, first aircraft in the batch) was rolled out in November 1952 and cleared for manufacturer's flight tests. After the rigging procedure and initial ground checks, the aircraft was dismantled and loaded on a barge, which took it down the Moskva River to Zhukovskiy, where the OKB had its flight test facility on the south side of LII's airfield. The factory's runway at Fili was rather short, and flying the bomber to Zhukovskiy from there would have been an excessive risk. The river transport operation was no simple matter, since the bulky load had to pass under numerous bridges in central Moscow. At Zhukovskiy the bomber was reassembled and readied for its maiden flight. Later, the static test airframe (c/n 0002) was also transported to Zhukovskiy, where TsAGI was located.

OKB-23 managed to meet the tough two-year development schedule. Once the obligatory taxi tests had been completed, on 20 January 1953 – barely a year and 10 months after the design bureau's establishment – the as-yet-unserialled M-4 prototype took off on its maiden flight in spite of foul weather. The cloud base was at 180–200 m (590–660 ft.), with a gusty wind and driving snow; still, Vladimir M. Myasishchev chose not to postpone the flight because the generals had arrived in force, and cancelling

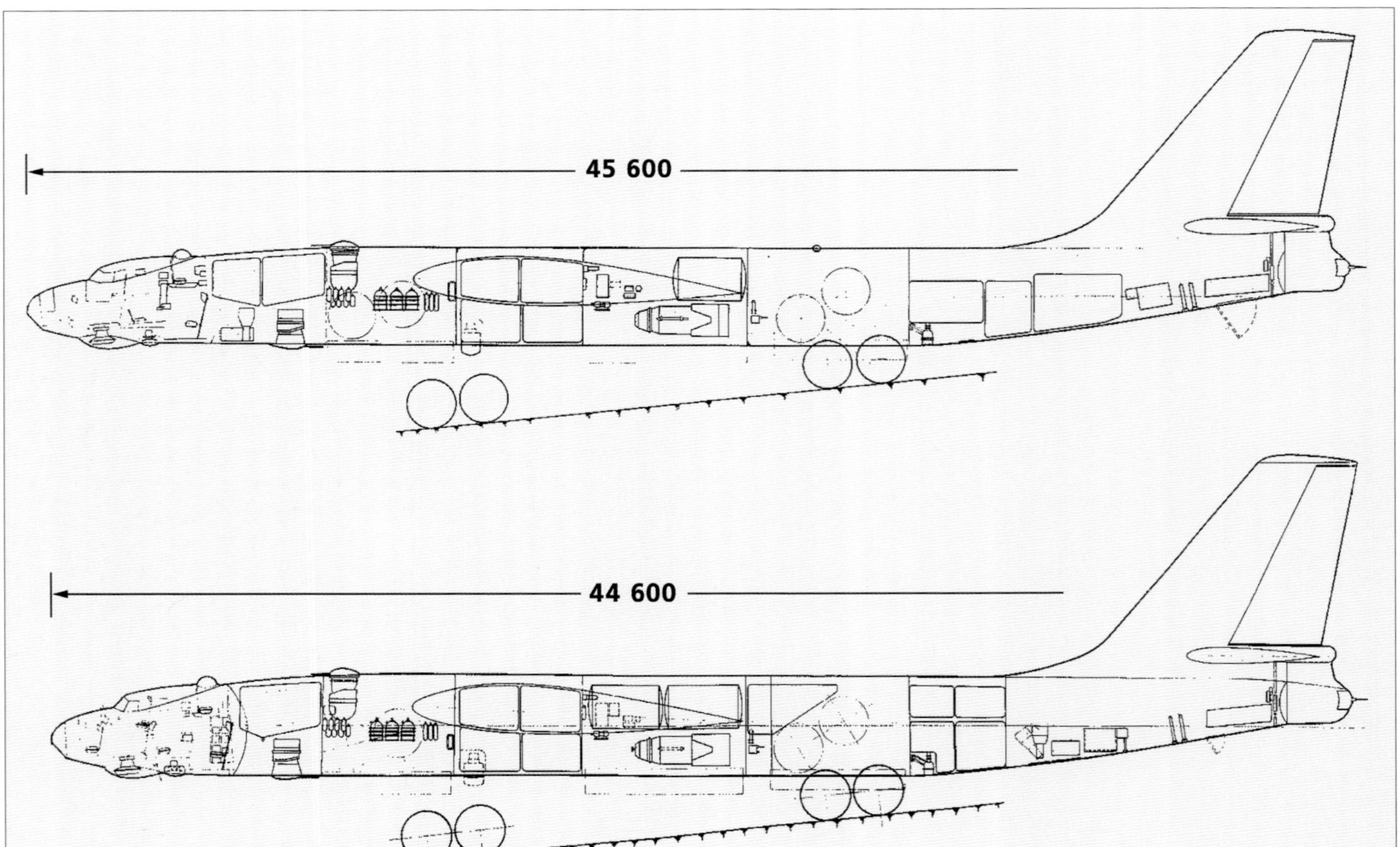

The M-4's forward fuselage (section F-2) in the process of assembly. Note the ejection hatches (three of the five are visible) and the cutout for the starboard beam blister.

the flight in front of them would mean losing face. The aircraft was flown by Myasishchev OKB test pilot Fyodor F. Opadchiy as captain and LII test pilot Aleksey N. Gratsianskiy as copilot; the crew also included navigator A. I. Pomazunov, flight engineer G. A. Nefyodov, radio operator I. I. Rykhlov, and test engineers I. N. Kvitko and A. I. Nikonov. The first flight lasted only 10 minutes, but it was nevertheless a major breakthrough for the entire Soviet aircraft industry and defence industry. A large crowd gathered to attend the first flight – including some who had hoped to see Myasishchev's new bomber crash.

In fact, the first flight could indeed have ended in a crash – the aircraft overreacted to a normal rudder input, jerking and rolling sharply; this abnormal behaviour was the reason why Opadchiy cut the flight short. At the time, the prototype was not yet fitted with test equipment sensors and recorders; these were installed before the second flight, and the phenomenon manifested itself again in the second flight at 3,000 m (9,840 ft.). The designers suspected that rudder overcompensation might be the cause of the trouble. To check this possibility, Opadchiy decided to switch off the hydraulic control actuators briefly in one of the flights, going to manual-control mode. No sooner had he done this when the guess was confirmed – the aircraft jerked sharply to one side; an immediate push of the actuator control button restored normal flight with a slight sideslip. It turned out that the culprit was a technician who had made an error in adjusting the controls. Once the cause of the problem was known, it did not take long to fix it.

Left: These cutaway drawings from project documents give a comparison of the first prototype ('aircraft M,' above) and the second prototype ('aircraft DM'). The latter's forward fuselage is 1 m (3 ft., 3 3/8 in.) shorter, and the fuel tanks and equipment in the fuselage have been rearranged.

Right: A riveter working on the rear fuselage of an M-4 (section F-5) sitting on a handling dolly at plant No. 23. The root section of the fin, built integrally with the fuselage, is visible on the left.

There were other minor spills too. Problems arose with the hydraulics (including the control surface actuators) and the structural strength of some airframe components. Despite the fact that the M-4's landing gear had passed exhaustive tests on the ShR test bed, the nose gear unit proved prone to shimmy oscillations. Moreover, on 25 March 1953 all four tyres on the main gear bogie blew and came off the rims as the aircraft touched down after its eighth flight; the cause was traced to a mismatch between the characteristics of the brake system and the antiskid unit. However, there were pleasant surprises as well; it was established that the M-4 had a top speed of 962 km/h (597 mph) and a technical range of 10,700 km (6,645 miles).

The second prototype (c/n 4300003 – that is, year of manufacture 1954, plant No. 23 [the first digit was omitted to confuse hypothetical spies], batch 00, third aircraft) was completed in December 1953, joining the first M-4, whose manufacturer's flight tests continued until 15 April 1954. This aircraft was briefly known in-house as the 2M but was soon redesignated DM (that is, *doob**lyor** samo**lyo**ta M* – second prototype of 'aircraft M').

The 'aircraft DM' differed from the first prototype in having a cranked wing trailing edge, the sweep being reduced on the inboard portions to increase wing area (less flaps) by 5.75 m² (61.8 sq. ft.) and flap area by 6.9 m² (74.2 sq. ft.) or 35%; the maximum flap setting was increased to 35°. The forward fuselage was shortened by 1 m (3 ft., 3⅜ in.), reducing overall length from 45.6 to 44.6 m (from 149 ft., 7⁹⁄₃₂ in., to 146 ft., 3²⁹⁄₃₂ in.). The fuselage fuel cells and equipment were rearranged. The second prototype had two fuselage fuel cells ahead of the wings instead of four, two optional extra tanks in the upper part of the bomb bay instead of one, an extra fuel cell in the mainwheel well, and eight small fuel cells in the rear fuselage instead of six larger ones, while the strike cameras were relocated from their position just aft of the flight deck to the rear fuselage. Additionally, 'aircraft DM' had a revised nose gear unit featuring the bogie-tilting mechanism from the outset; on the first prototype this feature was added after the 28th flight. The tilting mechanism increased the angle of attack 3° (from 7°30' to 10°30') immediately before liftoff. The redesigned nose gear unit functioned well and was recommended for production. The second prototype was considerably lighter than the first one. Knowing that the first aircraft was terribly overweight, the OKB urged the design staff to seek ways of making it lighter; a 50-ruble reward was promised for each kilogram of structural weight saved.

'Aircraft DM' was effectively the pattern aircraft for series production, featuring a full avionics and equipment fit. This included an RBP-4 panoramic ground-mapping radar (***rah**diolokatsi**on**nyy bombardi**ro**vochnyy prit**sel*** – radar bombsight, an advanced derivative of the Rubidiy-MM2) in a chin radome and a PRS-1 Argon gun-laying radar above the tail gunner's station; the targeting equipment further included PS-48 gunsights (*pritsel'naya **stahnt**siya* – sighting station, 1948 model) and an OPB-11S optical bombsight. Apart from the radar, the flight/navigation suite featured an ARK-5 Amur automatic direction finder (*avtoma**ti**cheskiy **rah**dio**kom**pas* – ADF), an MRP-48 ***Dya**tel* (Woodpecker) marker beacon receiver (***mar**kernyy **rah**diopri**yom**nik*), RV-2 *Kri**stall*** (Crystal) high-range and RV-17 low-range radio altimeters (***rah**diovy**so**tomer*), etc.; the aircraft was fitted with an AP-5 autopilot. The communications suite featured a 1-RSB-70M HF communications radio (***rah**dio**stahn**tsiya bombardi**rov**shchika* – bomber-type radio), an RSIU-3M UHF command radio (***rah**dio**stahn**tsiya istre**bi**telya **ool'**trakorotkovol**no**vaya* – fighter-type VHF radio) comprising an R-800A transmitter and an R-800B receiver, and an SPU-10 intercom (*samo**lyot**noye peregovornoye oostroystvo*). The bomber was equipped with an SRO-1 ***Ba**riy* (Barium) identification friend-or-foe transponder (*samo**lyot**nyy **rah**diolokatsi**on**nyy ot**vet**chik* – lit. 'aircraft-mounted radar responder').

The SPV-25 defensive armament suite consisted of six 23 mm TKB-495 (AM-23) cannons in three powered installations – a DB-33A dorsal barbette, a DB-34A ventral barbette, and a DB-35AM tail barbette (DB = *distantsi**on**no oopravl**ya**yemaya **bahsh**nya* – remote-controlled turret). The ammunition supply was 500 rounds per gun for the dorsal and ventral barbettes and 800 rpg for the tail barbette. The offensive armament comprised free-fall bombs of up to 9,000 kg (19,840 lb.) calibre; the maximum design ordnance load was 24 tons (52,910 lbs.). In theory the M-4's bomb bay could accommodate up to four 6,000 kg (13,230 lb.) bombs; in practice, however, it was impossible to cram more than two 6,000 kg bombs or one 9,000 kg bomb into the aircraft. Strike cameras were fitted for bomb damage assessment and recording the gunnery results.

The manufacturer's flight tests of 'aircraft DM' began in January 1954, involving 32 flights and lasting a full year. Again, Fyodor F. Opadchiy was captain on the first flight, with Myasishchev OKB test pilot Boris K. Galitskiy as copilot; later, Galitskiy would be heavily involved both in testing new versions of the M-4 family and in check-flying production M-4s. The original PS-48 gunsights borrowed from the Tu-4 proved unsatisfactory for a jet bomber and were eventually substituted with new (1953-model) PS-53 gunsights featuring VB-53M ballistic computers (*vychis**li**tel' ballis**ti**cheskiy*) as fitted to the production Tu-16.

Airmen are rather superstitious, and sometimes with good reason. Believe it or not, the first major trouble with the M-4 occurred in the 13th test flight. The mission was nothing out of the ordinary – the aircraft was to make level passes at different altitudes as the maximum speed was being determined. Now, the M-4's elevators were manufactured in two sections. As the bomber accelerated after reaching one of the assigned altitudes, test engineer I. N. Kvitko, sitting in the tail gunner's station, reported that the inboard and outboard sections of each elevator were deflecting differentially in a scissorlike fashion, which was abnormal. Fyodor F. Opadchiy throttled back and aborted the mission, landing safely at Zhukovskiy. There the engineers did not rack their brains too long about the cause of the incident, choosing to simply fix the elevator sections together by riveting metal plates over the joint.

That accomplished, the bomber took to the air again. Suddenly, as the aircraft accelerated at 3,000 m (9,840 ft.), a loud bang resounded through the airframe and the bomber started climbing of its own accord, shuddering all the while. At 10,000 m (32,810 ft.), Kvitko reported that one elevator had broken away completely, damaging the stabiliser and the rudder as it departed. Throttling back, Opadchiy ordered the crew to eject, but no one did.

As gently as they could, the pilots turned the crippled bomber homeward. After descending to 2,000 m (6,560 ft.), Opadchiy tried a simulated flare-out and discovered that even with one elevator gone, there was still sufficient elevator authority to control the aircraft. But the nightmare was not over yet; when the runway in Zhukovskiy was already in sight, the huge bomber suddenly started acting up again – the nose dipped abruptly. It was the captain's instant reaction that saved the aircraft and the test crew; not caring about the landing speed, Opadchiy thrust the throttles forward and the nose lifted immediately. Just as the captain throttled back the engines again, the M-4 touched down, and the 5,000 m (16,400 ft.) runway was long enough for the bomber to stop in time. Whew!

It turned out that the accident had been caused by the designers being overenthusiastic with weight-saving measures. As already mentioned, in its original form the first prototype had exceeded its design maximum takeoff weight by more than 25 tons (55,115 lbs.). The airframers had then set to work, trimming away excess weight wherever they could; as a result, the elevator skin ended up being too thin, failing in the 13th flight.

The systems and equipment designers were no less enthusiastic about cutting weight – again, sometimes to the detriment of safety. For instance, the avionics engineers justifiably complained about the antiquated electronic components, the electrical engineers equally rightly clamoured about the excessively heavy and inefficient generators and servo drives, etc.; however, while little could be done to lighten the equipment proper, the numerous mounting racks and brackets seemed to offer unlimited scope for saving weight. This approach was bound to end in trouble sooner or later – and it did. Several years later, disaster struck when a production M-4 was undergoing tests. A fuel line failed because some of the retaining clamps had been deleted to save weight, causing the pipe to flex and break. As fuel gushed into the innards of the airframe, a spark from a defective wire was enough to turn the entire aircraft into a blazing torch. None of the crew managed to eject . . .

In the course of manufacturer's flight tests, the first prototype made 46 flights (28 with the original nose gear unit and 18 with the bogie-tilting mechanism added), logging a total of 147 hours, 15 minutes. The tests, which lasted until 15 April 1954, were taking rather longer than expected, which meant the M-4 could not be submitted for state acceptance trials by the deadline stipulated by the government. Note that this deadline was postponed twice – first to December 1953 in accordance with Council of Ministers directive No. 2462-1019, dated 19 September 1953, and then to February–March 1954, in accordance with CoM directive No. 2970-1278, dated 18 December 1953. In November 1953 the new minister of aircraft industry, Pyotr V. Dement'yev, wrote thus to Council of Ministers chairman Gheorgiy M. Malenkov: *'The CoM directive tasked Myasishchev with building a long-range fast bomber powered by four 8,700 kgp [19,180 lbst] AM-3 engines and possessing the following performance: effective range, 9,500 km [5,900 miles]* [*sic!* – *Auth.*]*; top speed, 925 km/h [575 mph]; service ceiling over the target, 13,000 m [42,650 ft.]. The aircraft was to be submitted for state acceptance trials in May 1953. The flight test results necessitated a redesign of certain subassemblies of "aircraft M" in order to improve their structural strength and reliability. The redesign work caused the flight tests to be suspended between May and September 1953. . . . Due to the fact that plant No. 23 switched to Il-28 production in January 1953, the manufacture of Myasishchev bombers there was suspended.* [Actually, the plans to build the Ilyushin Il-28 *Beagle* tactical bomber at plant No. 23 were eventually cancelled; still, a lot of time was wasted because of this – *Auth.*] *On 19 September 1953 the plant was again assigned to producing the Myasishchev bomber, and the deadline for the state acceptance trials of 'aircraft M' was set at December 1953.'*

Yet, this revised schedule was not kept either. It was not until 30 March 1954 – 15 days before the completion of the manufacturer's flight tests – that the first prototype M-4 was formally submitted for joint state acceptance trials to the Red Banner State Research Institute of the Air Force named after Valeriy P. Chkalov (GK NII VVS – *Gosu**dar**stvennyy Krasnozna**myonn**yy nau**chn**o-is**sled**ovatel'skiy institu**t** **V**oyenno-vozdoosh**n**ykh sil*). (Note: Some sources erroneously quote the date as 30 April 1954 – that is, 15 days *after* the completion of the manufacturer's flight tests. 'Joint' means that the trials were conducted jointly by the manufacturer [the OKB] and the customer [the air force]. The 'Red Banner' bit means that the institute was awarded the Order of the Red Banner of Combat.) A state commission had been appointed well in advance (by Council of Ministers directive No. 2957-1271, issued on 17 December 1953) to conduct the trials. It was chaired by the chief of general staff, Air Marshal Sergey I. Rudenko, with Air Marshal Aleksandr A. Novikov – the recently appointed commander of the VVS's strategic bomber arm – as vice chairman.

The trials officially commenced on 15 April (concurrently with the ending of the manufacturer's tests). In reality, however, they did not begin until 4 May; this was because three days earlier the first prototype M-4 took part in the traditional May Day parade in Moscow. Escorted by four Mikoyan/Gurevich MiG-17 *sans suffixe* fighters (NATO reporting name *Fresco-A*) and followed closely by the first nine production Tu-16 *Badger-A*s, the bomber flew over Red Square to the tremendous jubilation of the watching public – and the amazement of the attending foreign diplomats, military attachés, and journalists. This public demonstration was of such importance to the Soviet political and military leaders that it was decided to postpone the trials for more than two weeks, giving the crew time to practice for the spectacular flypast over Red Square.

As a rule, new Soviet military hardware was kept a closely guarded secret for a long time. However, there is always an exception to the rule, and the M-4 was one. The purpose of this early debut was plain: being aware of the jet strategic bombers developed in the USA, the Soviet government wanted to show the Western world that the Soviet Union, too, now had such an aircraft capable of delivering a retaliatory strike against any aggressor. This goal was undoubtedly achieved: photos of the M-4 were circulated in the world press, and the Americans were justifiably alarmed. Allen Welsh Dulles, the then director of the US Central Intelligence Agency (CIA), wrote: *'There is evidence in 1954 that the Soviet Union is producing intercontinental heavy bombers comparable to our B-52s,'* commenting that *'this has led some in our nation to suppose that we are lagging behind in bomber development.'* This statement was somewhat exaggerated, since the Americans did not know that '25 Red' was one of only two M-4s flying at the time. Nor did they know about the M-4's range problems (more about this will be discussed later).

Front and three-quarters front views of the as-yet-unserialled first prototype M-4 during state acceptance trials. Note the short air data probe at the tip of the nose, which was part of the test instrumentation.

Rear and side views of the same aircraft, showing the strong tailplane dihedral, the pre-1955 placement of the national insign a (with stars on the rear fuselage), and the inordinately long forward fuselage with 'towel rail' aerials on both sides

Left: The first prototype makes the M-4's public debut at the 1954 May Day parade, with two of the four escorting MiG-17s visible.

Right: Close-up of the first prototype's forward fuselage. Note the aerial of the instrument landing system under the extreme nose and the gently elliptical shape of the beam blisters.

Far right: The first prototype's rear fuselage and tail gunner's station, with the faired spent-case / belt link collector below the tail turret

After its public debut at the 1954 May Day parade, the M-4 was allocated the NATO reporting name *Bison*. Heated discussions about the Soviets' new wonder weapon broke out in the Western press. Western experts rightly concluded that the aircraft was a high-altitude, long-range bomber. They theorised that the high-aspect-ratio wings had been selected to obtain good high-altitude performance but were a liability from a structural standpoint, having inadequate torsional stiffness, and that the wingtip fairings were balance booms meant to alleviate bending loads or torsional oscillations. The wingspan was estimated as 48–52 m (157 ft., 5$^{49}/_{64}$ in., to 170 ft., 7¼ in.), which was not too far from the truth. However, it was presumed that the four turbojet engines were rated at about 6,800 kgp (14,990 lbst) each, and that the aircraft had a maximum design speed of about Mach 0.95 and a takeoff weight of about 113.5 tons (250,220 lbs.). The weight and thrust estimates turned out to be wildly inaccurate. Little did the experts know that the Russians could create a turbojet delivering 8,750 kgp (19,290 lbst); nothing in the same league existed then on their side of the Iron Curtain.

Not only that, but for a while the West hadn't the foggiest idea who was responsible for the bomber, and one source misidentified the aircraft as the Ilyushin Il-38! (This designation was later found to belong to a shore-based antisubmarine warfare aircraft based on the Il-18 *Coot* four-turboprop airliner and code-named *May*.) In fact, a description of an official recognition model of the M-4 built by the US Navy's Office of Naval Research / Special Devices

Here the same aircraft is seen after 1954 with the tactical code '25 Red' added.

Center published in the 8 August 1955 issue of *Aviation Week* said that the tail surfaces *'have the geometry characteristic of Ilyushin's latest designs.'* It also stated that *'the Red bomber'* had *'external fuselage dorsal and ventral turrets, sighting blisters, radomes and antenna, indicating a lower state of the avionic art in Russia than here.'*

However, let's return to the *Bison*'s home pasture, so to speak. To hasten the completion of the state acceptance trials, these were held jointly by OKB-23, LII, and GK NII VVS. The greater part of the programme was performed by GK NII VVS project test pilot Col. Aleksandr V. Sarygin (captain) and Maj. V. D. Khromov (copilot). Other military pilots who flew the M-4 at this stage included Engineer-Colonel Kostryuk, who headed the institute's Department 1 (concerned with tests of heavy aircraft), Col. Ganechkin, Maj. Semyon M. Antonov, et al. Col. Selivanov was the M-4's project navigator during the state acceptance trials, while Col. Sviridenko was engineer in charge. Concurrently, plant No. 23 was tooling up for full-scale production of the Myasishchev bomber, which would supersede the Tu-4 on the assembly line.

Stage A of the trials involving the first prototype continued until 5 August 1954 (some sources say 30 July or even 30 August!) with 34 days' downtime, involving 29 test flights concerned mostly with performance and stability/handling tests. Together with the seven flights before and during the May Day parade, the aircraft logged 129 hrs., 58 min., in these flights. The results were not entirely encouraging. True, most of the target performance figures were met; in particular, the bomber showed a top speed of 950 km/h (590 mph) at 6,700 m (21,980 ft.) and a service ceiling of 12,300 m (40,350 ft.). On the other hand, maximum-range flights immediately showed that range was no more than 6,500 km (4,040 miles) with a full ordnance load and 9,800 km (6,090 miles) with 5 tons (11,020 lbs.) of bombs – provided that the aircraft

The first prototype taxies at LII's airfield with the flaps fully deployed

descended from 11,000 to 12,000 m (36,090 to 39,370 ft.) and entered the airfield's holding pattern with two engines shut down to save fuel. If the bomber returned to base with all four engines running, range with a 5-ton bombload did not exceed 9,620 km (5,975 miles); effective range with fuel reserves was more like 9,500 km (5,900 miles). The takeoff run, too, turned out to be longer than expected.

The state commission's protocol, which was endorsed on 25 August 1954, pointed out that the armament, the autopilot, the radar, the JATO boosters, and some other items had not been tested during manufacturer's flight tests and therefore not submitted for state acceptance trials. The concluding part of the protocol read as follows:

'1. The first prototype of the M-4 development aircraft designed by V. M. Myasishchev and powered by four AM-3A turbojet engines meets the specifications set forth in Council of Ministers directive No. 949-469 of 24.03.51 as far as the maximum speed of 918 km/h [570 mph] at 9,000 m [29,530 ft.] and the speed of 800 km/h

Left and above left: Two more views of the first M-4 taxiing after 1954. The lower photo shows well how far the area-increasing outer flaps extend and how the inboard flaps follow the contours of the engine housings.

Top: M-4 '25 Red' begins its takeoff run.

Above: '25 Red' a few seconds after becoming airborne. Note the almost-level position of the nose gear bogie, indicating that the aircraft was not yet fitted with a bogie-tilting mechanism to reduce the takeoff run.

Right: The first prototype M-4 in landing configuration

A fine study of the first prototype M-4 in cruise configuration.

[496 mph] in maximum-range flight mode are concerned. It fails to meet the specifications with respect to the following:

• maximum range with a 5,000-kg bombload (a range of 9,800 km [6,090 miles] was attained instead of the required 11,000–12,000 km [6,830–7,450 miles])

• cruise altitude in maximum-range flight mode (8,500–13,500 m [27,890–44,290 ft.] instead of the required 13,000–16,000 m [42,650–52,490 ft.])

• takeoff run at maximum TOW without rocket boosters (2,610 m [8,560 ft.] instead of the required 2,000 m [6,560 ft.].

2. The decision on whether the aircraft is suitable for air force service can be taken when the cannon and bomb armament and other systems not tested and verified on the first aircraft have passed their state acceptance trials on the second prototype.

3. The M-4 is a modern high-speed bomber. It can be mastered by average-skilled pilots flying the Tu-4 and Tu-16 and is capable

A different view of the same aircraft underscoring the sleek lines of the M-4, which are slightly spoiled only by the brazenly protruding engine jet pipes

The M-4's performance obtained in the state acceptance trials

Power plant	4 × Mikulin AM-3A
Takeoff thrust	4 × 8,700 kgp (19,180 lbst)
Length overall	47.665 m (156 ft., $4\frac{37}{64}$ in.)
Wingspan	50.526 m (165 ft., $9\frac{7}{32}$ in.)
Wing area	350 m² (3,767.3 sq. ft.)
Empty weight	73,595 kg (162,250 lbs.)
All-up weight:	
normal *	130,000 kg (286,600 lbs.)
maximum	181,500 kg (400,130 lbs.)
Bombload:	
normal	5,000 kg (11,020 lbs.)
maximum	24,000 kg (52,910 lbs.)
Fuel capacity:	
nominal	132,390 litres (29,125 imp. gal.)
actual (maximum)	123,600 litres (27,192 imp. gal.)
Top speed:	
at sea level	670 km/h (416 mph)
at 6,700 m (21,980 ft.)	947 km/h (588 mph)
at 9,000 m (29,530 ft.)	918 km/h (570 mph)
Unstick speed:	
with a 138-ton (304,230 lb.) all-up weight	245 km/h (152 mph)
with maximum AUW	278 km/h (172 mph)
Landing speed with a 100-ton (220,450 lb.) landing weight	220 km/h (136 mph)
Service ceiling	12,500 m (41,010 ft.)
Climb time with a 138-ton AUW:	
to 5,000 m (16,400 ft.)	5.4 minutes
to 10,000 m (32,810 ft.)	15.7 minutes
to service ceiling	35.0 minutes
Climb time to 10,100 m (33,140 ft.) with maximum AUW	29.2 minutes
Range with a 5-ton bombload at 800 km/h (496 mph)	9,800 km (6,090 miles)
Cruise altitude in maximum-range flight	11,000 m (36,090 ft.)
Takeoff run without JATO boosters: †	
with a 138-ton AUW	1,300 m (4,270 ft.)
with maximum AUW	2,610 m (8,560 ft.)
Takeoff distance to h=15 m (82 ft.) without JATO boosters: †	
with a 138-ton AUW	2,800 m (9,190 ft.)
with maximum AUW	n.a.
Landing run: ‡	
with brake parachutes	1,470 m (4,820 ft.)
without brake parachutes	2,285 m (7,500 ft)
Landing distance from h = 15 m: ‡	
with brake parachutes	2,990 m (9,810 ft.)
without brake parachutes	n.a.
Defensive armament	6 × 23 mm TKB-495 cannons

Notes:

* With a 5-ton (11,022 lb.) bombload and half-filled fuel tanks

† Engine speed 4,650 rpm

‡ With a 100-ton landing weight and a 220 km/h landing speed

of operating at altitudes up to its service ceiling and at speeds ranging from a minimum of 670 km/h [416 mph] to a maximum of Mach 0.85. Flight at speeds in excess of Mach 0.85 is not recommended due to considerable vibration of the entire aircraft.

4. The crew's working conditions with regard to physiology and hygiene (as regards workstation ergonomics, cabin temperature distribution, noise levels, and placement of ancillary equipment) are unsatisfactory at all altitudes since they cause excessive crew fatigue.

5. Overall structural strength has been partially checked but can be checked in full only after flights at maximum dynamic pressure and g-loads have been made. [The M-4 was designed for an operational limit of 2.7 g – *Auth.*]

6. The PT-3076-53 brake parachute system developed and manufactured by NIEI PDS has passed the trials successfully. The question of whether the M-4 shall be equipped [as standard] with the PT-3076-53 system is to be decided once the PT-3237-53 brake parachute system made of Kapron has passed its state acceptance trials. [PT = *para**shoot** tormoz**noy*** – brake parachute; NIEI PDS = *Naoo**ch**no-is**sled**ovatel'skiy eksperimen**tahl'**nyy insti**toot** para**shoot**no-de**sahnt**nykh sis**tem*** – the Paradropping-Systems Research & Experimental Institute; Kapron™ is a Soviet synthetic fabric similar to nylon – *Auth.*]

. . . 8. The defects and shortcomings detected during the state acceptance trials of the first prototype M-4 and set forth in this report's concluding section and lists 1 and 2 shall be rectified on the second prototype and subsequent production examples. [List 1 is an appendix to the state commission's protocol enumerating the most-critical defects with which the Ministry of Defence cannot accept the aircraft; less critical defects are entered in list 2 – *Auth.*]

Signed: GK NII VVS Department 1 Chief Engineer-Colonel Kostryuk 19.08.54

GK NII VVS Department 2 Chief Engineer-Colonel Titov 20.08.54

GK NII VVS Department 3 Chief Major-General Shelimov (aircraft maintenance service)'

The report also stated that the M-4's lateral stability deteriorated at speeds in excess of Mach 0.81–0.82, and that the aircraft displayed reverse roll reaction to rudder inputs at these speeds (that is, it rolled to starboard instead of to port when left rudder was applied). At high angles of attack with the landing gear and flaps up, the aircraft displayed longitudinal instability. The forces on the aileron control wheels were rather high at medium and high speeds. When the powered flight controls were switched off and the emergency manual mode was tried, the control runs became deformed, requiring much-bigger control inputs to achieve the desired reaction. Finally, in rainy weather the flight deck glazing was flooded with water to such an extent as to make landing all but impossible – forward visibility dropped virtually to zero.

The Myasishchev OKB took due notice of these findings. The deficiencies discovered in the course of the state acceptance trials were rectified on the prototypes and initial production aircraft, and appropriate changes were introduced on the production line. However, new problems cropped up as old ones were eliminated. For instance, during maximum-range flights it transpired that the aircraft's 1-RSB-70 communications radio had a range of only 3,000–3,850 km (1,860–2,390 miles) when working with ground-based ***Boon**ker* (bunker) transmitters and R-250 receivers; in practice this meant long-range bomber crews would have no radio contact with the ground most of the time!

Opposite page: Three views of the second prototype M-4 ('aircraft DM') at Zhukovskiy during initial flight tests, with flaps deployed and bomb bay doors open. Note the shorter forward fuselage.

Right: Another shot of the second prototype during initial tests, showing the increased-area flaps. The aircraft was totally devoid of markings at this stage.

Yet, the OKB disagreed with some of the state commission's findings, arguing that some of the more serious objections were unfounded. On 7 September 1954 Aleksey I. Shakhoorin, who had returned to MAP after Stalin's death in 1953 (he had been fired as people's commissar of aircraft industry in March 1946 for slow progress in jet fighter development), wrote as follows to the Council of Ministers Machinery Industries Bureau:

'State acceptance trials of the first prototype M-4 powered by AM-3 engines were completed on 30 July 1954. . . . The state commission holding the trials of the M-4 notes that the aircraft's performance meets the specifications set forth in Council of Ministers directive No. 949-469 of 24 March 1951, with the exception of maximum technical range and takeoff run without the assistance of rocket boosters.

Chief designer Comrade Myasishchev believes that GK NII VVS has incorrectly determined the aircraft's maximum range in the test flight on 21 July. He argues that the amount of unusable fuel is actually much smaller than the figure quoted by the air force, being 250 kg [551 lbs.] rather than 1,450 kg [3,196 lbs.], and that the amount of fuel required for taxiing to the hardstand after landing was not included into the 5% remaining after landing. As a result, the M-4's range ended up being 500 km [310 miles] less than the figure obtained during manufacturer's flight tests.

The 2,610 m takeoff run recorded during state acceptance trials cannot be regarded as a minimum figure either, because the first prototype M-4 lacks the increased-area flaps fitted to the second aircraft; besides, the takeoff technique involving nose gear bogie tilting has not yet been mastered by the crew, since a maximum takeoff weight mission has been flown only once to date. During manufacturer's flight tests, the second prototype demonstrated a shorter takeoff run of approximately 2,000–2,200 m [6,560–7,220 ft.].

. . . The delays in the manufacturer's flight test programme of the second prototype M-4 were caused by procrastination in resolving the issue of extending the AM-3 engine's time between overhauls from 100 to 150 hours.'

'Aircraft DM' at a later stage of the test programme, now wearing its c/n (4300003) on the nose and fin and a black antiglare panel, but still no national insignia

LII test pilot Aleksey N. Gratsianskiy, the M-4's copilot in the maiden flight

Myasishchev OKB test pilot Fyodor F. Opadchiy, who performed the M-4's first flight

Myasishchev OKB test pilot Boris K. Galitskiy, who tested the second prototype M-4

GK NII VVS test pilot Aleksandr V. Sarygin, who participated in the state acceptance trials

The fact that the M-4 fell short of the expectations as far as range and some other characteristics were concerned could well have led to the cancellation of the programme. However, just a few days after the beginning of the M-4's state acceptance trials, its competitor suffered a serious setback. Powered by the provisional Kuznetsov 2TV-2F–coupled turboprops, which had been installed to save time, the first prototype Tu-95 ('aircraft 95/1') took to the air on 12 November 1952, ahead of the Myasishchev bomber. On 11 May 1953 the '95/1' crashed on its 17th flight after suffering a catastrophic engine fire, killing four of the eleven crew members. This accident, whose cause was traced to a defective engine reduction gearbox, led to the decision to abandon the 2TV-2F and to definitely use the originally intended Kuznetsov TV-12 (NK-12) turboprop. Yet, the protracted testing of this engine caused considerable delays of the Tu-95 programme; the second prototype ('aircraft 95/2') sat engineless until December 1954, and its maiden flight did not take place until 16 February 1955. In contrast, the AM-3A turbojet was already in production and showed adequate reliability.

A unique photo of general designer Vladimir M. Myasishchev holding a model of the M-4

In this situation, it was decided to go ahead with the Myasishchev bomber. MAP drafted a Council of Ministers directive envisaging M-4 production not only in Moscow (at plant No. 23), but also in Kuibyshev (now renamed back to Samara) at plant No. 18, which had been earmarked to build the Tu-95. This draft was not endorsed, and eventually plant No. 18 did launch full-scale production of the Tu-95 *Bear-A* bomber in 1956. Still, on 19 September 1953, the M-4 was ordered into production at plant No. 23, which was to manufacture an initial-production batch of 11 bombers in 1954–55 – with the proviso that the aircraft be upgraded in order to obtain the desired performance as soon as possible. This was a political decision aimed at achieving real parity in the arms race against the USA. (Jumping ahead of the story, we may mention that by the time the second Tu-95 flew, M-4 production was well underway and the aircraft was undergoing evaluation at Engels-2 air base near Saratov in southern Russia.)

Before the onset of *perestroika* ('restructuring') – the policy of democratisation, openness, and 'new thinking' launched by the first and only Soviet president, Mikhail S. Gorbachov, the Soviet popular press frequently embellished the M-4 while trying to 'play down' the importance of its chief Soviet 'competitor,' the Tu-95. For example, Vladimir M. Myasishchev's former deputy V. K. Karrask wrote in one of his publications: *'The Tupolev OKB, which then held a monopoly in the field of heavy aircraft develop-*

ment, considered it impossible to create an equivalent to the qualitatively new Boeing B-52 in the conditions prevalent in the Soviet Union, proposing a turboprop aircraft instead. This aircraft (the Tu-95) was, in effect, an extension of the traditional [bomber design] concept, which had by then reached the limit of its resources; thus it could not match the B-52's capabilities in full.

After the war . . . V. M. Myasishchev developed a number of turbojet-powered aircraft projects. This work culminated in his offer to create a jet-powered strategic bomber having transcontinental range and able to compete with the Boeing B-52 in all aspects.'

However, this salesman's hyperbole was dashed when the actual performance figures of the *Bear* and the *Bison* had been disclosed, showing that the praise lavished on the M-4 by its proponents was somewhat exaggerated. For instance, although the M-4 was 57 km/h (35 mph) faster than the Tu-95, with an identical 5-ton (11,020 lb.) bombload, its range in these conditions was 5,000 km (3,105 miles) shorter; also, the M-4 had an operational g-limit of 2 and a never-exceed speed (V_{NE}) of Mach 0.9.

The tone of Karrask's publication quoted above echoes the intense rivalry between the Tupolev and Myasishchev bureaus as 'the Lion and the Unicorn were fighting for the crown' (the Soviet bomber crown, that is). Yet, no documentary proof has been found of Vladimir M. Myasishchev attacking Andrey N. Tupolev, whereas cases of the contrary were plenty. (Speaking of which, Myasishchev was one of Tupolev's disciples – one of the best disciples, it should be noted – but it was more probably the two men's personal qualities and character that explains this 'pecking order.') Defending the Tu-95, Andrey N. Tupolev railed against the M-4 in 1954; however, Mikhail V. Khroonichev, who was then CoM vice chairman, unexpectedly supported Myasishchev. In July 1954 he wrote to the new Soviet leader Nikita S. Khrushchov:

'With regard to chief designer Comrade Tupolev's memo concerning the main performance data of the "aircraft M" bomber designed by Comrade Myasishchev, I report the following:

. . . The actual flight performance of the "aircraft M" bomber is as follows:

	Manufacturer's flight tests report	Comrade Tupolev's memo
Top speed, km/h [mph]	962 [597.5]	925 [574.5]
Cruising speed, km/h [mph]	800 [497]	800 [497]
Takeoff run at MTOW, m [ft.]	2,970 [9,740]	3,250 [10,660]
Technical range, km [miles]	10,700 [6,645]	10,600 [6,580]
Effective range with one fuel top-off, km [miles]	12,650 [7,860]	11,500–12,000 [6,830–7,450]

As for Comrade Tupolev's comments about the "aircraft M" bomber stated in the text of the memo, the facts are as follows:

• Item 1 alleging that, in its current form, "aircraft M" is incapable of reaching any vital targets in the USA – this needs to be checked with the Ministry of Defence, taking into account the positioning of our airfields and the flight routes; however, according to chief designer Comrade Myasishchev's calculations, the "aircraft M" can reach aircraft factories located in the northern part of the USA

• Item 2 . . . – if the aircraft is refuelled in flight near the airfield [of origin], effective range will be 15,000 km [9,320 miles]; if refuelling takes place en route to the target, the range will be 17,000 km [10,560 miles], making it possible to strike at many of the USA's [political and industrial] centres

• Item 3 alleging that the takeoff [that is, field performance – Auth.] *problem remains unresolved – in actual fact, the takeoff run in maximum TOW (overload) configuration in standard conditions (ISA +15°C) is 2,970 m (this figure comes from the manufacturer's flight tests report). However, according to Comrade Myasishchev's calculations, the second prototype "aircraft M" and the first production aircraft are expected to have a takeoff run of 2,600 m [8,530 ft.] thanks to an improved flap design.*

Comrade Tupolev's claim that the takeoff run will exceed 4,000 m [13,120 ft.] at higher ambient temperatures is equally applicable to all long-range bombers, since engine thrust decreases in hot conditions and, according to Comrade Myasishchev's OKB, the takeoff run grows by 400 m [1,310 ft.] with every 10°C. This rule applies equally to turbojet and turboprop engines.'

The reference to 'range with one fuel top-off' in the above excerpt is purely hypothetical (the key word is 'if') because neither the M-4 nor the Tu-95 had in-flight refuelling capability yet. Here it is worth noting that even though the Tu-95's performance was fairly high and the *Bear* surpassed the M-4 in range, it did not meet its design specifications in full either! The draft of the state acceptance trials protocol prepared in 1956 read:

'the Tu-95 aircraft powered by NK-12 engines possesses the following performance:

a) maximum technical range with a 5,000 kg [11,020 lb.] bombload, full main fuel tanks, and a normal all-up weight of 156 tons [343,915 lbs.] does not exceed 12,970 km [8,055 miles], whereas the required figure is 15,000 km [9,320 miles]. With a full fuel load and a 172-ton [379,190 lb.] maximum TOW, range is 15,040 km [9,340 miles] instead of the specified 17,000–18,000 km [10 560–11,180 miles];

b) maximum speed is 890 km/h [552 mph] at 7,000 m [22,965 ft.] instead of the specified 920–950 km/h [570–590 mph] at 8,000–9,000 m [26,250–29,530 ft.];

c) service ceiling with an average all-up weight of 120 tons [264,550 lbs.] is 11,300 m [37,070 ft.] instead of the specified 13,000–14,000 m [42,650–45,930 ft.];

d) the takeoff run is 1,740–2,260 m [5,710–7,410 ft.] depending on takeoff weight instead of the specified 1,500–1,800 m [4,920–5,900 ft.].'

Anyway, in a transatlantic or transpacific mission, the Tu-95 gave the crew a chance of returning home safely – or at least of reaching an area under the control of friendly forces where the crew could bail out and be rescued. The M-4 did not. The obvious question is, Why? Why didn't Myasishchev achieve the required range? Given the extremely fuel-thirsty jet engines of the day, all the Myasishchev OKB could do to meet the stringent requirements was to give the M-4 an excellent lift/drag ratio provided by the bomber's 'clean' wings of extremely high aspect ratio (17.45). Yet, this was not enough, as it turned out. There were two possible approaches to resolving the M-4's range problem. One way was to increase the fuel capacity; however, this would increase the takeoff

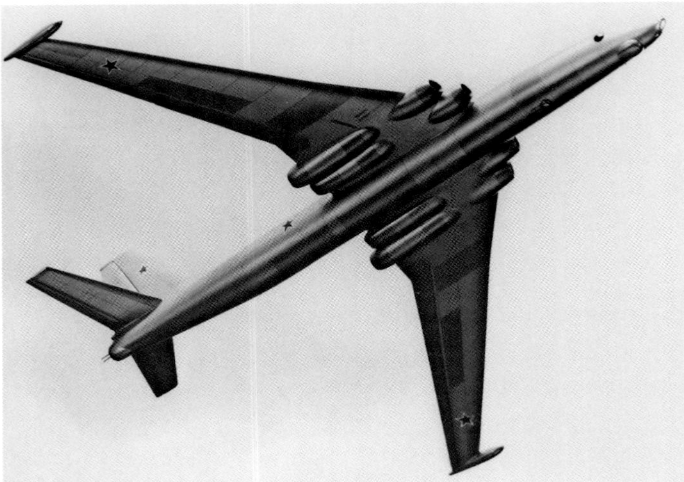

A reasonably accurate official recognition model of the M-4 built by the Special Devices Center of the US Office of Naval Research

weight (which was limited by the landing gear's strength) and the runway loading, so this option was ruled out. (As already mentioned, anticipating this problem, back at the ADP stage the OKB had considered using two underwing auxiliary landing-gear struts, which were to be fitted just outboard of the engines and jettisoned after takeoff, but this feature was ultimately discarded.) The other option was to use in-flight refuelling (IFR). Work on IFR systems in the Soviet Union was already underway but far from complete, and the M-4 entered production and service without IFR capability.

The bottom line was that even though OKB-23 had managed to design, build, and test-fly the new strategic bomber two months ahead of schedule (Stalin had given Myasishchev two years to complete the job), it did not succeed in either achieving the stipulated 12,000 km (7,450-mile) range or in beating the Americans – or its Tupolev competitor, for that matter. The first prototype XB-52 entered flight test on 15 April 1952, followed by the second prototype (YB-52) on 2 October of that year.

Some time after the state acceptance trials, the first prototype M-4 gained the tactical code '25 Red' (apparently derived from the programme's code number) writ large on the forward fuselage. (Note: Unlike Western military aircraft, which have serial numbers allowing positive identification, since 1955 Soviet/CIS military aircraft have two-digit tactical codes that are usually simply the number of the aircraft in its unit. Three-digit codes are usually allocated to aircraft in training units or development aircraft, in the latter case tying in with the machine's construction number (c/n), fuselage number (f/n or line number), or manufacturer's designation. Transport aircraft are an exception; some Soviet/Russian air force transports had three-digit codes, which were the last three digits of the aircraft's former quasi-civil registration.)

On 26 March 1955 the second prototype M-4 was, in turn, officially submitted for state acceptance trials, although its flights under the trials programme did not begin until 10 May because the aircraft was used for conversion training of long-range bomber crews and participated in that year's May Day parade. This time the trials also involved the upgraded first prototype and the 11th production M-4 (more on which will be said in chapter 3).

The second prototype was used for testing the M-4's offensive and defensive armament, although only conventional bombs were dropped at this stage. Between 9 September 1955 and 7 July 1956, the bomber's offensive and defensive armament underwent an upgrade involving the replacement of the powered barbettes and the PRS-1 gun-laying radar. Apart from bombs, 'aircraft DM' was used to check its suitability for antishipping strike; the plan was to attack incoming NATO warships before they came within striking distance of the Soviet shores. During this part of the programme, which ended in December 1956, the aircraft carried a complement of six 450 mm ($17\frac{45}{64}$ in.) calibre RAT-52 rocket-propelled torpedoes weighing 627 kg (1,382 lbs.) each, or six naval mines of various types – the 985 kg (2,170 lb.) Lira moored mine, the 1,300 kg (2,870 lb.) Serpey bottom mine, the 825 kg (1,820 lb.) APM floating mine, and the 1,100 kg (2,430 lb.) AMD-2M and IGDM bottom mines. (RAT = *re'ak**tiv**naya aviat**sion**naya tor**pe**da* – jet-powered air-droppable torpedo; Lira (pronounced ***lee**ra*) means 'Lyre'; APM = *aviat**sion**naya **pla**vayushchaya **mee**na* – air-dropped floating mine; AMD = *aviat**sion**naya **mee**na **don**naya* – air-dropped bottom mine; IGDM = *indooktsi**on**no-**ghid**rodina**mich**eskaya **don**naya **mee**na* – bottom mine with a combined induction/hydrodynamic detector. As for the nonsensical word *Ser**pey***, it is nothing more than an accidental anagram of the intended code name *Persey* (Perseus). Quite simply, a typist had made an error when typing out the document clearing the weapon for service, and no one had taken the responsibility to set her straight, believing that 'the Powers That Be know better'!) Eventually, however, mines and torpedoes were excluded from the M-4's weapons range.

As mentioned earlier, to reduce the M-4's takeoff run, the designers intended to use jet-assisted takeoff (JATO) boosters. The latter, known as SUM (***star**tovyy ooskori**tel**' Myasishcheva* – Myasishchev takeoff booster), were developed in house by OKB-23, utilising a liquid-propellant rocket motor designed by Aleksey M. Isayev and components from the V-300 surface-to-air missile created by Semyon A. Lavochkin's OKB-301 for the S-25 *Berkut* (Golden eagle; NATO code name SA-1 *Guild*) SAM system. The JATO booster was initially tested on a ground rig and later on a Tu-4 (the ShR-2 landing-gear test bed, which had been appropriately modified and redesignated *izdeliye* SUR) and one of the M-4 prototypes. Eventually, however, it was considered too complex to use, and the boosters were never installed on operational M-4s.

Later on, the first prototype served as a test and development aircraft with OKB-23; among other things, it was a propulsion test bed for the Dobrynin VD-7 turbojets used on other aircraft of the M-4 family, and a test bed for the fuel tanks' inert gas pressurisation system, preventing an explosion if the aircraft was hit by enemy fire. The second prototype served in a similar capacity as a systems test bed for the ill-starred Myasishchev M-50 *Bounder* four-jet supersonic bomber.

In 1957 a group of OKB-23 employees (Vladimir M. Myasishchev, Gheorgiy I. Nazarov, Vladimir M. Baryshev, Lazar' M. Rodnyanskiy, Fyodor F. Opadchiy, and Leonid L. Selyakov) received the prestigious Lenin Prize – one of the highest government awards in the Soviet Union – for their major contribution to the development of Soviet strategic aviation.

Chapter 3

A Herd of Wild *Bisons*: Version and Mission Variety

M-4 Long-Range Bomber (*izdeliye* 25)

Full-scale production of the M-4, powered by AM-3A engines with a takeoff rating of 8,750 kgp (19,290 lbst) and a maximum continuous rating of 7,000 kgp (15,430 lbst), began at aircraft factory No. 23 in 1954; the first production bomber – initially uncoded – took off in August of that year. The original plans envisaged building some 200 M-4s, but in the event the production run was quite small, only 32 production bombers being completed and delivered before the M-4 was superseded on the assembly line by a more advanced version – the 3M (*see below*) – in 1956.

According to some sources, batch 0 comprised three aircraft – the two prototypes and the static-test airframe, while subsequent production batches consisted of a single aircraft each. However, it would be more appropriate to say the production-standard M-4 had no production batches as such and the aircraft were numbered consecutively. Anyway, the first production M-4 was c/n 4300104 (that is, year of manufacture 1954, meaning of first zero unknown, plant No. 23, first production aircraft, and fourth airframe built). Thus, the final M-4 built was c/n 6303235.

Originally, when the bombers were in their natural-metal factory finish, the c/n was stencilled on both sides of the fuselage nose and the fin. It disappeared from the nose when the M-4s received a so-called 'antiflash' colour scheme (albeit a partial one) as protection from the effects of a nuclear explosion – the undersurfaces, fuselage sides, and rudder were given a coat of special AS-2SP high-gloss white paint to reflect the flash of the explosion as much as possible.

Busy scene at Zhukovskiy, with probe-equipped M-4A '85 Red' (c/n 4300104) in the foreground and the M-4-2 tanker prototype ('71 Red,' c/n 4300205) beyond. The Tu-4 in the background is the *izdeliye* SUR test bed for the M-4's SUM JATO booster (still with the experimental bicycle landing gear from its previous role as the *izdeliye* ShR-2 test bed). The third *Bison-A* must be c/n 4300306.

Outwardly, production M-4s from the second production aircraft (c/n 4300205) onward differed from the second prototype in lacking the bulged fairing under the tail turret, housing a spent-case / belt link collector (saving the belt links was deemed impractical, and they were simply ejected), and in having a different aerial arrangement on the communications radio. A second pair of wing fences were added outboard of the existing one. Not so obvious was the fact that production *Bisons* had a smaller AoA, with the nose gear bogie tilted for takeoff (9°30' in rotated condition instead of 10°30'); additionally, a revised bogie actuator that tilted the bogie more slowly was tested on the fifth production bomber (c/n 5300508) and soon was introduced on later production aircraft. These measures were meant to prevent the bomber from pitching up into a stalled position immediately after takeoff – a situation where ejection would be impossible (this phenomenon had resulted in several crashes early in the M-4's service career).

Also, production bombers featured a complete avionics and equipment suite; in particular, the second prototype's AP-5 autopilot was replaced with the later AP-15 model. Conversely, the number of weapons options was reduced by excluding guided bombs, naval mines, and torpedoes from the weapons range – albeit not before a comprehensive test programme was undertaken with the latter two kinds of weapons (more about this below).

As already mentioned, the first prototype M-4 had been dismantled and shipped to Zhukovskiy prior to its maiden flight. However, in order to expedite delivery of production bombers, it was desirable to avoid this complicated procedure if at all possible. (Quite apart from the time and effort involved in dismantling, loading, unloading, and reassembling the aircraft, the long way down the meandering Moskva River from Fili to Zhukovskiy was time consuming. Moreover, it took the barges through the centre of Moscow, necessitating large-scale security measures; this was a time when everybody was paranoid about espionage.) The problem was that plant No. 23 had a runway no more than 1,500 m (4,920 ft.) long. The solution was to reduce the bomber's takeoff weight by filling just enough fuel for one hour's flight, allowing the M-4 to be flown out of Fili. Vladimir M. Myasishchev gave the go-ahead for this experiment, which was flawlessly performed on 16 December 1954 by a factory test crew under Boris K. Galitskiy on M-4 c/n 4300104. (An alternative account says the first production aircraft was delivered to Zhukovskiy by road in pieces and reassembled there, and the first M-4 to be flown out of Fili was the *second* production aircraft.) All subsequent *Bisons* were ferried to Zhukovskiy in this fashion for predelivery flight tests before they were handed over to Long-Range Aviation (DA – ***Dahl'****nyaya avi****aht****siya*), the strategic-bomber arm of the Soviet Air Force. This technique saved both time (as much as a month) and money.

Almost immediately, the OKB and the factory set about improving the bomber. A longer bomb bay was introduced from the third production aircraft (c/n 4300306) onward. In 1955, plant No. 23 started manufacturing M-4s powered by RD-3M or RD-3M-500A turbojets, which were advanced versions of the AM-3A developed by Aleksandr A. Mikulin's deputy Prokofiy F. Zoobets; the '500' denoted a time between overhauls (TBO) extended from 200 to 500 hours. The existing *Bison* fleet was progressively reengined by 1957. Besides being more fuel efficient, the RD-3M offered higher thrust, with a normal takeoff rating of 9,500 kgp (20,940 lbst) and a 10,500 kgp (23,150 lbst) contingency rating, which allowed takeoff to be continued in the event of an engine failure. The new power plant also increased range with a 5-ton (11,020 lb.) bombload to 10,500 km (6,520 miles). (In passing, it may be noted that when Mikulin fell from favour at the top level and was removed from office as head of OKB-300 in 1955, the use of his initials as an engine designation became taboo, and the AM-3 was promptly rechristened RD-3 for *re'ak****tiv****nyy* ***dvig****atel'* – jet engine.)

The tremendous efforts of OKB-23 and other aircraft industry enterprises involved bore fruit: the Soviet Union became the world's first nation to field a new-generation strategic bomber.

Opposite page: The first production M-4 (c/n 4300104) still had the spent-case / belt link collector fairing under the tail turret. Here the bomber is still unserialled.

Top and above: The second production M-4 (likewise still unserialled) introduced a cleaner rear fuselage without the said fairing. The aircraft is seen here at Zhukovskiy during tests.

Right: The forward fuselage of M-4 c/n 4300205, showing the miscellaneous aerials and the still-empty platform for a star tracker on top of the nose

Left: M-4 c/n 4300104 flies with the bomb bay doors open.

Below: The first (*foreground*) and second production M-4s at OKB-23's flight test facility in Zhukovskiy. Note the photo calibration markings on the rear fuselage of the nearest bomber (aft of the insignia applied as per pre-1955 standard) for use in test programmes.

Even though the M-4 had entered flight test later than its turboprop competitor (the Tu-95) and the B-52, it beat both of them to service entry, achieving initial operational capability (IOC) in early 1955. Indeed, the M-4 was remarkable in that it was way ahead of many aircraft of simpler design in terms of service entry.

The first few production aircraft were used for conversion training of service pilots, which took place at the Myasishchev OKB's flight test facility in Zhukovskiy. To expedite the state acceptance trials, the 11th production M-4, coded '24 Red' (c/n 5301114) –the final aircraft of the aforementioned initial-production lot – was seconded from the Air Force, participating in stage B of the trials alongside the two prototypes. Between 21 February and 30 March 1955, it logged 52 hrs., 24 min., in 16 flights, and then a further 13 hrs., 8 min., in 13 flights between 23 May and 14 June 1955. Among other things, it made several night flights, as well as flights to verify the avionics and the bomb armament. Up to 25 July, the three aircraft made 99 test flights between them under stage B. These revealed that the performance of the production aircraft differed little from that of the prototypes (for better or for worse).

Several additional test programmes were held in November and December 1955. M-4s c/ns 4300104, 4300205, 5301215, and 5301316 performed a number of high-speed takeoffs with a TOW of 100–130 tons (220,450–286,600 lbs.) and flap settings of 20°–35°; the aircraft handled best with the flaps at 20°–25°. M-4 c/n 5301215 was also used for testing the fuel jettison system; fuel dumps showed that there was no risk of the fuel being ignited by the engine efflux, and the CG stayed within the prescribed limits.

The concluding part of the stage B trials report, which was endorsed on 11 August 1956, read:

'1. The M-4 high-speed long-range bomber designed by V. M. Myasishchev and powered by four AM-3A turbojet engines meets the specifications set forth in CoM directive No. 949-469 of 24.03.51 as far as the maximum speed of 918 km/h [570 mph] at 9,000 m [29,530 ft.] and the speed of 800 km/h [496 mph] in maximum-range flight mode are concerned. It fails to meet the specifications with respect to the following:

- *Maximum range with a 5,000 kg bombload (a range of 9,800 km [6,090 miles] was attained instead of the required 11,000–12,000 km [6,830–7,450 miles])*
- *cruise altitude in maximum-range flight mode (8,500–13,500 m [27,890–44,290 ft.] instead of the required 13,000–16,000 m [42,650–52,490 ft.])*

• *Takeoff run at maximum TOW without rocket boosters (2,610 m [8,560 ft.] instead of the required 2,000 m [6,560 ft.]). The booster has failed its state acceptance trials.*

2. The M-4 is suitable for service with first-line air force units as a long-range bomber.

Takeoff in the M-4 in adverse weather, at night, or in a cross-wind is unsafe because of the vigorous pitch-up [caused by the tilting bogie], which causes the aircraft to reach near-critical AoAs, and because of the too-low unstick speed, which does not enable effective operation of the flight controls.

The windshield wipers have not been submitted for state acceptance trials, making takeoff and landing in snowy or rainy weather impossible.

Operation of the aircraft and its combat application are hampered by the numerous defects discovered in the course of the state acceptance trials and listed in this report. These shall be eliminated on all existing M-4s, the scope and schedule of the work being agreed by MAP and the air force.

3. The functioning of the M-4's bomb armament is unsatisfactory because of the unreliable operation of the RBP-4 [radar] and OPB-11R bombsight. Providing all components of the bomb armament are serviceable, the system ensures acceptable bombing accuracy with the OPB-11R and high bombing accuracy with the RBP-4.

4. State acceptance trials of the bomber's special equipment under a separate special test programme are being conducted by the air force's special test range jointly with the Ministry of Medium Machinery and have not yet been completed.

(Note: The equipment in question was associated with the carriage of nuclear weapons. The test range mentioned here is the 71st Test Range at Bagerovo AB, in the eastern part of the Crimea Peninsula, 14 km (8.7 miles) west of the city of Kerch, which had overall responsibility for testing Soviet nuclear weapons. Actually the designation '71st Test Range' was something of a misnomer because, in addition to an instrumented test range with a target for the bomb drops, the establishment included three Air Force units. The latter were the 35th OSBAP (*otdel'nyy smeshannyy bombardirovochnyy aviapolk* – Independent Composite Bomber Regiment) performing the test drops, the 513th IAP (*istrebitel'nyy aviapolk* – fighter regiment) providing protection, and the 647th SAPSO (*smeshannyy aviapolk spetsiahl'novo obespecheniya* – Composite Special Support Air Regiment) with a mixed bag of utility, transport, and reconnaissance aircraft, which was tasked with air sampling in the wake of nuclear tests, aerial photography, and transport/liaison duties. Only instrumented dummy bombs were dropped at Bagerovo; nobody in their right mind would test a live nuke in the Crimea, which was, to use a common Soviet cliché, *the all-Union resort*! The Ministry of Medium Machinery (MSM – *Ministerstvo srednevo mashinostroyeniya*) was responsible for the Soviet nuclear programme in all of its aspects, civil and military.)

5. The M-4's cannon armament has passed its trials successfully and, once the dangerous concentration of explosive blast gases in the ventral barbette has been eliminated and the field-of-fire limiter on the dorsal barbette [which prevents 'self-inflicted injuries to the bomber's vertical tail – *Auth.*] *works reliably, it can be accepted for air force service.*

6. The M-4's mission equipment meets the air force's specific operational requirement as regards the complement and types of equipment.

7. The AP-15 autopilot has passed its state acceptance trials successfully, conforming to the air force's specific operational requirement No. 3 226 of 19 April 1947.

8. The in-flight refuelling system with which all M-4 aircraft are to be equipped, pursuant to CoM directive No. 2970-1278 of 18 February 1953, has been installed on two pairs of aircraft and is currently due to be submitted for state acceptance trials after passing manufacturer's flight tests on one pair of M-4s.

According to MAP's OKB-23, a single fuel top-off on the outbound leg of the flight (en route to the target) can extend the M-4's range by 2,900 km [1,800 miles]. . . .

11. The PT-3076-53 brake parachute system, made of silk, and the PT-3237-53 Srs 2 brake parachute system, made of Kapron, have passed their state acceptance trials successfully. Production M-4s shall be equipped with the PT-3237-53 Srs 2 brake parachute system, which has higher performance than the PT-3076-53.'

Tests under various programmes as part of the state acceptance trials continued on several aircraft even after the above report had been signed. In particular, M-4 c/n 4300306 was used for testing the envisaged nuclear weapons and the heat-insulated bomb bay, equipped with an environmental-control system maintaining the correct temperature right up to the aircraft's service ceiling. This feature was necessary because nuclear munitions are sensitive to subzero temperatures and may malfunction after having had a cold soak at high altitude – the chain reaction might not go ahead, leading to nondetonation. Also, a system performing the prerelease procedure (disengaging the multistage safety system, arming the detonator, and activating the barometric sensors to explode the nuke at the preset altitude) and releasing the bomb was fitted.

First, the aircraft underwent manufacturer's flight tests, logging 25 hours in eight flights in which the nuke's control system and the bomb bay's heat insulation were tested, bombing accuracy was assessed, and the strength of the MBD6-M gantry-type bomb rack was checked. (MBD6-M = *mostovoy bahlochnyy derzhahtel'* – gantry-type rack, group 6 (a reference to the weight class of munitions to be carried), optimised for 'aircraft M'; that is, M-4. By comparison, a similar bomb rack for the Tu-95 was designated MBD6-95.) Between 12 June and November 1956, the aircraft made a further 15 flights under the state acceptance trials programme, logging 42 hrs., 26 min., with good results. Later, the M-4 was readied for an additional round of state acceptance trials as a nuclear bomb carrier in a configuration called *nositel* I ('carrier'; that is, 'phase I nuclear weapons delivery vehicle'); at this stage its 'antiflash' protection features were tested in March 1957.

In 1957, two M-4s (c/ns 5301821 and 5301922) were involved in additional armament tests. In September–October of that year the first production M-4 (by then coded '85 Red') was used for testing the crew rescue system. The aircraft made 19 flights under this programme, in which 13 dummies and eight live parachutists were ejected; the flights showed that safe ejection was possible at speeds up to 650 km/h (403 mph).

On 27 December 1958 the Council of Ministers issued directive No. 1404-688, ordering a special test programme to be held in

M-4 '24 Red' (c/n 5301114) used in stage B of the state acceptance trials

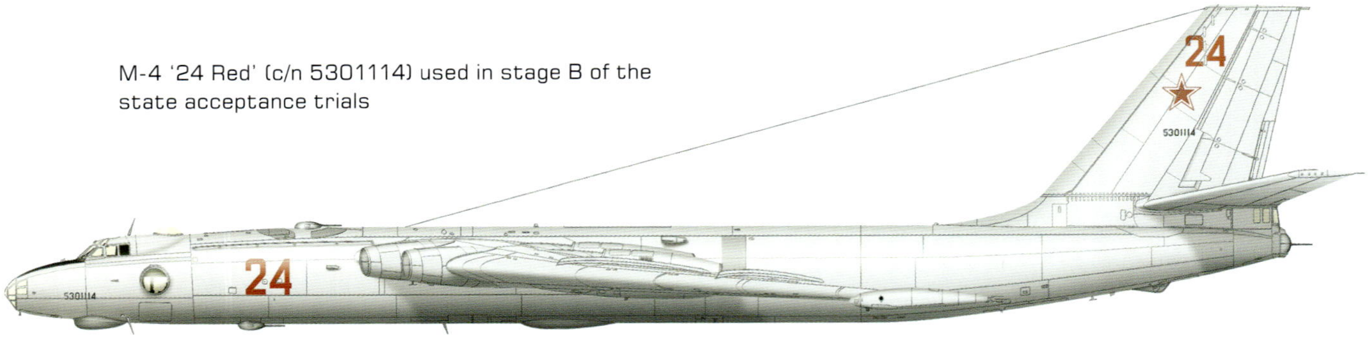

Top and above: A midproduction M-4 (c/n 5301114) seen during state acceptance trials of the bomber's armament

Top right: M-4 c/n 5301114 wore the tactical code '24 Red' on the nose and tail in huge characters.

Above right: This view illustrates the bulge on the bomb bay doors enabling carriage of a FAB-9000 bomb plus long-range fuel tanks.

Right: A three-quarters rear view of the same aircraft with the flaps partly deployed. Note the lack of star insignia on the rear fuselage.

5301114
24
24

5301114
24
24
5301114

24
24
5301114

Top: Uncoded M-4 c/n 5301215 was used for fuel jettison system tests. The outer-wing fuel jettison valves are in operation here.

Above: This photo of the same bomber shows different fuel jettison valves in operation (in the inner wings and the rear fuselage). Note the photo calibration stripes on the fuselage.

order to determine the M-4's behaviour at high angles of attack. In keeping with this directive, M-4 c/n 6303033 was seconded from the air force, arriving at Zhukovskiy on 18 December 1959. In February–April 1960, this aircraft made nine test flights to verify the automatic critical flight mode indication system (ASKR – *avto**maht** signali**z**ahtsiï kri**ticheskikh** re**zhim**ov*), warning the pilots of dangerous AoAs.

The tests were not altogether without incident. One day in 1956, a hydraulic reservoir failed just as an M-4 captained by Myasishchev OKB test pilot Mark L. Gallai, with B. M. Stepanov as copilot and A. A. Titov as flight engineer, was about to take off. The massive cast-iron reservoir burst like a fragmentation bomb, inflicting great damage; quite apart from the obvious failure of the backup hydraulic system, the fragments severed the rudder control rods and the No. 3 engine's fuel line, punching numerous holes in the lower fuselage skin as they went. (Noticing this, a radio operator watching Gallai's takeoff from the control tower at Zhukovskiy uttered in bewilderment, *'Something's just shot out of 'em!'*) As soon as the bomber became airborne, it veered sharply to one side because of the inoperative rudder; Gallai had to wait until he had gotten up to about 15 m (50 ft.) before he could cancel the yaw with opposite bank. As he did, the No. 3 engine flamed out due to fuel starvation, and test engineer S. A. Sokolov, sitting in the tail gunner's station, reported a serious fuel leak from the lower fuselage: *'The fuel is gushing like a fountain!'* To top it all, the rudder kept flapping and causing the bomber to yaw left and right, which the pilots counteracted with opposite bank.

Climbing to a few hundred metres, with extreme caution the pilots turned the stricken bomber around for an emergency landing. Care had to be exercised in handling the throttles, because when the engines were throttled back, the exhaust plume changed its shape and could ignite the pall of kerosene mist stretching for hundreds of feet behind the aircraft – and then the chances of sur-

One of the last M-4s built, '01 Red' (c/n 6302831), at the LII airfield. The dark objects on to of the wings are probably rolled-up canvas covers.

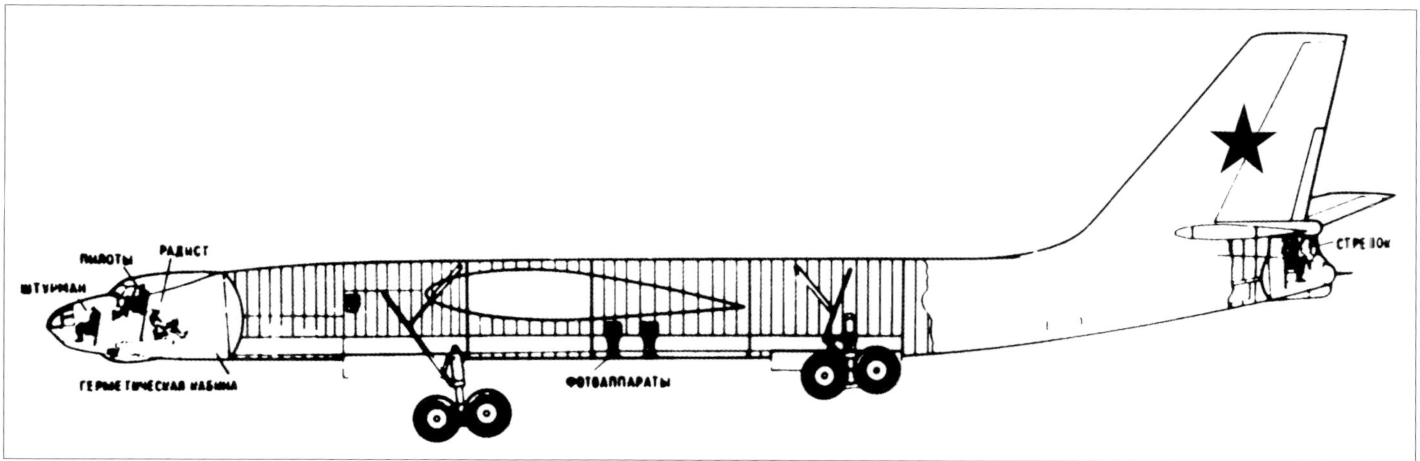

A cutaway drawing from the project documents depicting the layout of the 'aircraft 25F' (VM-25F) photoreconnaissance aircraft. Note the cameras installed in the bomb bay.

vival would be very slim indeed. As the M-4 touched down, Titov shut down the nos. 2 and 4 engines to reduce the fire hazard; the No. 1 engine could not be shut down just yet because it drove a hydraulic pump, providing hydraulic power for the brakes. Gallai ordered it shut down as the aircraft lumbered to an almost complete halt on the hardstand, but even that was a bit too early, as it turned out. Still under momentum, the huge bomber flattened the chocks placed in front of the wheels by the ground crew and crawled on relentlessly, refusing to stop until it bumped into the stripped-out hulk of another aircraft lying on the verge of the hardstand. The wing leading edge was slightly damaged in the collision, which made Gallai really angry, but his anger evaporated when he realised the true extent of the damage. The test crew's experience and skill saved the day.

Originally known simply as the *Bison* in NATO parlance, the initial 'glass-nosed' bomber became the *Bison-A* when new versions appeared and were identified by the West, necessitating the addition of suffix letters. The Western popular press sometimes referred to the M-4 by the name ***Molot*** (Sledgehammer). The origin is obscure but the name is a fitting one – from a Western viewpoint anyway, since the shape of the M-4 in plan view is not unlike Thor's hammer!

In the autumn of 1959 the Myasishchev OKB decided to set a series of world records in the M-4. To this end, the aircraft involved – variously reported as the second prototype ('aircraft DM') or the first production machine – was lightened as much as possible by removing all nonessential equipment. On 16 October the bomber took off from Zhukovskiy with captain Nikolay I. Goryainov and copilot Anatoliy S. Lipko at the controls; the crew also included A. S. Iksanov, V. S. Koozovlyov, S. A. Sokolov, and V. V. Kolosov. On that day the aircraft established two class C-1 payload-to-height records, lifting payloads of 5 and 10 tons (11,022 and 22,045 lbs.) to an altitude of 15,313 m (50,239 ft.).

Two weeks later, on 30 October, the same machine set five class C-1 world speed records, averaging 1,028.664 km/h (638.92 mph) over a 1,000 km (621-mile) closed circuit with payloads of 5, 10, 15, 20, and 25 tons (11,022, 22,045, 33,070, 44,090, and 55,115 lbs.). Actually, the payload suspended in the bomb bay weighed 27 tons (59,520 lbs.); the above figures are the incremental values used in establishing such records. This time Lipko was the captain, with Myasishchev OKB test pilot Boris I. Yumashev as copilot; the crew of eight also included navigator V. I. Milyutin, radio operator L. N. Goosev, I. M. Korchagin, A. A. Moshin, V. P. Glushkov, and V. V. Kolosov.

Now, to have the records officially registered, the aircraft type and engine type had to be duly entered in the documents submitted to the FAI (*Fédération Aéronautique Internationale* – International Aeronautics Federation). Using the true designations was out of the question because they were classified. Hence, as was customary in Soviet times, the aircraft was entered for the record

Top: The crew of the M-4 that set world records as the '103M.' *Above*: The same airmen at a reception in the Kremlin, where they were handed government awards for their achievement.

Provisional specifications of the VM-25F	
Length overall	48.75 m (159 ft., $11\frac{19}{64}$ in.)
Wingspan	53.0 m (173 ft., $10\frac{39}{64}$ in.)
Wing area:	
overall	355 m^2 (3,821.19 sq. ft.)
less leading-edge root extensions	308 m^2 (3,315.28 sq. ft.)
Empty weight	68,000 kg (149,910 lbs.)
Fuel load	53,000 kg (116,840 lbs.)
Takeoff weight:	
option A	105,000 kg (231,480 lbs.)
option B	120,000 kg (264,550 lbs.)
Top speed at 9,000 m (29,530 ft.)	910 km/h (565 mph)
Cruising speed:	
option A	750–800 km/h (465–496 mph)
option B	760–800 km/h (472–496 mph)
Flight altitude over the target at half range:	
option A	16,000–16,500 m (52,500–54,130 ft.)
option B	16,200–16,700 m (53,150–54,790 ft.)
Technical range:	
option A	7,000–7,500 km (4,350–4,660 miles)
option B	5,000–5,500 km (3,100–3,420 miles)

attempts under a fictitious designation, 103M; unlike the 'pseudonym' of the 3M bomber prototype involved in a different record-breaking flight (*see below*), this designation was seemingly picked randomly. In fact, there were people in the West who took this designation at face value, and some Western publications referred to the M-4 as the 103M.

VM-25F High-Altitude Reconnaissance Aircraft (*izdeliye* 25F, Project)

In parallel with the M-4 bomber, the Myasishchev OKB undertook development of a dedicated photoreconnaissance (PHOTINT) version provisionally designated VM-25F (***f**otorazvedchik* – photoreconnaissance aircraft). Like the baseline 'aircraft M,' the aircraft was to be powered by AM-3A engines. However, it had a larger wingspan but nevertheless a lower structural weight and takeoff weight (two versions with different TOWs were considered – see the table on this page). To save weight, the fuel capacity was reduced and the defensive armament was limited to a single tail barbette with two 23 mm cannons. The crew was reduced to five – two pilots, a navigator, a radio operator, and a tail gunner (it seems that the pilots would have to take on the flight engineer's functions as well). The former bomb bay housed a battery of vertical and oblique cameras for day or night photography shooting through camera ports in the bay doors. The aircraft was not built.

Provisional specifications of the VM-25D	
Fuselage length (length overall?)	45.6 m (149 ft., $7\frac{17}{64}$ in.)
Height on ground	11.35 m (37 ft., $2\frac{55}{64}$ in.)
Wingspan	49.84 m (163 ft., $6\frac{13}{64}$ in.)
Wing area:	
overall	330 m^2 (3,548 sq. ft.)
less leading-edge root extensions	283 m^2 (3,043 sq. ft.)
Bombload:	
normal	5,000 kg (11,020 lbs.)
maximum	24,000 kg (52,910 lbs.)
Top speed at 9,000 m (29,530 ft.)	920–940 km/h (570–583 mph)
Cruising speed	800 km/h (496 mph)
Flight altitude over the target	12,800 m (42,000 ft.)
Technical range with a 5-ton (11,020 lb.) bombload	10,500 km (6,520 miles)

VM-25D Long-Range Fast Bomber (*izdeliye* 25D, Project)

This projected version differed from the basic M-4 mainly in featuring a 6 m^2 (64.5 sq. ft.) increase in flap area and having provisions for carrying air-to-surface missiles. Once again, the power plant consisted of four AM-3A turbojets. The aircraft was to feature in-flight refuelling capability, which most likely accounts for the *D* suffix to the designation (***d**ahl'niy* – long-range). The nose gear unit incorporated a 'jump strut' increasing the angle of attack for takeoff.

Left: Coded '26,' this display model depicts a projected derivative of the M-4 powered by two Dobrynin VD-5 turbojets (note the larger air intakes). This aircraft was not 'aircraft 26' (VM-26) – this designation applied to a totally different project.

Opposite page: The first project configuration of the 2M (VM-28) bomber with two VD-5 turbojets in the wing roots and two more in underwing pods. Note how the wings are positioned farther forward on the fuselage as compared to the M-4. The outrigger struts are still positioned at the wingtips in this version.

Provisional specifications of the bomber with two VD-5 engines	
Fuselage length (length overall?)	45.6 m (149 ft., 7 17/64 in.)
Height on ground	11.35 m (37 ft., 2 55/64 in.)
Wingspan	49.84 m (163 ft., 6 13/64 in.)
Wing area:	
overall	330 m² (3,552.09 sq. ft.)
less leading-edge root extensions	283 m² (3,046.19 sq. ft.)
Empty weight	63,800 kg (140,650 lbs.)
Takeoff weight	160,000 kg (352,730 lbs.)
Fuel load	89,500 kg (197,310 lbs.)
Bombload:	
normal	5,000 kg (11,020 lbs.)
maximum	24,000 kg (52,910 lbs.)
Top speed at 9,000 m (26,250 ft.)	900–920 km/h (560–570 mph)
Cruising speed	800 km/h (496 mph)
Flight altitude over the target	12,200 m (40,030 ft.)
Service ceiling over the target	13,800 m (45,280 ft.)
Technical range with a 5-ton (11,020 lb.) bombload	11,850 km (7,360 miles)

Long-Range Fast Bomber with Two VD-5 Engines (Project)

Another projected derivative of the 'aircraft M' under consideration in 1953 featured a more efficient power plant – two Dobrynin VD-5 axial-flow turbojets developed by OKB-36 in Rybinsk (Yaroslavl' Region), with a takeoff rating of 13,000 kgp (28,660 lbst) and a maximum continuous rating of 11,000 kgp (24,250 lbst). Again, the engines were placed aft of the wing torsion box, and the wing root sections were redesigned to accommodate them. Since the M-4 had the in-house designation VM-25 (*izdeliye* 25), the authors of some publications mistakenly labelled its twin-engined derivative as the VM-26 (*izdeliye* 26). In reality, the latter designation applied to a projected medium military transport/troopship aircraft with two Kuznetsov TV-2F turboprops dating back to 1951–52; no separate designation is known for the bomber with two VD-5s.

Like the M-4, the bomber had a crew of eight and a defensive armament suite comprising six 23 mm cannons in three powered barbettes. Similarly, there were provisions for JATO boosters – two large booster rockets delivering 9,000 kgp (19,840 lbst) apiece were to be mounted under the wing roots.

A model of the twin-engined version was tested in TsAGI's T-102 wind tunnel. Eventually, however, this version was abandoned in favour of a projected version with four VD-5s – the VM-28 (*see below*).

Long-Range Fast Bomber with AL-7 Engines (Project)

One more power plant option considered by OKB-23 in 1953 at the project stage was a version of the M-4 powered by Lyul'ka AL-7 turbojets – apparently the original AL-7 *sans suffixe* nonafterburning version dating back to 1952, with a takeoff rating of 6,500 kgp (14,330 lbst). Although much less powerful than the mighty AM-3, the Lyul'ka engines were expected to give the bomber a range in excess of 13,500 km (8,385 miles). This version was not proceeded with.

VM-28 High-Altitude Strategic Bomber (2M – Second Use of Designation, *izdeliye* 28, Project)

Back in August 1951, when the Myasishchev OKB was working on the advanced development project of the SDB bomber, it became clear that the aircraft's maximum range would be attained at altitudes ranging from 8,500 to 14,000 m (27,890–45,930 ft.). However, insufficient wing lift prevented the later 'aircraft M' from reaching really high altitudes where it would be safe from the potential adversary's air defences; in this respect the Soviet bomber was markedly inferior to the B-52.

Yet, on 29 March 1952 – long before the M-4 had even made its first flight – the Council of Ministers issued directive No. 1049-542, ordering OKB-23 to develop a high-altitude, long-range bomber powered by four Dobrynin VD-5 turbojets. The prototype was to commence manufacturer's tests in December 1953 and be submitted for state acceptance trials in May 1954. Thus, the work on the two types went on simultaneously, with about one year's lag for the new machine, which was allocated the in-house designation 2M (that is, second iteration of the 'aircraft M'), which had previously been briefly allocated to the second prototype M-4 before the latter was redesignated 'aircraft DM.' The project was also known as the VM-28 or *izdeliye* 28.

In accordance with the directive, the 2M was to carry 5 tons (11,020 lbs.) of bombs over a 12,500 km (41,010-mile) distance; the 'customer' specified the maximum flight altitude over the target as 17,300 m (56,760 ft.) and the top speed as 940–970 km/h (580–600 mph). Calculations showed that to achieve this performance, the aircraft was to have a takeoff weight of some 185 tons (407,850 lbs.), including a maximum fuel load of 107 tons (235,890 lbs.), part of which would be carried in drop tanks. Moreover, to attain the stipulated altitude the bomber would need a wing area of 450 m² (4,838.7 sq. ft.). To avoid the technical difficulties associated with building a wing that large, the designers decided to make do with 400 m² (4,301 sq. ft.) and provide the wings with powerful high-lift devices.

The appearance of the 2M changed considerably as the work progressed; the aircraft was similar in layout to the M-4, with the exception of the engine placement, which was the object of special studies. Four different arrangements were considered. **Version 1** had two of the engines buried in the wing roots, just like on the

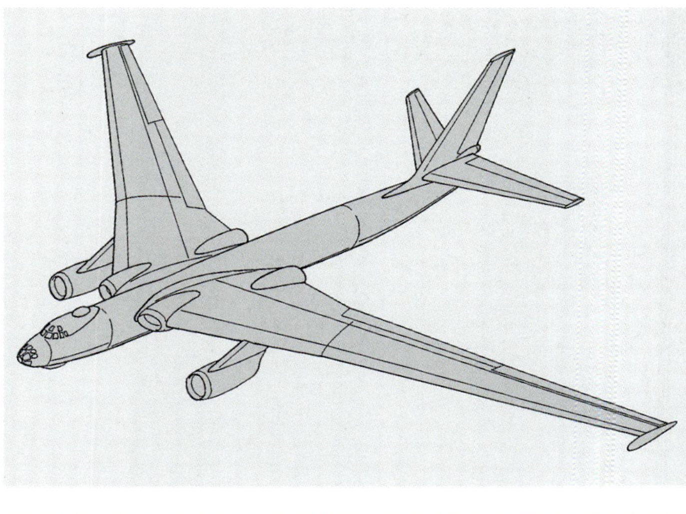

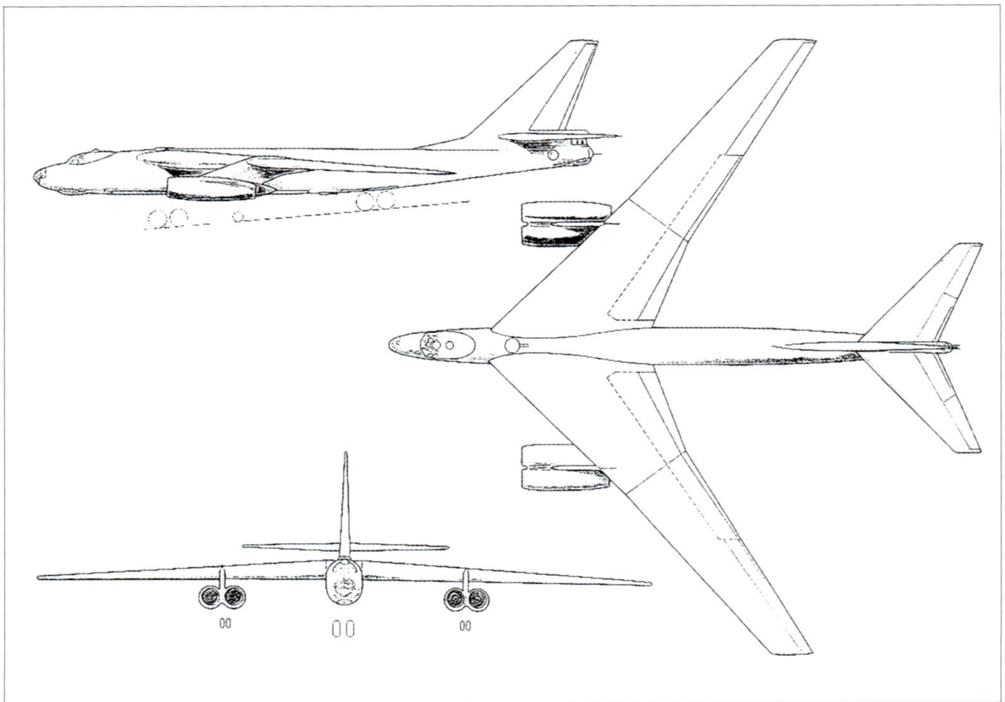

Left: A three-view drawing from the project documents, showing the second version of the 2M (VM-28), with paired engine nacelles housing the outrigger struts. The enlarged wings are clearly visible.

Below: A cutaway drawing of the same project version. Note the underwing drop tanks, the repositioned ventral cannon barbette located aft of the mainwheel well, and the redesigned rear pressure cabin seating two gunners, with lateral observation blisters in similar manner to the Tu-95.

twin-VD-5 bomber project described earlier, while the other two were carried under the wings on sharply swept pylons fairly close to the fuselage; the outrigger struts retracted into M-4 style wing-tip fairings.

Version 2 had the engines arranged in side-by-side pairs in B-52-style nacelles, which were carried on pylons located in line with the wing trailing-edge kink; the pylons' leading edge continued across the top of the wings to form boundary layer fences, limiting spanwise airflow. To be precise, the engine nacelles were more like those of the East German Baade 152 V4 short-haul airliner, since the engines were installed with a perceptible 'toe-in' and the gaps between the engine jet pipes accommodated fairings housing the outrigger struts. The main wing tanks occupying the entire wing torsion box held 69,300 kg (152,780 lbs.) of fuel; a further 19,000 kg (41,890 lbs.) was carried in three fuselage tanks, the first two of which were rigid extra tanks suspended in the upper part of the bomb bay. Additionally, there were provisions for two cigar-shaped drop tanks carried on integral pylons under the wings at half span; these could hold between 3,100 and 9,500 kg (6,830–20,940 lbs.) of fuel each.

Version 3 had four individual underwing engine nacelles carried on pylons at 25% and 61% of the wingspan. This time, all four pylons formed boundary layer fences.

Version 4 likewise had four individual underwing engine nacelles, but these were carried on pylons at 25% and 50% of the span, with cigar-shaped outrigger strut fairings protruding beyond the leading edge well inboard of the wingtips. This version was selected as the final one for further development and reached the

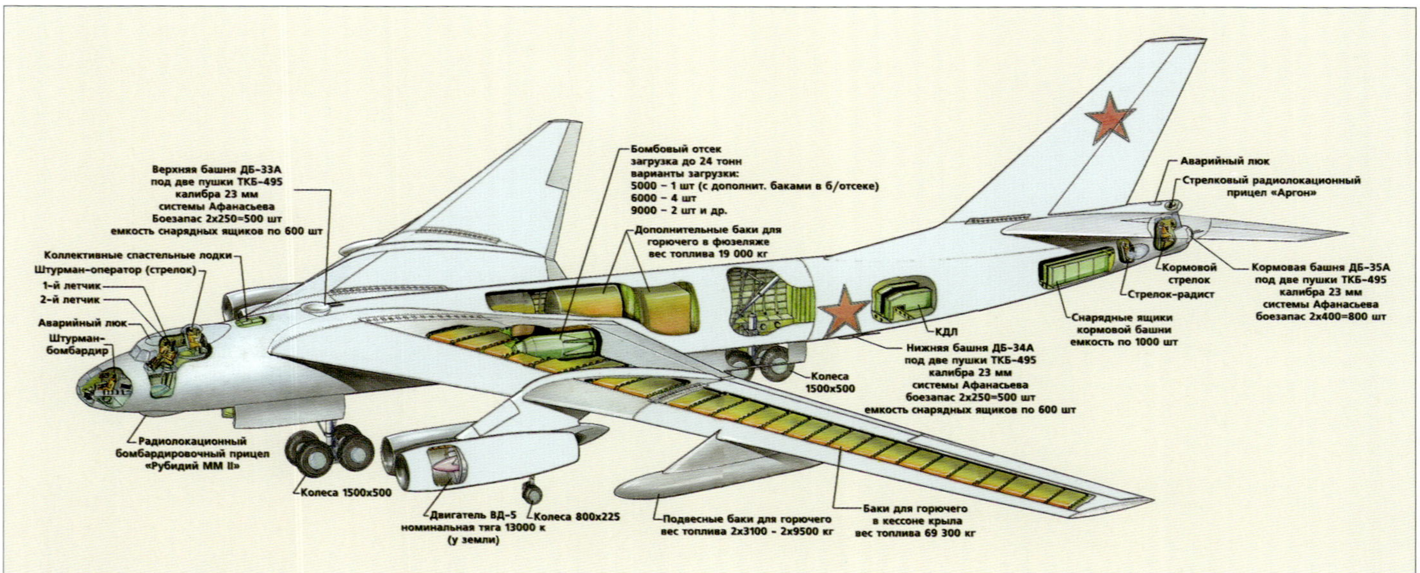

Right: The penultimate version of the 2M (VM-28) project. The outer engine nacelles are conformal *à la* B-47, and the outrigger struts retract into these nacelles in the same fashion. The sweep angles of the wings and tail surfaces are indicated.

Below and below right: A desktop model showing a similar configuration of the 2M with four pylon-mounted engines (the outer engines are not conformal). The outrigger struts are, unfortunately, missing.

Bottom and bottom right: Another model showing the final version of the 2M with four pylon-mounted engines placed closer together and outrigger strut pods located well inboard of the wingtips. The nose gear unit does not have twin wheels – the rear pair of wheels was 'lost' due to poor retouching of the original image.

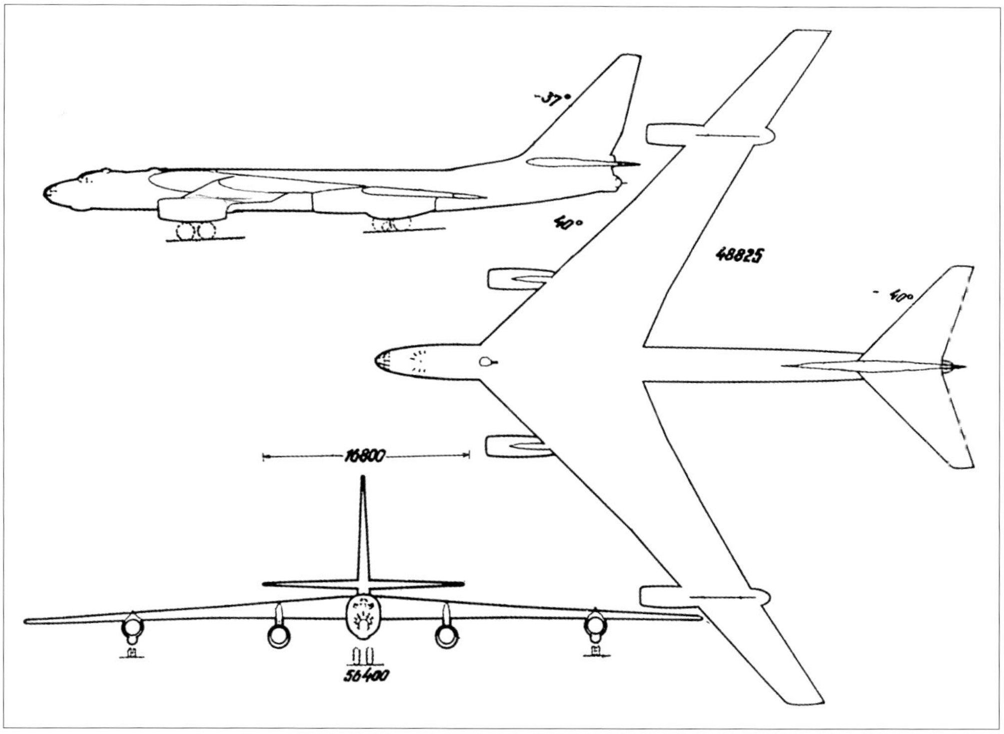

full-size mockup stage. The Air Force's mockup review commission examined the mockup in December 1952 (some sources say January 1953) and generally approved; however, some alterations had to be introduced in accordance with the critical remarks of the commission. For example, the outer engines were moved outward to 70% of the span and housed in conformal nacelles; the latter also accommodated the outrigger struts, so that the aircraft looked like a B-47 gone wrong.

The 2M's proportions differed from those of the M-4 – the wings were located farther forward along the fuselage, and so was the nose gear unit. The wings and horizontal tail had 40° sweep at quarter chord; the vertical tail was swept back 37° at quarter chord. To meet the requirement that the maximum ordnance load be increased from 24 to 27 tons (from 52,910 to 59,520 lbs.) the fuselage cross-section was changed from circular to oval, making for a more capacious bomb bay. The bomber had a crew of six; the two pilots, the navigator / bomb aimer sitting ahead of them in the glazed nose, and the radar operator / dorsal gunner sitting beneath an astrodome occupied the forward pressure cabin, which lacked the M-4's distinctive lateral sighting blisters. The gunner / radio

Left: An artist's impression of the 2M in the penultimate version. It does not conform to the actual project, having M-4-style forward beam blisters and a forward-mounted ventral barbette.

Below left: A wind tunnel model of the 2M in the penultimate version

Below: Another wind tunnel model, showing the 2M in the final version. The humps on top of the outer wings are attachment points for suspending the model in the wind tunnel.

operator working the ventral barbette and the tail gunner (a.k.a. defensive fire commander) sat in a new two-seat rear pressure cabin, which had lateral-sighting blisters in the manner of the Tu-16 and Tu-95. (It should be noted that the OKB presented two versions of the rear pressure cabin to the mockup review commission – there was also an M-4-style single-seat tail gunner's station.)

The weapons options included 250 kg (551 lb.) FAB-250 M46, 5-ton (11,020 lb.) FAB-5000 M46, and 9-ton (19,840 lb.) FAB-9000 M46 high-explosive bombs (*foogahsnaya* ***avia******bom****ba*); 6,000 kg (13,230 lb.) BRAB-6000 armour-piercing bombs (*brone****boy****naya* ***a****vi****abom****ba*); or, on antishipping missions, AMD-2M, Desna, and A-2 naval mines and type 45-36AV torpedoes. (The latter designation means 450 mm (17⁴⁵⁄₆₄ in.) calibre, 1936 model, *[tor****ped****a] aviats****ion****naya, vy****sot****naya* – air-droppable torpedo for high-altitude attacks.) For self-defence the aircraft was provided with six 23 mm cannons, the ventral barbette being relocated to the rear fuselage underside; the tail turret was aimed by means of a new *Ks****enon*** (Xenon) gun-laying radar enclosed by a larger fairing. The ammunition supply was 125 rpg for the dorsal and ventral barbettes and 400 rpg for the tail barbette (some sources say 250 and 400 rpg, respectively).

A scratch-built display model of the 2M in the penultimate project configuration

A scratch-built display model of the 2M in the final configuration

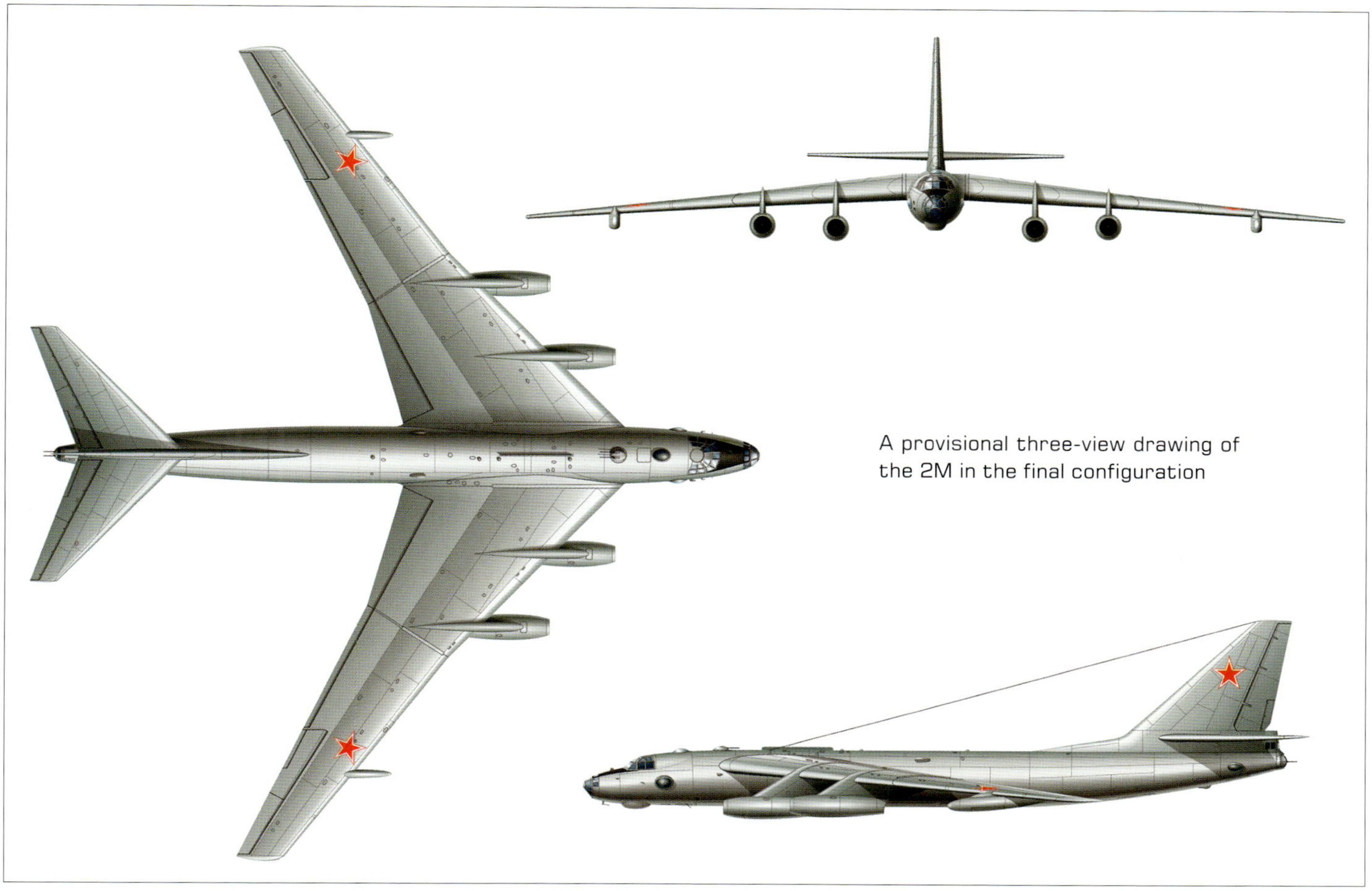

A provisional three-view drawing of the 2M in the final configuration

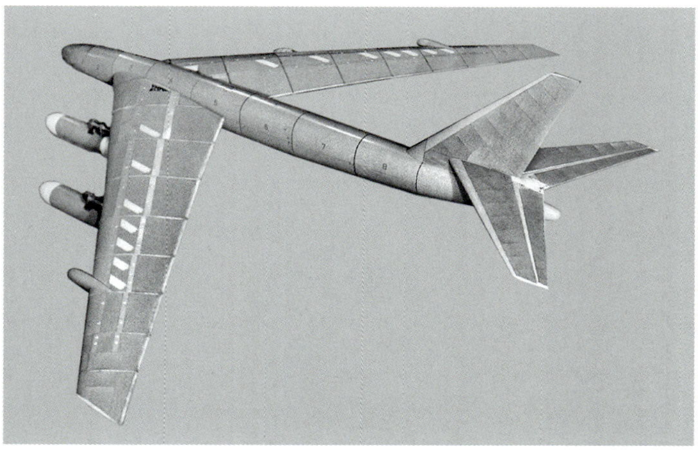

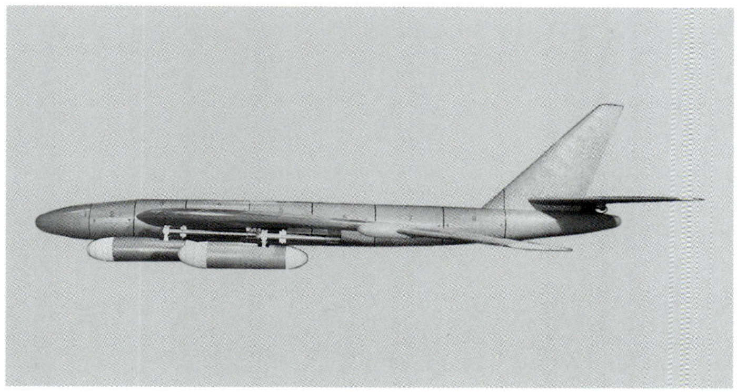

Above and above right: Two views of a scaled-strength model of the 2M in the final configuration, with a very schematic reproduction of the engine pylons

Right: A partial wind tunnel model of the 2M in the same configuration. The outrigger strut fairings are not reproduced on this model.

The dimensions of the 2M varied from version to version; in one project version, the wingspan was 48.5 m (159 ft., 1 29/64 in.), the overall length 50.0 m (164 ft., 1/2 in.), and the wing area was estimated at 320–340 m² (3,440–3,656 sq. ft.). In another version, the aircraft was to have a wingspan of 52.92 m (173 ft., 7 15/32 in.) and a fuselage length

Left: A full-size wooden mockup representing the final configuration of the 2M (VM-28) bomber. Note that there are no sighting blisters on the forward fuselage sides.

Below left: A separate mockup of the rear pressure cabin, seating two gunners, with beam blisters for the forward one. Note the two-pronged radome of the gun-laying radar, which tips up to form an escape hatch.

Below: A mockup of the alternative single-seat tail gunner's station, with high-set lateral blisters looking like half-closed eyes

Bottom: A cutaway drawing of the 2M's ultimate configuration

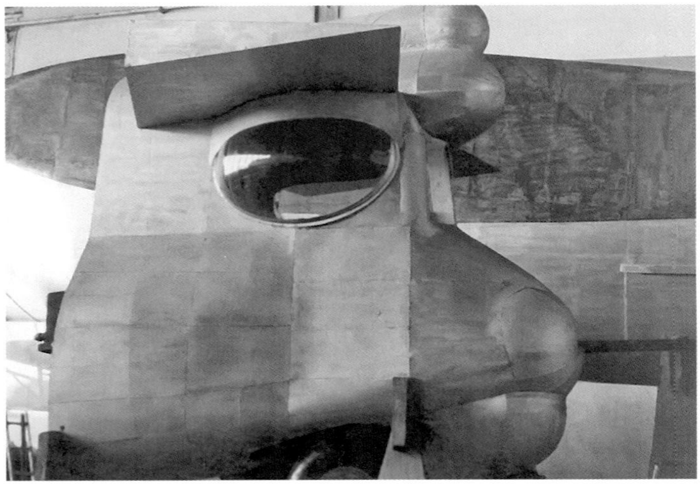

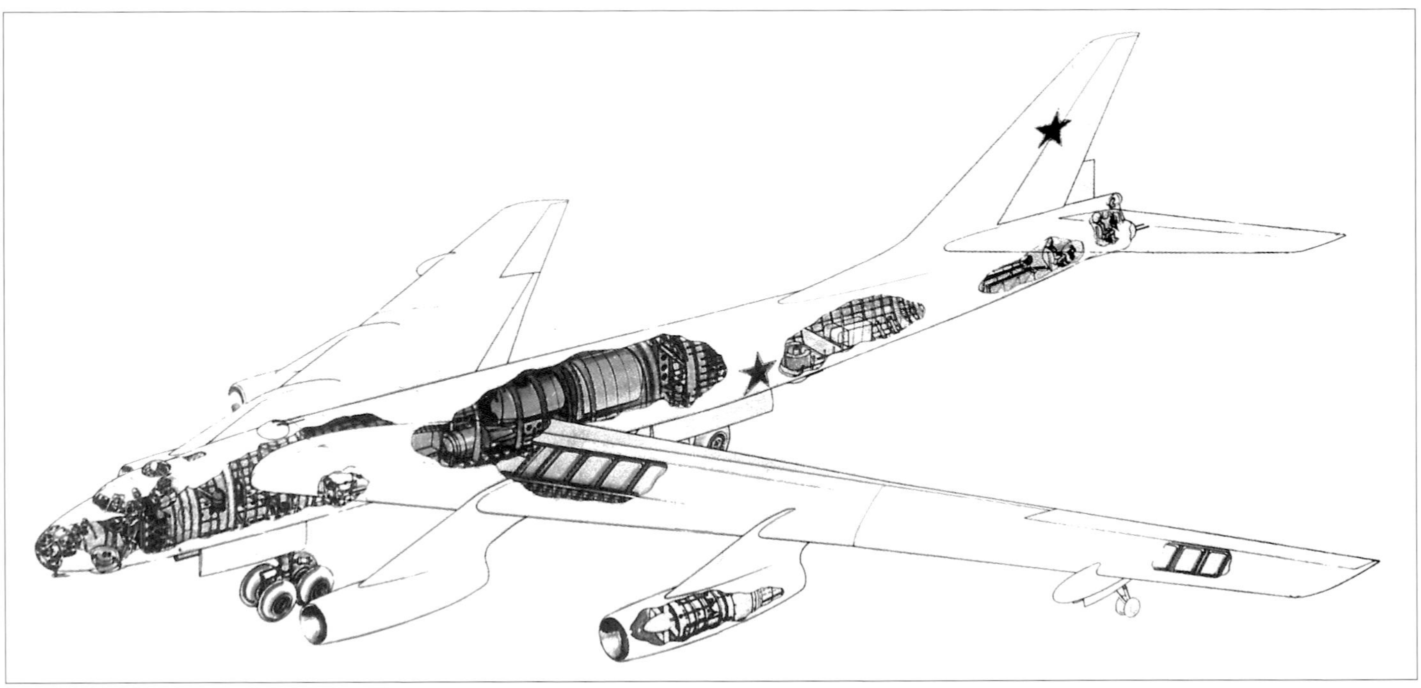

Above: The copilot's instrument panel and engine instrument panel of the 2M mockup

Above right: The captain's instrument panel on the same mockup

Right: The gunner / radio operator's station of the 2M mockup, showing one of the beam blisters and the gunsight in it

of 47.6 m (156 ft., 2¹⁄₆₄ in.). Details of two further versions of the VM-28 (2M) are given in the table below; please note that the numbering does not necessarily match the versions listed above.

In September 1953, the final version of the project was again inspected by the mockup review commission, and the government authorised construction of the 2M (VM-28) prototype in parallel with the manufacture of the aforementioned 11-ship initial production batch of M-4 bombers. Council of Ministers directives required the 2M to begin state acceptance trials first in May 1954, then in June 1955, but the aircraft was never built. This was because the Myasishchev OKB had offered a new version of the M-4 bomber – the 3M, which was to yield the required results at much less cost. Another reason was the protracted development of the VD-5 engine (it first ran on 9 February 1953) and the fact that OKB-36 was already working on the more promising VD-7 turbojet, whose development had been officially sanctioned by a Council of Ministers directive dated 6 July 1954. The Myasishchev OKB had selected this engine for the 3M.

Hence, in October 1954 the government pulled the plug on the 2M project. A new Council of Ministers directive issued on 16 October 1954 said:

'Due to the inadequate range (7,700 km) [4,780 miles] of the 2M high-altitude, long-range bomber designed by Comrade Myasishchev and considering that the Ministry of Aircraft Industry has received new assignments involving modification of the M-4 bomber with a view to obtaining a range of up to 14,000 km [8,700 miles], as well as construction of an experimental compound bomber with a range of 13,000 km [8,075 miles] and a service ceiling of 14,000–15,000 m [45,930–49,210 ft.], . . . the proposal to terminate all further design and construction work on the 2M high-altitude, long-range bomber put forward by the Ministry of Aircraft Industry, the Ministry of Defence, and the Ministry of Agricultural Machinery is hereby accepted.'

The reference to the Ministry of Agricultural Machinery (MSKhM – *Ministerstvo sel'skokhozyaistvennovo mashinostroyeniya*) may sound a bit wild but is explained by the fact that after the Second World War, this ministry was responsible not only for ploughs and such but also for the production of various weaponry and ordnance. (Agricultural machinery, indeed! This reminds us of a Russian joke dating back to the days of Sino-Soviet military confrontation in the 1970s, where a 'Soviet peaceful tractor

Provisional specifications of the VM-28 (2M)

	Version 1	Version 2
Fuselage length	48.825 m (160 ft., 2¹⁵⁄₆₄ in.)	47.6 m (156 ft., 2¹⁄₆₄ in.)
Wingspan	56.4 m (185 ft., ¹⁵⁄₃₂ in.)	55.8 m (183 ft., ⁵⁵⁄₆₄ in.)
Wing area, m² (sq. ft.):		
overall	429 (4,617.72)	429 (4,617.72)
less leading-edge root extensions	390 (4,197.93)	390 (4,197.93)
Flight altitude over the target, m (ft)	17,000 (55,770)	16,000 (52,500)
Takeoff weight, kg (lbs.)	146,000 (321,870)	180,000 (396,825)
Empty weight, kg (lbs.)	74,000 (163,140)	74,000 (163,140)
Fuel load, kg (lbs.)	66,500 (146,600)	100,000 (220,460)
Technical range with a 5-ton (11,020 lb.) bombload, km (miles)	7,680 (4,770)	10,510 (6,527)
Takeoff run, m (ft.)	1,000 (3,280)	1,600 (5,250)
Service ceiling, m (ft.)	17,300 (56,760)	16,600 (54,460)
Cruising speed at 9,000 m (29,530 ft.), km/h (mph)	940–970 (580–600)	940–970 (580–600)

ploughing a field' is fired upon by six Chinese artillery batteries, then returns fire and destroys them!) Actually the reference to MSKhM in this particular Council of Ministers directive is odd for a quite different reason, since this ministry no longer existed in 1954, having been amalgamated with three other ministries on 5 March 1953 to form the Ministry of Machinery.

M-4P Long-Haul Airliner ('M-6P,' M-29, *izdeliye* 29, Project)

In the mid-1950s, passenger air traffic in the Soviet Union was booming, sparking a requirement for new medium-haul and long-haul fast airliners to satisfy the needs of Aeroflot Soviet Airlines, the sole national air carrier. Such aircraft needed to have adequate seating capacity and adequate range to serve domestic and international routes efficiently – that is, fly nonstop point-to-point services without the refuelling stops necessary for the piston-engined types that formed the backbone of Aeroflot's fleet.

The Tupolev OKB was first past the post in this respect, developing the Tu-104 *Camel* twin-turbojet, medium-haul airliner as a derivative of the Tu-16. In 1956 it received a new assignment, commencing development of the 'aircraft 95T' military transport and the 'aircraft 95P' airliner, based on the Tu-95.

However, the Myasishchev OKB was not left out either – and took the same approach, developing airliner and transport derivatives of the M-4 bomber at its own initiative in 1954. At OKB-23 the airliner and military transport projects, which were designed with maximum commonality in mind, received the blanket designation ***tema*** (subject) 29 or M-29, though some publications referred to it as VM-29. The commercial version received the parallel designation M-4P (*passa**z**heerskiy [samo**lyot**]* – airliner). O. A. Sidorov was the project chief; a lot of work on the M-4P was performed by Yu. P. Bobrovnikov.

An important comment must be made here. Originally the M-29 was to be powered by AM-3A or AM-3M turbojets, which by then had reached an adequate reliability level; later they were to be replaced by fuel-efficient VD-7 turbojets, which would improve operating economics and give longer range. This caused some researchers to assume that the aircraft was based on the 3M bomber (a.k.a. M-6), which was designed around the same VD-7 engines from 1954 onward, and thus to refer to the projected airliner version as the M-6P. However, the 3M had a later programme code (Subject 36), and the M-29 cannot have been developed from it 'in reverse.'

The M-4P shared the bomber's wings, tail surfaces, landing gear, and power plant arrangement. These were mated to a new and larger fuselage that turned the aircraft from a midwing monoplane to a low-wing monoplane. The fuselage had an unusual 'double-bubble' cross-section with a larger upper lobe; this allowed the task of maximising usable internal volume to be reconciled with the need to retain maximum structural commonality with the bomber, thereby cutting costs. (The Boeing Company had used the same principle when deriving the C-97 Stratofreighter and the Model 377 Stratocruiser airliner from the B-50 Superfortress bomber.) The upper and lower lobe diameter was 3.7 m (12 ft., 1⁴³⁄₆₄ in.) and 3.5 m (11 ft., 5⁵¹⁄₆₄ in.), respectively, making five-abreast seating possible; only the upper lobe (the flight deck and cabin) was pressurised, with the unpressurised lower lobe accommodating the wheel wells, avionics/equipment bays, and baggage compartments.

It should be noted that the approach taken by the two design bureaus was rather different. While the Myasishchev OKB opted for a low-wing layout with an all-new fuselage from the outset, the 'aircraft 95T' and 'aircraft 95P' were midwing designs with a slim fuselage similar to that of the Tu-95 bomber. Eventually, however, the Tupolev OKB arrived at the same solution, evolving the Tu-114 *Rossiya* (Russia; NATO reporting name *Cleat*) four-turboprop, long-haul airliner and the projected Tu-115 military transport, which had a larger fuselage and a low-wing layout.)

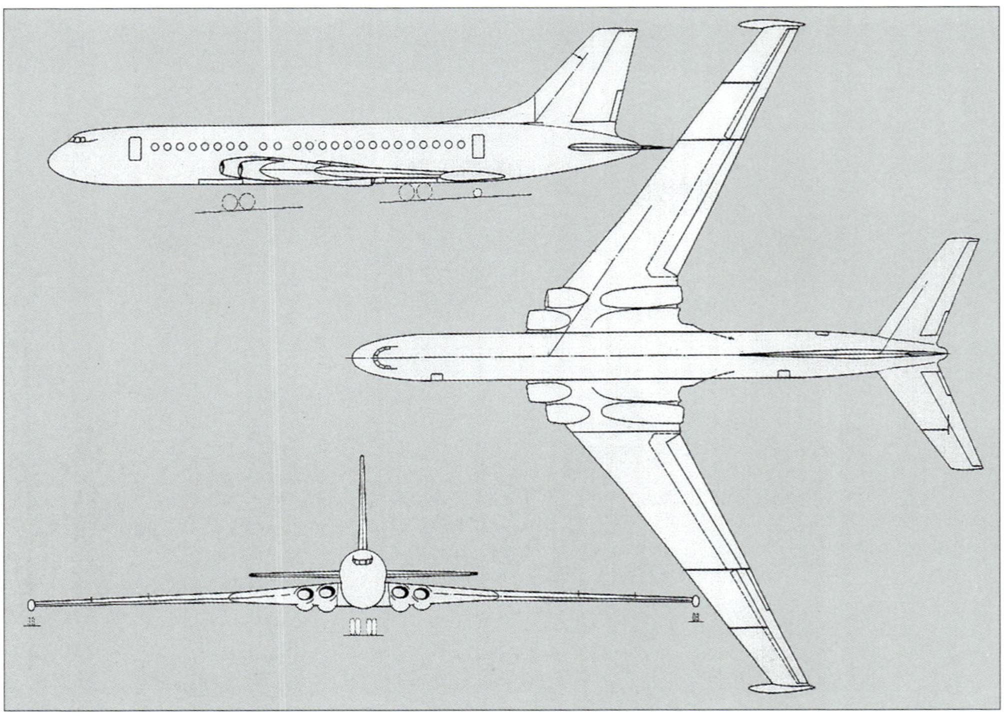

Left: A provisional three-view drawing of the M-29 (M-4P) airliner, showing a single passenger deck, a stepped flight deck windscreen, and no navigator's station glazing

Above right: A display model of the M-29 in a similar configuration, except that the cabin windows at the rear (in the galley and baggage compartment) are located higher. The symbolic pre-1958-style Soviet registration CCCP Λ-1956 (SSSR L-1956 in Cyrillic characters) alludes to the year of design.

Right: A different model of the M-29 featuring a reshaped parabolic nose with a Boeing Stratoliner–style integrated flight deck windscreen and double-deck passenger accommodation

The M-4P was offered in two versions. The airline variant accommodated 100–130 passengers with a high degree of comfort (one of the layouts had two cabins seating 68 and 38, respectively, with lavatories in between and coat storage fore and aft, opposite the portside entry doors). There was also a long-range VIP transport version with sleeper compartments for 15 to 30 passengers fore and aft, a conference room amidships, and a buffet at the front. The aircraft had a five-man flight crew – captain, copilot / flight engineer, navigator, radio operator, and flight technician.

According to the PD project, the airliner was to have a takeoff weight of 158–162 tons (348,320–357,140 lbs.), a payload of 16–24 tons (35,270–52,910 lbs.), and a seating capacity of 80–130. The aircraft would have a range of 5,400–7,000 km (3,354–4,347 miles), a cruising speed of 800 km/h (496 mph), and a take-

off and landing run of 1,300–1,400 m (4,265–4,590 ft.), with brake parachutes to reduce the landing run.

In March–April 1955, the M-4P project was submitted for review to MAP, the air force and the Ministry of Civil Aviation (MGA – *Ministerstvo grazhdahnskoy aviahtsii*). On 12 August 1955 the Council of Ministers issued directive No. 1561-868, followed up by MAP order No. 571 on 29 August, which officially tasked OKB-23 with developing an airliner version of the *Bison*. These documents contained more-stringent requirements, as detailed in the table on this page. The stipulated maximum range would allow the M-4P to fly nonstop from Moscow to New York (7,500 km / 4,658 miles), to say nothing of shorter services such as Moscow–Magadan (6,000 km / 3,726 miles), Moscow–Beijing (5,800 km / 3,602 miles), and Moscow–Delhi (4,400 km / 2,733 miles).

Also, the bicycle undercarriage inherited from the bomber entailed operational limitations because of the very wide track of the outrigger struts placed at the wingtips; few airports had runways and taxiways of sufficient width. This prompted a study of a new undercarriage layout (unfortunately, no details are available).

The M-4P project was terminated because MAP favoured the competitors' designs – the Tu-114, the Antonov An-10 *Cat*, and the Ilyushin Il-18 *Coot* medium-haul airliners. The prejudiced attitude of MAP's top management to Myasishchev (and very probably lobbying by his chief competitor, Tupolev) was also a factor. On

Specifications of the M-29 (M-4P) as per CoM directive

	Option A	Option B
Payload, tons (lbs.)	30 (66,140)	5–15 (11,020–33,070)
Seating capacity	170–180	100–120
Range with one-hour fuel reserves, km (miles)	3,500–4,000 (2,174–2,484)	7,500–8,000 (4,658–4,968)
Maximum speed, km/h (mph)	900–950 (559–590)	900–950 (559–590)
Cruising speed, km/h (mph)	800–825 (496–512)	800–825 (496–512)
Takeoff/landing run, m (ft)	1,400–1,500 (4,590–1,640)	1,400–1,500 (4,590–1,640)

Right: A drawing from the M-4P's project documents, showing the standard airline version. The seating capacity is 80 with four-abreast seating throughout, 100 with five-abreast seating, and 130 with six-abreast seating; the plan view shows a 106-seat version (68+38).

Below right: This drawing from the ADP documents shows the VIP ('special') version with sleeper accommodation for 30 passengers.

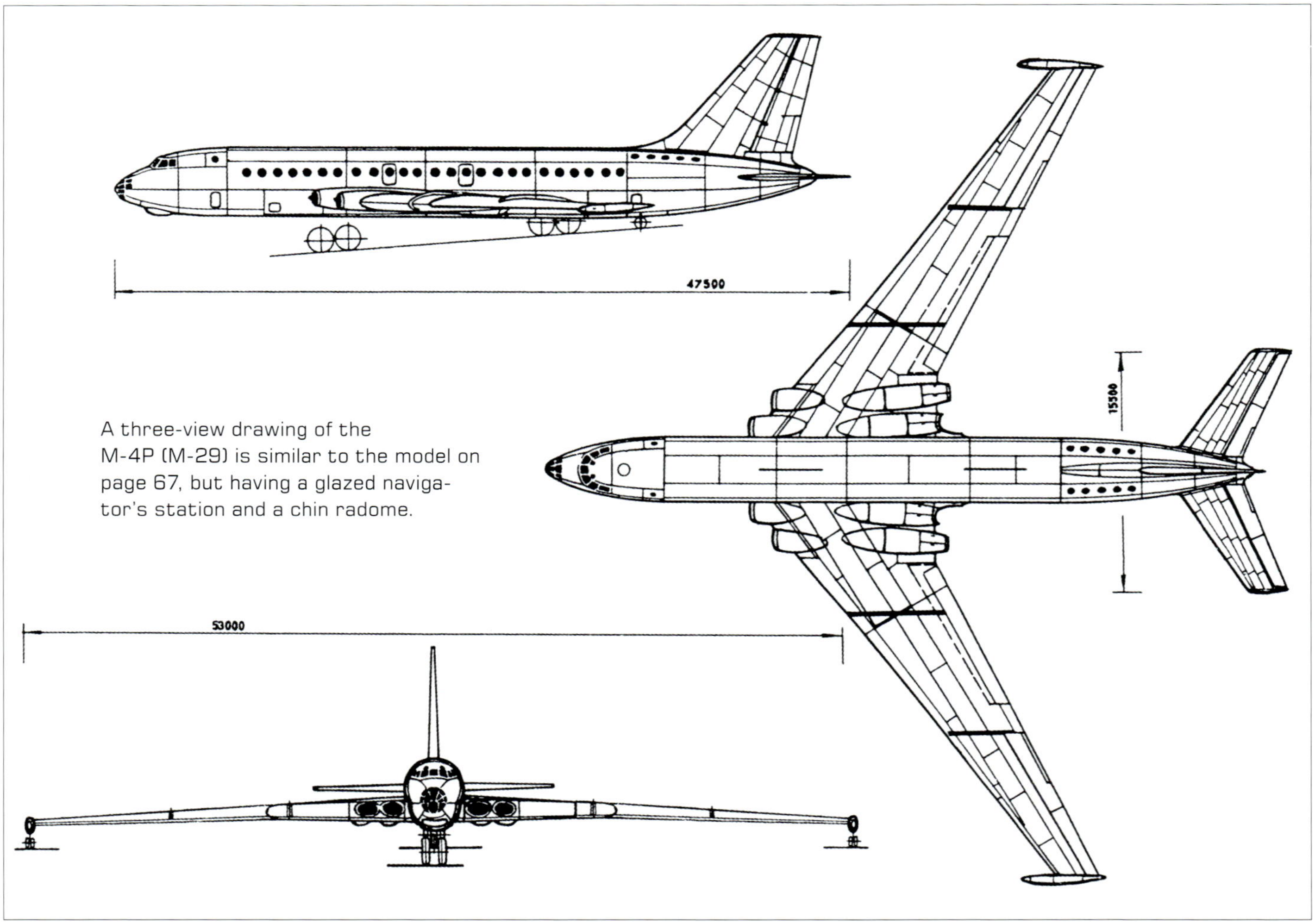

A three-view drawing of the M-4P (M-29) is similar to the model on page 67, but having a glazed navigator's station and a chin radome.

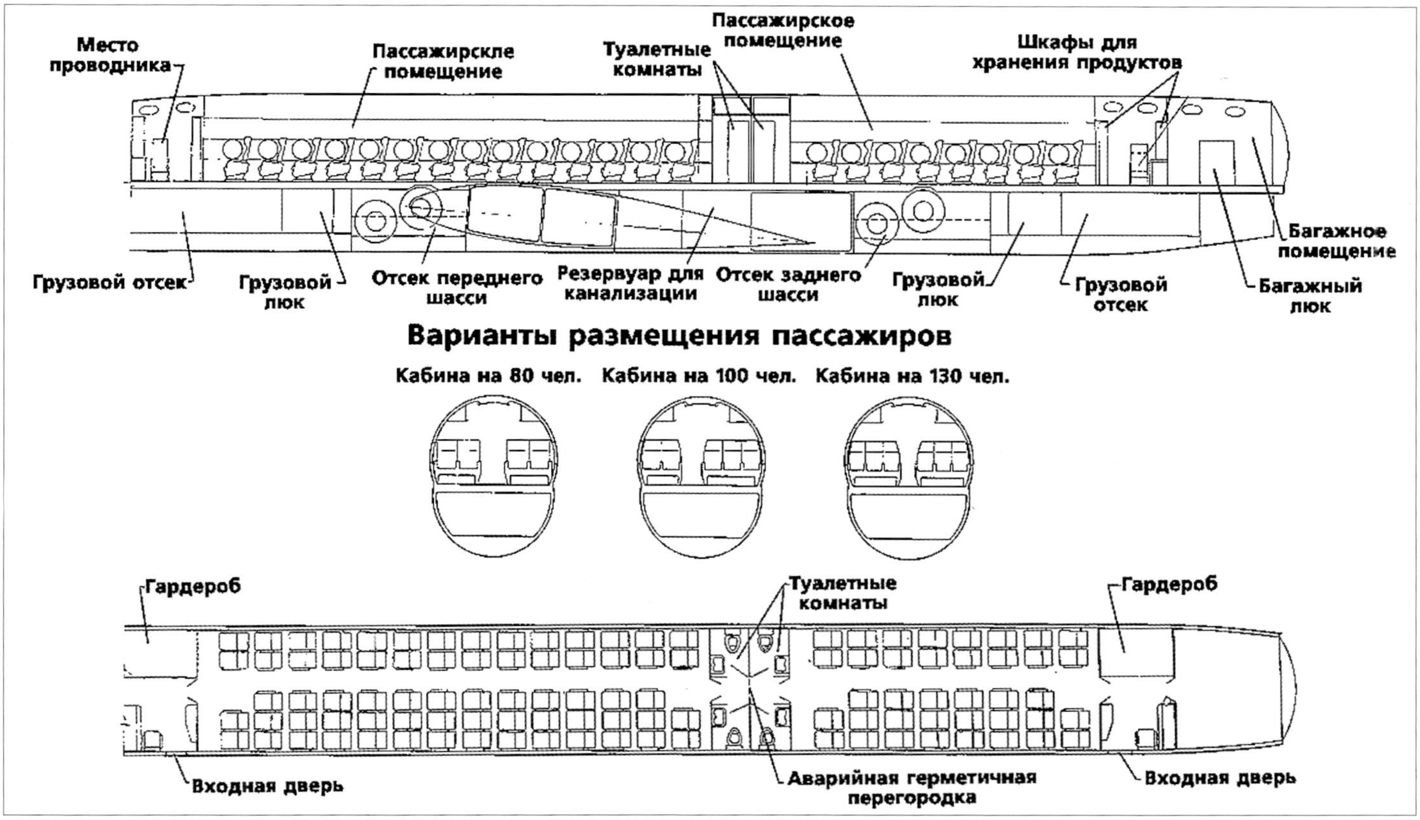
Место проводника
Пассажирскле помещение
Туалетные комнаты
Пассажирское помещение
Шкафы для хранения продуктов
Багажное помещение
Багажный люк
Грузовой отсек
Грузовой люк
Отсек переднего шасси
Резервуар для канализации
Отсек заднего шасси
Грузовой люк
Грузовой отсек
Варианты размещения пассажиров
Кабина на 80 чел.
Кабина на 100 чел.
Кабина на 130 чел.
Гардероб
Туалетные комнаты
Гардероб
Входная дверь
Аварийная герметичная перегородка
Входная дверь

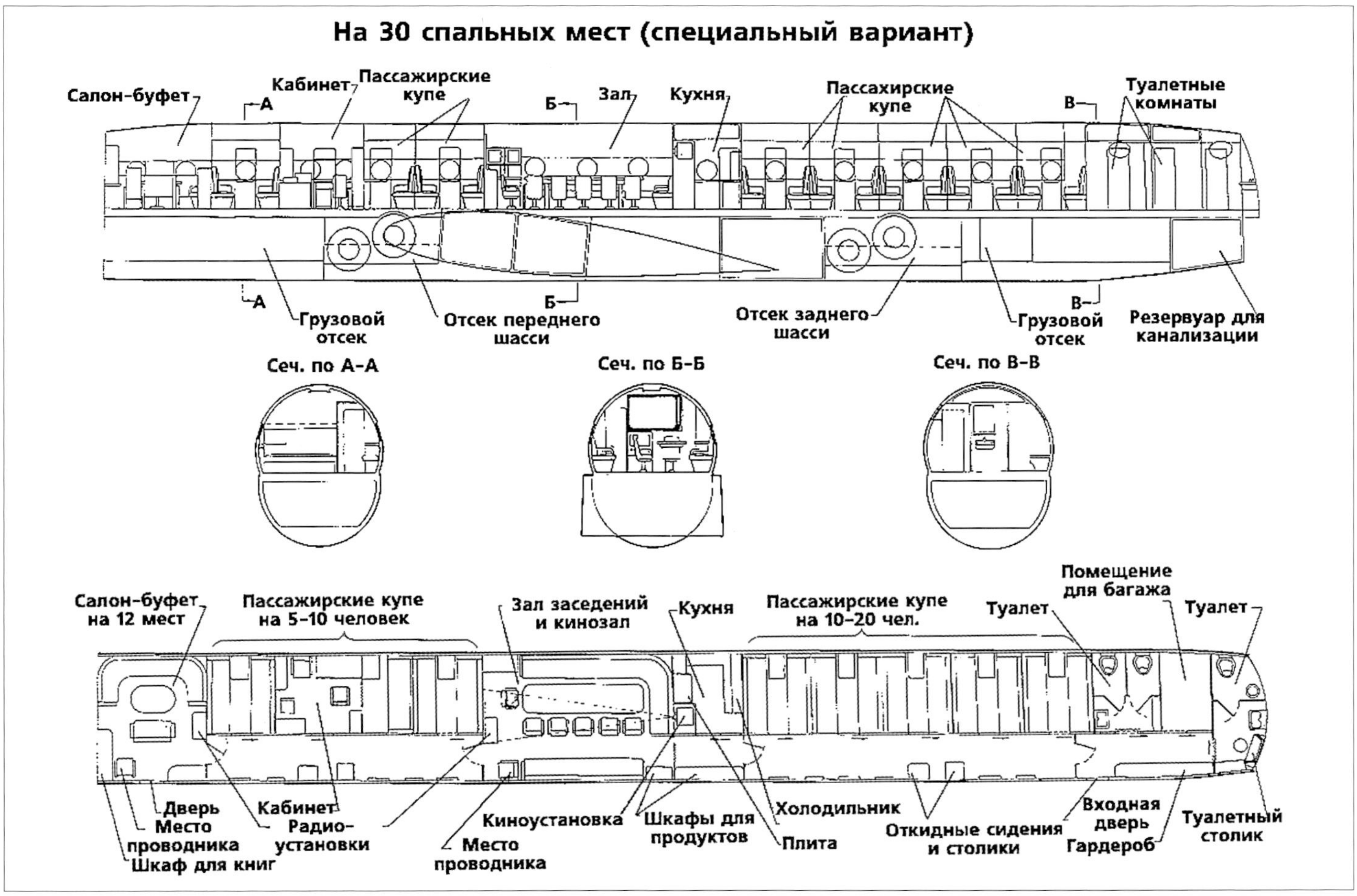
На 30 спальных мест (специальный вариант)
Салон-буфет
Кабинет
Пассажирские купе
Зал
Кухня
Пассахирские купе
Туалетные комнаты
А
Б
В
Грузовой отсек
Отсек переднего шасси
Отсек заднего шасси
Грузовой отсек
Резервуар для канализации
Сеч. по А-А
Сеч. по Б-Б
Сеч. по В-В
Салон-буфет на 12 мест
Пассажирские купе на 5-10 человек
Зал заседений и кинозал
Кухня
Пассажирские купе на 10-20 чел.
Помещение для багажа
Туалет
Туалет
Дверь
Место проводника
Шкаф для книг
Кабинет
Радио-установки
Место проводника
Киноустановка
Шкафы для продуктов
Плита
Холодильник
Откидные сидения и столики
Входная дверь
Гардероб
Туалетный столик

Left: Artist's impression of an Aeroflot M-4P registered CCCP Λ-1961 and having a 'solid' nose with a parabolic radome.

Right: A map from the project documents, showing the M-4P's possible destinations on flights from Moscow, including New York, Magadan (in the Soviet Far East), Beijing, and Delhi

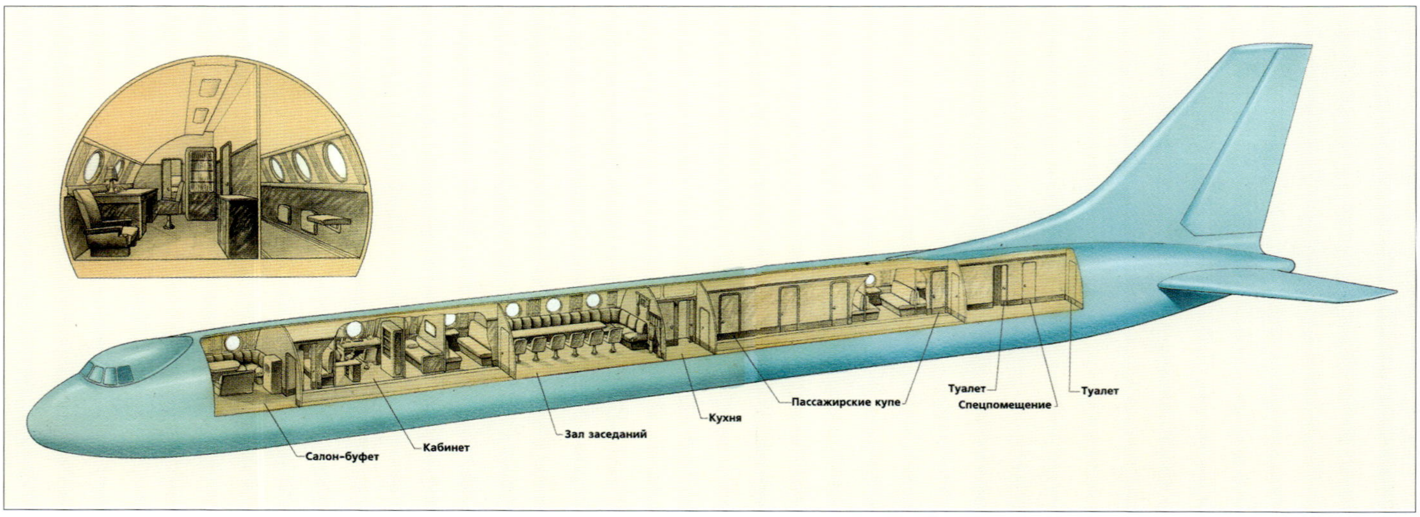

Above: The M-4P's VIP version with (*left to right*) a café compartment, a study for the 'main passenger' (top-ranking VIP), a conference hall, a galley, passenger compartments, and two lavatories with a 'special compartment' in between for carrying diplomatic mail

Left: An artist's impression of an M-4P with the registration CCCP-60931, under the system introduced in 1958. Note that the outrigger strut fairings are located well inboard of the wingtips.

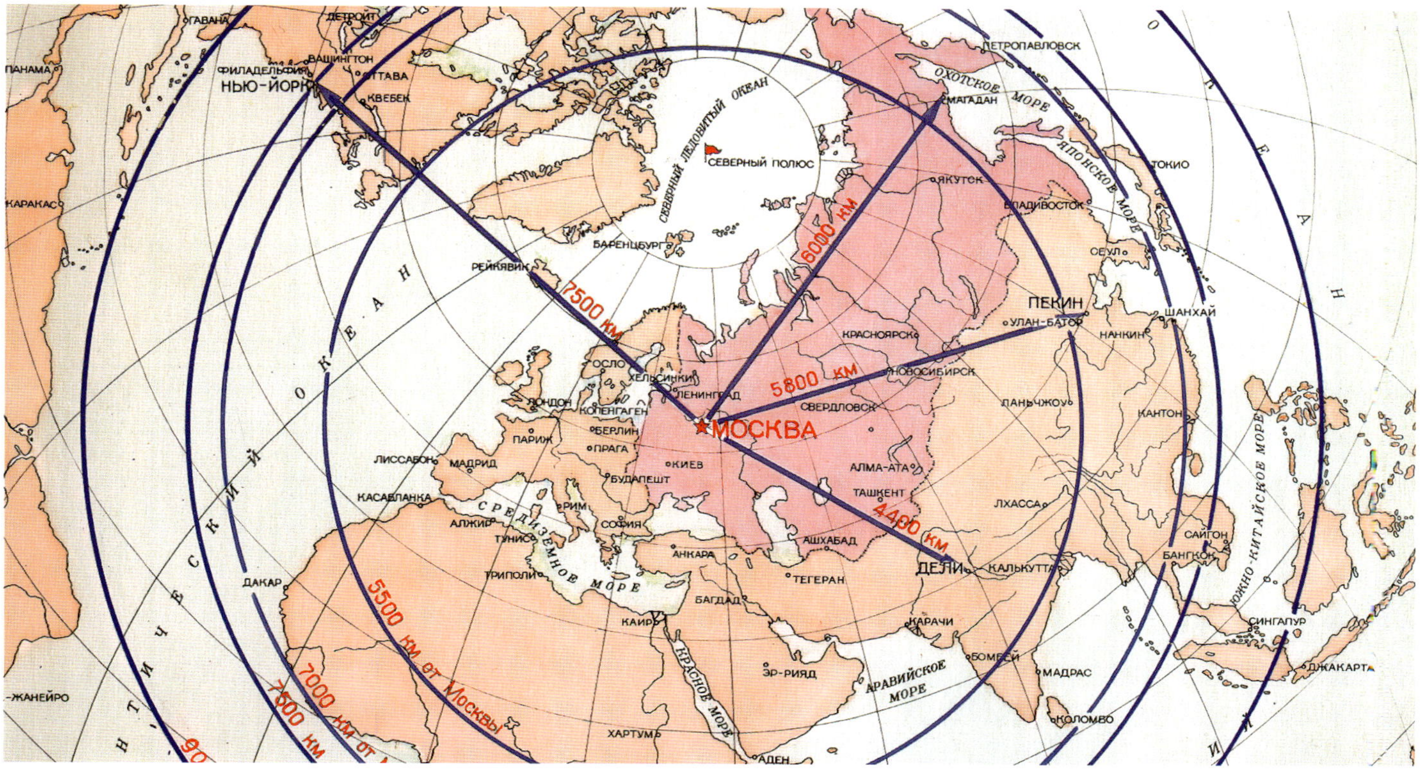

30 November 1955 the Council of Ministers issued directive No. 1966-1055 (followed up by MAP order No. 766 on 9 December), cancelling the project on the pretext of expediting development of other airliners amenable for military transport / troopship operations.

M-4T Military Transport Aircraft ('M-6D,' M-29, *izdeliye* 29, Project)

Also, in the mid-1950s the need arose for new heavy military airlifters to satisfy the needs of the Soviet Union's rapidly developing airborne forces (VDV – *Vozdooshno-desahntnyye voyska*). Hence, OKB-23 developed a military transport version of the M-29 project, known as the M-4T (***trahnsportnyy [samolyot]*** – transport aircraft). Some sources indicate the transport version was to be designated M-6D (*desahntnyy [samolyot]* – troopship).

Though basically similar to the M-4P airliner, the military transport version had a different fuselage design. It featured a strengthened cargo floor and a large cargo hatch closed by clamshell doors in the forward fuselage underside, in similar manner to the Douglas C-124 Globemaster II. The hatch was provided with a two-section folding ramp for loading wheeled or tracked vehicles; alternatively, loads secured to paradroppable platforms could be lifted through the hatch if the ramp was removed. A second hatch was provided in the rear fuselage underside for paradropping cargo or personnel.

The cargo hold was pressurised; it could accommodate up to 200 troops with their equipment and personal firearms or various items of weaponry (e.g., artillery pieces, mortars) and military vehicles to a total weight of 40 tons (88,200 lbs.). Two of the non-paradroppable payload options were five ZiS-151 6×6 army lorries, or three GAZ-63 4×4 army lorries and two 85 mm field guns. In the medical evacuation role, the aircraft could transport 180 stretcher patients and 15 medical attendants. Provision was made for a remote-controlled cannon barbette and gun-laying radar in the fuselage tail cone.

The M-4T project was cancelled together with the M-4P in November 1955.

Long-Range Minelayer/Torpedo-Bomber Version of the M-4

Separate state acceptance trials of the M-4's envisaged antishipping armament took place in late 1956, using the second prototype ('aircraft DM'), which by then had gained the unusual three-digit tactical code '101 Red.' Between 1 August and 10 October, the aircraft made 12 test flights from Zhukovskiy and from Kirovskoye AB near Simferopol' on the Crimea Peninsula; from the latter location, sorties were flown at a naval test range on the Black Sea. The test crew included project test pilot Vavilonov, navigator Makarov, radar operator Kryukov, and radio operator / gunner Zulutdinov.

For each mission the bomber was alternatively armed with six RAT-52M torpedoes or six naval mines of assorted types (IGDM, APM, Lira, Serpey, or AMD-2M). The munitions were carried internally on KD4-M racks (*kassetnyy derzhahtel'* – 'cassette-type rack'; that is, bomb cassette, optimised for 'aircraft M'). With 5% fuel reserves, maximum range in minelayer/torpedo-bomber configuration at a cruising speed of 800 km/h (496 mph) was 8,450 km (5,250 miles); the service ceiling over the target did not exceed 13,100 m (42,980 ft.), increasing to 15,300 m (50,200 ft.) at the end of the mission, after the torpedoes had been dropped.

The RAT-52 deserves special mention. It was conceived as a homing torpedo, but the guidance system was considered too complicated and was deleted in the production version. The torpedo weighed 627 kg (1,382 lbs.) complete with the parachute system and

had a 243 kg (535 lb.) warhead. Before dropping the torpedo, the navigator set the travel depth at 2–8 m (6.5–26 ft.), charged the torpedo's capacitors, and began the run-in to the target. At the proper moment, the bombsight automatically dropped the weapon. One second later, a small propeller-shaped drogue parachute deployed and the torpedo descended vertically. The main parachute deployed at 500 m (1,640 ft.), reducing the descent speed. It separated after the torpedo entered the water; then, special foreplanes were brought into play to turn the torpedo horizontally and were jettisoned immediately afterward. Next, the solid-fuel rocket motor fired and the torpedo accelerated to 58–68 kts. (107–130 km/h); by comparison, conventional wet-heater torpedoes could not travel faster than 40–45 kts. (74–83 km/h). Time from drop to impact was only about 35 seconds, which left the target no time for evasive action.

The RAT-52's chief shortcoming was the rocket motor's short burn time, resulting in a range of only 550–600 m (1,800–1,970 ft.), which took the bomber uncomfortably close to the target (within range of the ship's air defences). On the other hand, the torpedo could be dropped from any altitude between 1,500 m (4,920 ft.) and the aircraft's service ceiling at a speed of up to 800 km/h, which was of particular importance for jet torpedo-bombers. Live drops at the Soviet navy test range near Feodosiya showed a 'kill' probability of 17%–38% in a single-torpedo attack.

The trials were deemed successful; weapons loading was found to be convenient, and the weapons' separation from the aircraft was reliable and safe. Still, the antishipping version of the M-4 did not enter production because of the changing attitude toward this class of combat aircraft; torpedo-bombers had become outmoded by then, now that antishipping cruise missiles, which allowed the attack to be made from a safe distance, were coming on the scene.

M-4-2 (M-4-II) Refuelling Tanker (*izdeliye* 35)

In-flight refuelling was a sore spot for the Soviet air force. While fighters and tactical bombers had to do without (until the 1980s,

Opposite page: Front and rear views of the second prototype M-4 (c/n 4300003) during state acceptance trials of the antishipping armament in late 1956

Top and above: Two more views of the second prototype M-4 during the same trials, showing the unusual tactical code '101 Red' and the absence of the star insignia on the fuselage. A camera pack appears to be fitted instead of the tail guns.

Right: This picture of M-4 '101 Red' at the moment of takeoff shows how the nose gear's bogie-tilting feature worked.

A curious shot from the trials of the M-4's antishipping armament, showing a boom lift similar to the SPO-15 but based on a forklift truck being used to examine the rudder of '101 Red'

anyway), extending the range of strategic bombers by using IFR immediately became a high-priority task in the postwar years.

As already noted, the air force was dissatisfied with the M-4's short range, which was only 9,800 km (6,090 miles) instead of the required 11,000–12,000 km (6,830–7,450 miles) – without bombs. The fact that the competing Tu-95 boasted a range in excess of 13,000 km (8,075 miles) certainly did not help. To add offence to injury, the Tupolev OKB was already using the wingtip-to-wingtip IFR system developed by LII test pilots Igor' I. Shelest and Viktor S. Vasyanin in 1948 on the Tu-4 and Tu-16 bombers. (The refuelling technique was as follows: The tanker deployed a hose stabilised by a drogue parachute from one wingtip, and the receiver aircraft placed its opposite wingtip over the hose. Then the hose was rewound until the fitting at the end engaged an underwing receptacle. The receiver aircraft accelerated so that the hose

An IL-28 in formation with an M-4 to check the optimum position of the receiver aircraft with respect to the tanker.

formed a loop, rotating the receptacle and opening a valve, whereupon fuel transfer could begin.)

Hence, on 17 September 1953, MAP issued order No. 444, tasking the Myasishchev OKB with developing an IFR system for heavy aircraft; two M-4s equipped for in-flight refuelling (a tanker and a receiver aircraft) were to be submitted for state acceptance trials in the fourth quarter of 1954. Five versions of a system with a rigid fuel transfer boom (similar to Boeing's 'flying boom') were considered. Eventually, however, it was OKB-918, based in the town of Lyubertsy near Moscow, that developed the IFR system for the M-4; it was led by Semyon M. Alekseyev, former head of the defunct OKB-21 in Gor'kiy (the design office of the Gor'kiy aircraft factory), which had created the aforementioned I-215 fighter. This entity was tasked with developing aviation equipment, including IFR systems, and later became famous as NPP *Zvez**da*** (*naa**och**no-proiz**vod**stvennoye predpri**ya**tiye* – 'Star' Scientific & Production Enterprise).

The IFR system proposed by OKB-918 was a probe-and-drogue system similar to the one invented by the British company Flight Refuelling Ltd. As in the case of the wingtip-to-wingtip system, a modified bomber would serve as the tanker; its bomb bay accommodated a hose drum unit (HDU) – a powered drum with a retractable hose guide frame from which a fuel transfer hose about 50 m (164 ft.) long, terminating in a rigid stabilising drogue with a valve, was unwound. This assembly was called KAZ (***kom**pleksnyy agre**gaht** zaprah**v**ki* – integrated or self-contained refuelling module); the tanker could be easily reconverted to bomber configuration by removing the HDU. The flight engineer assumed the role of the refuelling system operator; he was assisted by the tail gunner, whose workstation commanded an unobstructed view of the rear hemisphere. The IFR system had the code name ***Ko**nus* (Cone), referring to the conical shape of the drogue.

The drogue's shape and weight were carefully chosen to provide optimum stabilisation in the turbulent slipstream. The HDU's drum drive and the fuel transfer pumps were hydraulically operated; the drum featured an emergency brake. The delivery rate was about 2,000 litres (440 imp. gal.) per minute.

The receiver aircraft featured a telescopic probe tipped with a special adapter atop the nose. The probe was 'fired' pneumatically into the drogue, locking into place, whereupon fuel transfer began.

To verify the system, OKB-918 converted an Il-28 bomber built by the Voronezh aircraft factory No. 64 ('01 Red,' c/n 2402101) into a makeshift 'tanker trainer.' An experimental winch emulating the HDU was installed in the bomb bay, paying out a 5 mm (3/16 in.) steel cable with a drogue of 640 mm (2 ft., 1 3/16 in.) diameter to a point 42 m (137 ft.) beyond the bomber's tail. Initially a 36 kg (79 lb.) unstabilised drogue was used; after the first four flights it was replaced with a drogue incorporating a stabilising device. The aircraft worked with the tenth-production Gor'kiy-built MiG-19 *Farmer-A* ('10 Red,' c/n N59210110), converted by LII in late 1957. This fighter had no fewer than four dummy refuelling probes (one ahead of the windshield and three on the port wing) because the optimum location had to be determined experimentally. Later, a second Il-28 mentioned in LII reports as 'Il-28LL No.418' was converted into a 'tanker' as part of the development effort – this time with a real HDU. It featured an experimental steerable drogue of

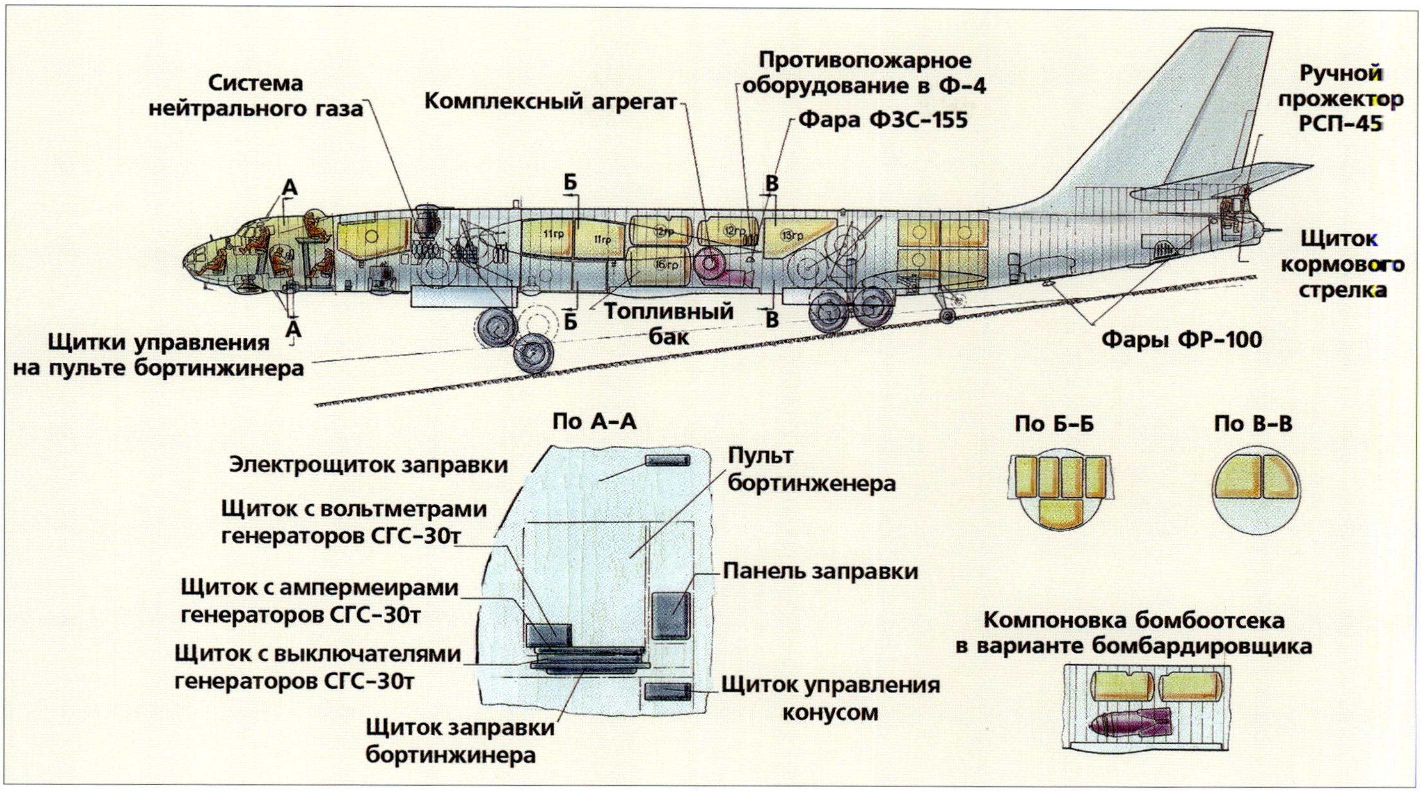

A drawing of the M-4-2 tanker from the ADP documents. *Clockwise from top left*: inert-gas pressurisation system; HDU; fire suppression equipment in fuselage section F-4; FZS-155 light; RSP-45 handheld searchlight; tail gunner's circuit breaker panel; FR-100 lights; fuel tank; circuit breaker panels at the flight engineer's station. Section A-A, *clockwise from top right*: flight engineer's station; refuelling management panel; drogue control panel; flight engineer's refuelling control panel; panel with SGS-30T generator switches; panel with SGS-30T generator ammeters; panel with SGS-30T generator voltmeters; refuelling circuit breaker panel. The scrap view shows the bomber configuration.

600 mm (1 ft., 11⅝ in.) diameter, incorporating an annular rudder of the same diameter 100 mm (3 15/16 in.) wide mounted at 60 mm (2 23/64 in.) from the base and having deflection limits of ±14° in the horizontal and vertical planes; the drogue was controlled by means of a joystick in the gunner's station. The hose had a diameter of 65 mm (1 13/16 in.) and was 23 m (75 ft.) long.

A third *Beagle* coded '08 Red' was converted into a receiver aircraft known as the Il-28LL, with a fixed refuelling probe offset to port on top of the navigator's station, making several contacts with the Il-28 'tanker' as part of the system's trials. Though not fitted with IFR equipment, a further Il-28 serialled '211 Blue' also contributed to the development of the probe-and-drogue IFR technique. In the autumn of 1954 it was used in simulated refuelling sessions, flying in line astern formation with an M-4 (emulating a tanker) in order to determine the optimum position in which the receiver aircraft would be least affected by the tanker's wake turbulence. LII test pilot Arkadiy P. Bogorodskiy made two flights with the M-4, maintaining a horizontal separation of 100, 80, 50, 30, 20, and 10 m (approximately 330, 260, 160, 100, 65, and 30 ft.). The first sortie was flown at 8,850 m (29,040 ft.), and the second at 8,300 m (27,230 ft.); the Il-28 approached at 20–25 m (65–80 ft.) below the lead aircraft's flight level, then slowly climbed until the buffet from the M-4's wake vortex was felt. Each of the tried relative positions of the two aircraft was maintained for five minutes to emulate fuel transfer and was filmed from a chase plane. Later, similar flights were made with the two M-4 prototypes emulating the tanker and the receiver aircraft. It was established that the receiver's optimum position was 8–12 m (26–40 ft.) below the tanker and 5–30 m (16–100 ft.) behind it.

The hose drum unit of the M-4-2 tanker (less hose and drogue). Note the signal lights on the drogue guide frame, indicating horizontal separation (too far / OK / too close) to the pilots of the receiver aircraft.

This research allowed the parameters of the equipment to be optimised. The drogue had a diameter of 0.85 m (2 ft., $9\frac{15}{32}$ in.) and a 60° cone; the hose had an inner diameter of 80 mm ($3\frac{5}{32}$ in.) and a length of 45 m (147 ft., 7 in.). The telescopic probe had a 1 m (3 ft., $3\frac{3}{8}$ in.) extensible section whose tip protruded 1.6 m (5 ft., 3 in.) beyond the bomber's nose when fully extended.

Apart from the IFR system as such, there were other complex issues to deal with; these included ensuring the rendezvous between the tanker and receiver aircraft far away from the base and in various visibility conditions, ensuring reliable contact between the two and sufficiently rapid transfer of large amounts of fuel, etc. Engineer Lazar' M. Rodnyanskiy was in charge of the programme at OKB-23. The programme was known in-house as *tema* 35, the tanker version of the *Bison-A* receiving the service designation M-4-2 (sometimes rendered as M-4-II).

The ADP of the probe-and-drogue IFR system was presented for review to MAP and the air force on 16 April 1954, well in advance of any government directives. It envisaged the possibility of using a converted Tu-4 as the tanker (in fact, 12 Kuibyshev-built Tu-4s were converted into single-point hose-and-drogue tankers as early as 1952). The receiver aircraft was to be fitted with extra tanks, increasing the fuel capacity by 12,500 litres (2,750 imp. gal.) and the all-up weight to 194 tons (427,690 lbs.); this would give a 1,200 km (750-mile) range increase on one top-off. On 22 May 1954, a full-size mockup of the KAZ HDU and the IFR probe was presented to the mockup review commission chaired by air force commander in chief Pavel F. Zhigarev, which approved it.

It deserves mention that alternative arrangements of the IFR system were explored as a result of the commission's findings. One of them was placing the HDU in the M-4 tanker's rear fuselage, thereby reducing the required hose length by 10–12 m (32 ft., $9\frac{45}{64}$ in., to 39 ft., $4\frac{7}{16}$ in.); the use of the Shelest/Vasyanin wingtip-to-wingtip system was also considered.

Detail design began next and was completed in July 1954. In addition to OKB-918, there were other subcontractors. In particular, MAP's OKB-383 designed the Model 1019 pumping station catering for the HDU; OKB-279 and OKB-476 were responsible for hydraulics and electrical components, respectively, while the Rubber Industry Research Institute (NIRP–*Naoochno-issledovatel'skiy institoot rezinovoy promyshlennosti*) developed a rubber hose that was adequate for the mission. The Myasishchev OKB built five test rigs to verify the HDU, the IFR probe, the automatic fuel controls, other fuel system components, and the electrics.

Real-life tests of the IFR system on the *Bison* began in the autumn of 1954. For starters, the first prototype M-4 ('25 Red') was fitted with a partly functional HDU as a 'tanker demonstrator';

Above left: A ground rig for testing the contact between the tanker's drogue and the receiver aircraft's IFR probe

Above: A separate rig was built for testing the M-4-2's transferable fuel tanks and the HDU's delivery pumps.

Left: M-4 '71 Red' (c/n 4300205) after conversion as the first prototype M-4-2. For security reasons, the last three digits of the c/n have been removed.

The first prototype M-4-2 in flight with the drogue partly deployed; a lot more of the hose needs to be paid out before the receiver aircraft can make contact. The tanker sports photo calibration markings on the rear fuselage; the faired FR-100 lights illuminating the rear fuselage as a visual reference during night refuelling are visible below them.

the HDU could unreel and rewind the hose but was not connected to the bomber's fuel system, which meant fuel transfer was impossible. Next, the second production M-4 (c/n 4300205) was converted into the first real prototype of the M-4-2 tanker in January 1955, featuring a fully functional HDU and a transfer tank with a delivery pump. By then the aircraft had gained the tactical code '71 Red' writ large on the nose and tail; in accordance with its new test mission, the tanker sported photo calibration markings in the form of crisscross lines oriented horizontally on the rear fuselage and vertically on the fin. Four FR-100 lights (***fa**ra roo**lyo**zhnaya* – taxi light, which is what the thing originally was) were installed in the rear fuselage – two in ventral fairings and two flush with the skin of the tail gunner's station to illuminate the underside of the tanker's fuselage and the horizontal tail at night; the drogue also had six small lights around the circumference to assist aiming.

Concurrently, the first production M-4 (c/n 4300104) – by then coded '85 Red' – was retrofitted with a refuelling probe. (Annoyingly, on many photos of these two aircraft, the construction numbers boldly stencilled on the nose have been retouched by military censors so that the aircraft appear to be marked '4300' or '43001' instead of their proper c/ns!) Together with the tanker demonstrator, the two aircraft participated in manufacturer's flight tests of the IFR system. These necessitated certain changes to the HDU; the length of the hose was increased to 56 m (183 ft., 8 in.) and several versions of the drogue were tried. The final version of the drogue weighed 220 kg (485 lbs.) and had an annular stabiliser. The HDU featured a set of 'traffic lights' coming on after contact had been made, to show the pilots of the receiver aircraft if the horizontal separation was OK. A yellow light meant 'too close, throttle back sharply'; a combination of yellow and green meant 'still too close, throttle back.' A green light meant 'OK, maintain this distance'; a combination of red and green meant 'too far, increase rpm,' and a red light meant 'risk of disengagement, increase rpm sharply.'

Making contact with the tanker was immediately found to be difficult because the airflow around the receiver aircraft's nose affected the drogue, making it sway. Experienced Myasishchev OKB test pilots – Mark L. Gallai, Fyodor F. Opadchiy, and B. M. Stepanov – were unable to 'hit the tanker' at first. Conversely, the young test pilot Nikolay I. Goryainov managed it on the first try – albeit after lengthy training on the abovementioned Il-28LL. On 11 July 1955 he performed the first successful refuelling of an M-4 by a sister aircraft; closing in on the tanker, he maintained the same speed for a while, then moved the bomber's nose close to the drogue and 'speared' it with the probe. Actually this first contact was a dry run, with no fuel transfer; in his second flight in the IFR-capable M-4, Goryainov made 10 consecutive contacts with the drogue.

After making 14 flights with the receiver aircraft, the tanker demonstrator was replaced by the actual tanker prototype. In a test flight on 7 December 1955, the tanker transferred 6,800 litres (1,496 imp. gal.); in another flight on 22 December it was 11,000 litres (2,420 imp. gal.). In the course of the manufacturer's flight tests, the first M-4-2 logged 63 hours, 21 minutes in 32 flights, while the receiver aircraft made 36 flights totalling 77 hours, 7 minutes. Altogether, the three aircraft made 102 flights at this stage (including 30 formation flights), logging 227 hours and transferring 120 tons (264,550 lbs.) of fuel, including 8 tons (17,640 lbs.) during night contacts.

On 15 February 1956 the Council of Ministers issued directive No. 247-159, ordering the probe-and-drogue IFR system into production. The system was to be fitted to production 3M bombers.

On 29 April 1956 – a year later than intended – the two aircraft were turned over to GK NII VVS for state acceptance trials. Actually, stage A of the latter began after a further five-month delay on 27 September 1956 but was halted soon afterward due to numerous technical problems that had to be rectified. Dangerous situations occurred frequently during the trials. The long hose could create a whiplash effect if the pilots made the wrong manoeuvre, and there were several cases when the hose broke; at best, the drogue remained impaled on the probe, and at worst, the remnant of the hose whipped around the receiver aircraft's horizontal tail, which could cause damage to the elevators and loss of control. If the two

The first M-4A (c/n 4300104) is shown here after apparently being retired. Note the cruciform photo calibration markings. All ejection hatch covers are missing, including the radome (the radar antenna is mounted on the pilots' ejection hatch cover, which is jettisoned together with the radome!). The offset IFR probe required the SP-1 star tracker to be shifted to the centreline; note also the drogue illumination light at the base of the probe. The white markings around the ventral cannon barbette were probably meant to identify skin areas requiring protection from the muzzle flames.

aircraft broke contact spontaneously, the receiver aircraft was doused with kerosene and the crew couldn't see anything for a few moments. In one of the flights, Nikolay I. Goryainov overshot and hit the tanker literally – the drogue collided with the aircraft's windshield; luckily the latter remained intact. (It may well be added at this point that 'hitting the tanker' was a lot easier on the later 3M and 3MD bombers, which had a longer and more pointed nose; consequently the airflow around the nose was much smoother, making the hose and drogue more stable.)

On 26 July 1956 the M-4's newly acquired IFR capability was demonstrated *urbi et orbi* during the traditional Aviation Day air event held annually at Moscow's Tushino airfield. The flypast included the tanker and receiver prototypes, which flew in front of the crowd in a simulated refuelling formation.

Later, two more *Bison-A*s were modified for the purpose of holding stage B of the state acceptance trials with the production version of the system; the drogue's diameter was increased to 0.95 m (3 ft., 1$\frac{1}{32}$ in.), and the length of the probe was increased by 1.25 m (4 ft., 1$\frac{7}{32}$ in.). M-4 '97 Red' (c/n 5301619) became the second prototype M-4-2, gaining similar photo calibration markings on the rear fuselage only, while M-4 '63 Red' (c/n 5301518) was fitted with an IFR probe. The manufacturer's flight tests, which were held between May 1956 and February 1957, involved 40 flights, in which fuel was transferred in 19 flights. These included two long-range missions in which the receiver aircraft was crewed by captain Nikolay I. Goryainov, copilot Ivan A. Laptev, navigator V. Zhuravlyov, flight engineer G. Nefyodov, and test engineer L. G. Gladoon. The first mission was flown on 26 September 1956 with one top-off, the aircraft flying eastward from Zhukovskiy via Chelyabinsk, Omsk, Novosibirsk, and Irkutsk to Yerofey Pavlovich Township in the Amur Region before turning back. Taking on 35.1 tons (77,380 lbs.) of fuel between Omsk and Novosibirsk, the bomber stayed aloft for 14 hrs, 13 min., and covered a distance of 11,500 km (7,150 miles).

The second mission, involving two fuel top-offs, was flown on 8 February 1957 from Zhukovskiy to the Shantar Islands in the Sea of Okhotsk and back. Receiving 35 tons (77,160 lbs.) of kerosene near Omsk from an M-4-2 captained by B. M. Stepanov on the outbound leg, the bomber flew on eastward while the tanker returned to Zhukovskiy for a refill and a crew change. On the way home, the bomber rendezvoused with the tanker – now captained by Fyodor F. Opadchiy – again to fill up another 24 tons (52,910 lbs.) of fuel over Lake Chany in the same area. As a result, the range attained was an impressive 14,500 km (9,006 miles), the aircraft staying aloft for 17 hours. Stage B, which ended in June 1958, again revealed a number of shortcomings, with unreliable operation of the HDU's servo system and the IFR probe being the most serious ones.

Starting in 1958, the surviving M-4s were progressively relegated to the tanker role (several aircraft had been lost in accidents by then). Together with the abovementioned prototypes, a total of 27 *Bison-A*s were converted into M-4-2s; these tankers worked with 3M and Tu-95 bombers. Most of the Tu-22 *Blinder* medium bombers were retrofitted with refuelling probes as well, while Tu-22M2 *Backfire-B* 'swing-wing' medium bombers / missile strike aircraft had them from the outset; however, the probes were removed from the *Backfires* in compliance with SALT (Strategic Arms Limitation Treaty) II, signed on 19 July 1979, which imposed a limit on the bombers' range.

The M-4-2 was powered by four Zoobets RD-3M-500A turbojets rated at 9,500 kgp (20,940 lbst); all armament was deleted. The maximum takeoff weight was 184 tons (405,640 lbs.), and the transferable fuel load was 41.6 tons (91,710 lbs.). Fuel transfer took place up to 2,500 km (1,552 miles) from the base, at 8,000–9,500 m (26,250–31,170 ft.) and 470–510 km/h (292–316 mph) indicated airspeed; at 2,250 litres (495 imp. gal.) per minute, the fuel could be transferred within 20 minutes.

M-4 Development Aircraft with IFR Capability ('M-4A')

As noted above, M-4 '85 Red' (c/n 4300104) served as the first receiver aircraft / test bed for the probe-and-drogue IFR system developed by OKB-918. A telescopic refuelling probe was installed ahead of the flight deck windscreen, slightly offset to port; it featured a drogue illumination light and a locking adapter providing a secure and fuel-tight connection between the probe and the drogue. The fuel transfer line ran along the port side of the forward fuselage, from the probe to a spot just aft of the forward crew cabin, to connect with the forward fuselage fuel cells, the external fuel line conduit passing above the port observation/sighting blister. Later, M-4 '63 Red' (c/n 5301518) was converted in the same fashion; these two aircraft are sometimes called M-4A in the aviation press, though this designation is clearly unofficial.

However, operational M-4 bombers were never fitted with IFR probes. The *Bison* did not receive actual IFR capability until the rather different 3M bomber entered service.

M-4R-2 (M-4R-II) Refuelling Tanker

In the early 1960s, some M-4-2s received an upgraded avionics suite that included a DISS-1 Doppler speed / drift sensor (***dop**lerovskiy izmeri**tel' skorosti i [oogla] snosa*), an additional ARK-42 ADF, an RSBN-2S Svod (Dome) short-range radio navigation system (***rah**diotekh**nich**eskaya sis**tem**a **bli**zhney navi**gah**tsii* – SHORAN), and an R-802V (aka RSIU-5V) command radio, replacing the older RSIU-4V. Such aircraft were unofficially designated M-4R-2 (M-4R-II), the R denoting ***rah**dio'oborudovaniye* (avionics).

3M Long-Range Bomber (M-6, *izdeliye* 36)

By the mid-1950s it had become obvious that the Soviet air force would have to field both the M-4 and the Tu-95, rather than one of them as originally planned. The two bombers complemented each other; the *Bison* offered higher speed and a higher war load, while

The same aircraft in flight, bearing the code '85 Red'; the c/n has been altered to '4300' by military censors. The fuel pipeline conduit leading from the IFR probe to the fuselage tanks is clearly visible. This photo has often been incorrectly captioned as the first prototype.

Left: M-4-2 '71 Red' (c/n 4300205) leaves contrails across the sky as it refuels M-4A '85 Red' (c/n 4300104) during a high-altitude test flight. This view illustrates the optimum curvature of the hose (which is connected to the drogue by a spherical joint) during contact.

Below left: Here, contact is still stable but the hose is nearly straight and horizontal separation is at the limit.

Bottom left: The second prototype M-4-2 ('97 Red,' c/n 5301619) refuels M-4A '85 Red.' The hose is strongly curved but it's still OK.

Below and bottom: Views from the navigator's station of the M-4A as the aircraft approaches the drogue and 'fires' the telescopic probe into it

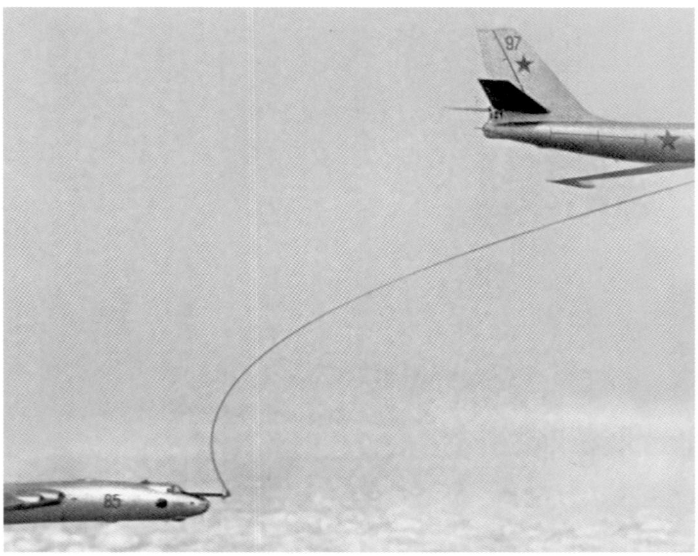

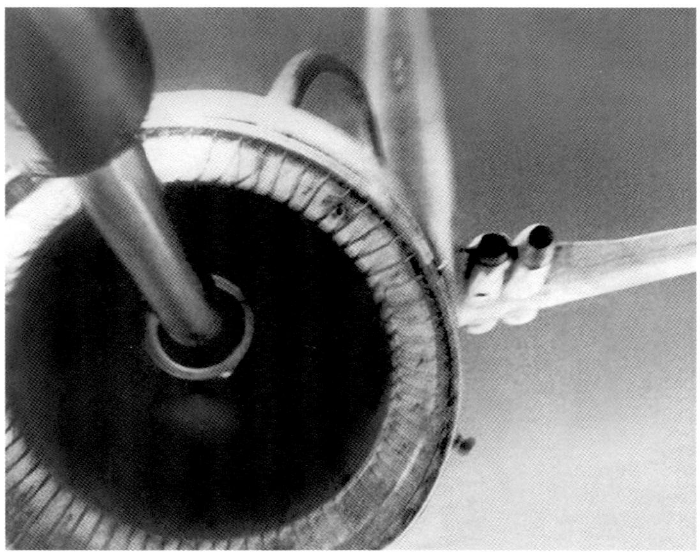

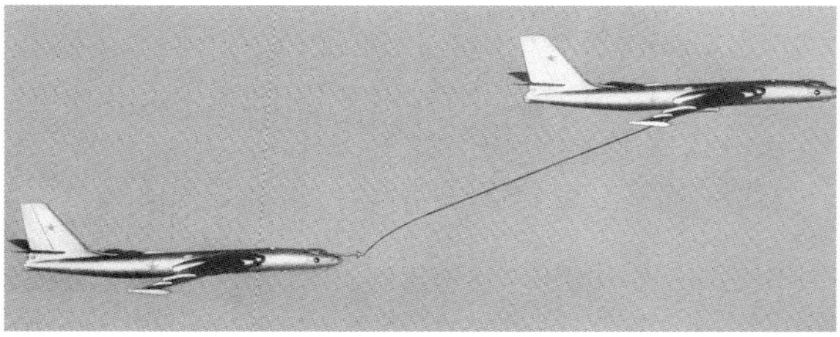

Top: An overall view of the M-4-2 and the M-4A making contact at maximum separation

Above and below: Two more views of '71 Red' refuelling '85 Red.' The M-4A stays well below the tanker's wake vortex.

Right: Here '85 Red' is seen from the tail gunner's station of the tanker as it closes in on the drogue.

Above: '77 Red' (c/n 5302124), a 'production' M-4-2 tanker serving with the 1230th APSZ (Air Refuelling Regiment) at Engels-2 AB. The flush aerials in the fin for the RSBN-2S short-range radio navigation system show this is an updated M-4R-2. Note the 'antiflash' colour scheme, with the c/n on the tail only and the much-smaller tactical code.

Left: M-4 '10 Red' (c/n 5300407) rests between missions with a loose canvas cover on the flight deck. This aircraft was later recoded '64 Red' and refitted as an M-4-2 tanker.

Below left: 1230th APSZ M-4-2 '62 Red' is refuelled by a TZ-22 fuel bowser – a 22,000-litre (4,840 imp. gal.) tanker semtrailer towed by a KrAZ-258B1 6×4 conventional tractor unit – at Engels-2 AB.

Top right: '61 Red' (c/n 6303134), the next-to-last *Bison-A*, also became an M-4-2, like all surviving sister ships. It served on in this guise until retired in the mid-1990s.

Above right: Another view of M-4-2 '61 Red.' Note how the outrigger struts run on the grass outside the paved taxiway, which was built with tricycle-gear aircraft in mind.

Right: Head-on view of an M-4-2 at Engels-2 AB

61
6303134
61

61

the *Bear* had longer range. (This goes to show that the Soviet Union had not succeeded in creating a true equivalent of the Boeing B-52 as far as the combination of speed, service ceiling, and range was concerned. On the other hand, the Soviet aircraft industry was then in the lead as far as development of powerful jet engines was concerned; it was the lack of a suitable turbojet rated at more than 6,000 kgp (13,200 lbst) for takeoff that forced the Boeing Company to use eight 5,000 kgp (11,000 lbst) Pratt & Whitney J-57s for the B-52.)

In April 1954, Minister of Aircraft Industry Pyotr V. Dement'yev reported thus to the Council of Ministers' Presidium:

'Until recently, the general belief was that a bomber powered by state-of-the-art turbojet engines cannot achieve a range of more than 9,500–10,000 km [5,900–6,210 miles]; therefore, [aircraft designers] strove to attain a range of 14,000–15,000 km [8,700–9,320 miles] by using turboprop engines. To this end, Comrade Tupolev has envisaged a power plant of four Kuznetsov TV-12 turboprops for his "aircraft 95" bomber prototype, which is to give the aircraft an effective range of 16,000–16,500 km [9,940–10,250 miles]. . . .

The Ministry of Aircraft Industry believes that creation of bombers with a 14,000 km range should be possible using turbojet engines as well. It has tasked the TsAGI and TsIAM research institutes with reconsidering current views [sic! – *Auth.*] *and finding practicable means of maximising the range of turbojet-powered heavy bombers jointly with the chief designers [of the aircraft and engine design bureaus].*

Building on experience accumulated both in country and abroad, as well as on research done by TsAGI, TsIAM, and the [aircraft and engine] OKBs, it has been established that it is currently possible to improve the fuel efficiency of turbojet engines significantly, reducing specific fuel consumption by 20%–25%, . . . and to improve the aerodynamics of aircraft, thereby substantially increasing range.

In accordance with these conclusions, chief designer Comrade [Vladimir A.] Dobrynin has offered to evolve the fuel-efficient VD-7 engine with a maximum takeoff rating of 11,000 kgp [24,250 lbst] and a cruise SFC of 0.85 kg/kgp·hr. from the existing VD-5 engine.

According to Tupolev OKB calculations, if the high-altitude, long-range bomber currently under construction [this must be the second prototype Tu-95 – *Auth.*] *is equipped with four VD-7 engines, the aircraft will have the following performance:*

- *effective range: 13,500–14,000 km [8,385–8,700 miles]*
- *service ceiling above the target: 13,500–14,000 m [44,290–45,930 ft.]*
- *top speed: 920–950 km/h [571–590 mph].*

Comrade Myasishchev has likewise performed calculations as to the possibility of installing four VD-7 engines on the 'aircraft M' bomber instead of the current AM-3s. The reengined aircraft would have the following performance:

- *effective range: 13,500–14,000 km*
- *service ceiling above the target: 13,500–14,000 m*
- *top speed: 950–1,000 km/h [590–621 mph].*

Additionally, Comrades Tupolev and Myasishchev propose equipping their aircraft with an in-flight refuelling system, which would extend their range to 16,000–17,000 km [9,940–10,560 miles].'

True enough, the fuel efficiency of the AM-3A and RD-3M turbojets left a lot to be desired. Hence, OKB-23 had closely followed the VD-7 turbojet's development progress almost from the inception of the 'aircraft M' bomber project. When the VD-7 commenced bench running in mid-October 1954, Vladimir M. Myasishchev firmly made up his mind to secure this engine for his

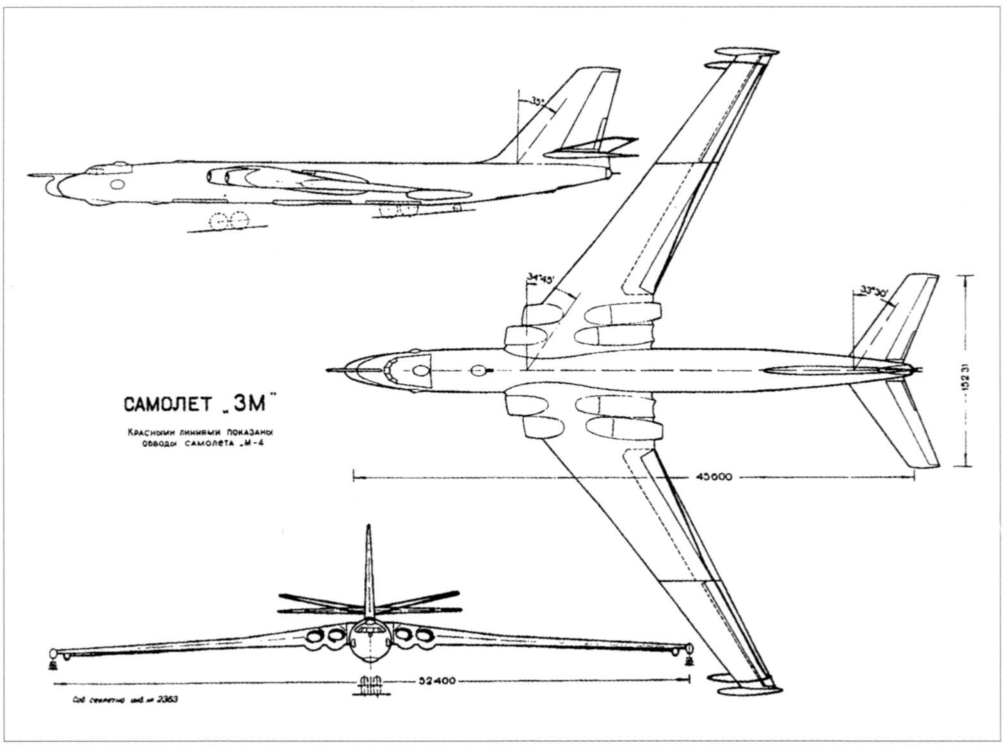

A three-view drawing of the 3M bomber from the project documents; the views are superimposed on those of the M-4 for the sake of comparison.

Right: A subscale mockup of the 3M's forward fuselage, showing the elongated 'thimble' nose topped by a centreline IFR probe and the blended radome. The latter has a curved joint line, and the beam blisters are elliptical, not circular. Note the star tracker mounting platform to port of the dorsal blister.

Below right: A full-size mockup of the 3M's forward fuselage – oddly, with no IFR probe! The radome joint line is now kinked 90° at the rear.

Bottom right: The bombsight and the radarscope (with the customary rubber sunblind) at the navigator's station of the full-size mockup

bomber. As the prototype engines accumulated hours on the test rigs, proving their viability and allowing the VD-7's reliability to be estimated, the airframers at OKB-23 started designing new wing root sections and engine bays tailored to the Dobrynin turbojet for the M-4.

The programme was formally launched on 6 July 1954, when the Council of Ministers let loose with directive No. 1428-642, ordering development of the M-6 bomber powered by VD-7 turbojets as a massive upgrade of the M-4. (Klimov VK-9 turbojets and Mikulin AM-13 coupled turbojets offering comparable overall performance were envisaged as alternative power plants in case the VD-7 did not come up to scratch. For a while, the Myasishchev OKB even considered installing TV-12 turboprops, which would give the M-6 a range in excess of 14,000 km.) A specific operational requirement envisaging a 40% increase in range and an improvement of field performance over the M-4 was endorsed on 27 August 1954.

The reengining programme was coded *tema* 36, the aircraft itself bearing the manufacturer's designation 3M (that is, third iteration of the 'aircraft M') and the provisional service designation M-6.

The new designation was justified. There was a lot more to it than simply installing new engines – despite the superficial similarity to the M-4, structurally the 3M (M-6) was a rather different aircraft. The different dimensions of the VD-7 brought about a revision of the bomber's central structural assembly consisting of the 23 m (75 ft., 5 33/64 in.) centre fuselage (built integrally with the wing centre section) and the inner wings with the engine bays and inlet ducts. Known colloquially as the *krest* (cross) for its cruciform shape, this huge subassembly having a 20 m (65 ft., 7 13/32 in.) span was outfitted separately during construction and delivered to the final assembly line with all systems and equipment in place.

Most of the airframe subassemblies attached to this 'cross' also underwent a major revision. The 3M featured new outer wings, giving greater span (the landing-gear track was increased accordingly) and a larger wing area. Wing aerodynamics were improved by eliminating the trailing-edge kink, which created vortices and hence drag; the wing camber was changed and both boundary layer fences on each wing were moved inboard considerably. The area-increasing trailing-edge flaps were built in one piece on each side instead of two sections. The dihedral tailplanes gave way to new stabilisers featuring longer span (15.2 m / 49 ft., 10 27/64 in.) and zero dihedral, and the stabiliser incidence adjustment range was changed.

Top, top left, and centre left: The first prototype M-4 was transported from plant No. 23 to Zhukovskiy down the Moskva River aboard a barge in semidismantled condition and under wraps.

Centre right: The first prototype 3M was transported from plant No. 23 to Zhukovskiy in the same fashion.

Above: Here the first prototype 3M is about to be unloaded from the barge. The zero-dihedral tailplane is very apparent. Interestingly, the aircraft is much less heavily wrapped up than the M-4 prototype had been.

Opposite page, top: The uncoded first prototype of the 3M (c/n 5320101) on a slush-covered runway at Zhukovskiy in March 1956

Right, above right, and centre right: Three views of the first prototype at the OKB's flight test facility. The reshaped nose with a chin fairing aft of the radome incorporating part of the navigator's station glazing is clearly visible. Note the cruciform photo calibration markings on the fuselage.

5320101

Left: OKB test pilot Nikolay I. Goryainov boards the 3M prototype before a test flight. The ventral blister aft of the entry hatch encloses the forward antenna for the PR-1 ECM/ELINT equipment.

Above: The first prototype deploys the triple brake parachutes on landing at Zhukovskiy.

The fuselage nose ahead of the flight deck glazing was extended by 900 mm (2 ft., 11⁷⁄₁₆ in.) and recontoured, receiving a thimble-like shape; this section was unpressurised. The distinctive nose glazing of the navigator's station and the chin radome vanished, giving place to a larger radome blended smoothly into the nose contour. The navigator / bomb aimer now sat aft of the radar instead of above and ahead of it; thus he had no upward visibility and very limited forward visibility, having to make do with three small windows on each side of the nose and a three-piece ventral sighting window built into a teardrop fairing that also incorporated his ejection hatch. Unlike the modified probe-equipped M-4s described above, the IFR probe was located on the centreline.

The structural-weight issue was addressed as well. Operational experience with the M-4 made it possible to revise the structural-strength standards used in bomber design. It turned out that the engineers had played safe, assuming the operational loads would be higher than they actually were, and incorporating excessive strength into some airframe components, which ended up being overweight. Now, armed with this knowledge, the structural-strength department made new calculations, shaving off no less than 6,500 kg (14,330 lbs.) as compared to the M-4; the landing gear alone accounted for 1,800 kg (3,970 lbs.) of this weight, but the main part of the weight saving was distributed between the wing spars and stringers, fuel cells, and the lighter engines.

Updates were made to the bomber's systems and equipment. For instance, the flight control system incorporated yaw dampers. An active/passive ECM suite was added, comprising an SPS-2 active jammer (*stahntsiya pomekhovykh signahlov* – lit. 'interference signal emitter'), three chaff dispenser modules in the mainwheel well for jamming ground-based air defence (AD) radars, and two more chaff dispensers in the rear fuselage for jamming the fire control radars of enemy fighters.

The avionics included an AP-15 autopilot; main and backup ARK-5 ADFs, RV-2, and RV-17 radio altimeters; an SPI-1 long-range radio navigation (LORAN) system (SPI = *samolyotnyy priyomoindikahtor* – aircraft-mounted receiver-indicator device); an SRZO-2 *Khrom-Nikel'* (chromium-nickel; NATO code name *Odd Rods*) IFF interrogator/transponder (*samolyotnyy rahdiolokatsionnyy zaproschik-otvetchik*) and an SRO-2 Khrom IFF transponder; an SP-50 *Materik* (Continent) automatic-landing-approach system (*[apparatoora] slepoy posahdki* – blind landing equipment); and a DAK-DB remote celestial compass for long-range bombers (*distantsionnyy astrokompas dahl'nevo bombardirovshchika*). The navigation/attack radar was the same RBP-4 as used on the M-4 – the Soviet avionics industry could not offer anything else at the time.

The maximum ordnance load initially remained unchanged at 24 tons (52,910 lbs.) but was later reduced to 18 tons (39,680 lbs.) for the simple reason that 6,000 kg (13,230 lbs.) bombs had disappeared from the Soviet air force inventory. The crew was reduced to seven; the flight engineer / gunner was replaced by an ECM officer (the copilot doubled as a flight engineer) and the radio operator was eliminated, his functions being taken over by the dorsal gunner. The reduction of the crew and the relocation of the navigator's station allowed the forward pressure cabin to be made shorter.

The redesign increased the bomber's lift/drag ratio at the optimum angle of attack from the M-4's 17.47 (an impressive figure as it is) to 18.5 – an unmatched result for an aircraft in this category.

Oddly enough, with the new bomber the Air Force chose 3M as the service designation, not the M-6 designation, which would be a logical follow-on to the M-4. Some OKB-23 documents, however, did refer to the bomber as the M-6 as per the CoM directive.

The first prototype 3M was built jointly by plant No. 23 and the OKB's colocated experimental production facility (the latter manufactured the new nose section, horizontal tail, flaps, and main gear unit). For whatever reason, in the case of the 3M there was no batch 0 or even batch 1 – production inexplicably began with batch 2.

A series of photos from the 3M production line at plant No. 23

Top: The starboard inner-wing torsion box, showing the figure-eight root sections of the spars accommodating the engine inlet ducts

Above: The centre fuselage on a dolly, showing the integral wing centre section, the nosewheel well, and the rescue raft access hatch

Top right: The rudder assembly jig

Above right: The forward fuselage buildup area; note the preassembled flight deck glazing frame awaiting installation

Right: The main gear unit's oleo strut being machined on a huge turning lathe

Thus, the first 3M received the c/n 5320101 (that is, year of manufacture 1955, plant No. [2]3, batch 2, first aircraft in the batch). The meaning of the last two digits is unknown, but they were 01 on all aircraft in batch 2; another peculiarity of this batch is that it consisted of 10 aircraft instead of the usual five. (Some documents give the c/n of this particular aircraft in abbreviated form as '201' – only the batch number and the number of the aircraft in it.)

The first prototype was completed on 30 October 1955. Since flight-cleared VD-7 turbojets delivering a stable thrust and stable rpm could be counted on the fingers of one hand at the time, the aircraft had to be fitted with two VD-7s to starboard and two AM-3A engines to port as a stopgap measure allowing initial flight tests to begin. In this unique lopsided configuration, the 3M took to the air on 27 March 1956, piloted by Mark L. Gallai (captain) and Nikolay I. Goryainov (copilot); the VD-7s were adjusted to limit their rating so as to avoid any asymmetrical thrust, thus providing a normal throttle position and coherent flight instrument readings. You have to give credit to Vladimir M. Myasishchev, who, with typical resourcefulness, devised this unconventional solution in order to speed up the trials.

The maiden flight nearly ended in a crash – immediately after liftoff, the 3M started inexorably increasing the angle of attack, even with the elevators fully down, and Gallai managed to keep it from stalling only by throttling back the engines, thereby reducing this tendency. Attaining an altitude of 500 m (1,640 ft.), the pilots turned the bomber around and landed with difficulty – the 3M would not touch down until the brake parachutes were deployed; the long runway at Zhukovskiy saved the day. As a result, revisions were made to the control system.

The first prototype made 13 flights with the provisional mixed power plant, logging 28 hrs., 38 min., by 15 June 1956. Later it was refitted to the intended configuration, making its first flight with four VD-7s on 26 July. The manufacturer's flight test programme continued until 8 January 1957, involving 35 flights with a total time of 106 hrs., 37 min. Originally, the aircraft had no tactical code but was eventually coded '60 Red.'

Meanwhile, the factory, which was building M-4s at the time, gradually started switching to 3M production. Aircraft c/n 5320201 was the static-test airframe, which was completed in June 1955 and transferred to TsAGI on 1 July. The first preproduction aircraft (c/n 6320301, or 203 for short) – or perhaps the second prototype after all? – was the first 3M to be rolled out with four VD-7s, completing manufacturer's flight tests in September 1956. That same year, Vladimir M. Myasishchev was 'promoted' to general designer, and in 1957 he received the prestigious Hero of Socialist Labour title, the civilian equivalent of Hero of the Soviet Union.

State acceptance trials of the 3M were to begin on 30 January 1957 but had to be postponed because of persistent problems with the VD-7 turbojets. This engine was much more fuel efficient than the RD-3M but was prone to surging at takeoff power. This even caused MAP to issue a grounding order, banning flights of the 3M with VD-7 engines. It took almost a full year to rectify this problem, which was caused by auto-oscillation of the compressor blades, and the concluding stage of the trials could not begin until 16 January 1958. The trials involved three batch 2 aircraft (c/ns 5320101, 6320301, and 6320401); the first aircraft was used for performance and handling tests, the second one for equipment and armament tests, and the third one for IFR system verification. GK NII VVS pilots Yuriy V. Sookhov, Semyon M. Antonov, and Neverov were the project test pilots; Nikolay N. Belyayev and Sergey G. Dedukh also flew the aircraft. As already mentioned, a welcome side effect of the 3M's new nose shape was that the IFR procedure became easier, since the tanker's drogue was more stable in the less turbulent airflow.

Between September and December 1957 the fourth preproduction 3M (c/n 6320601; originally uncoded, later coded '24 Red') was used

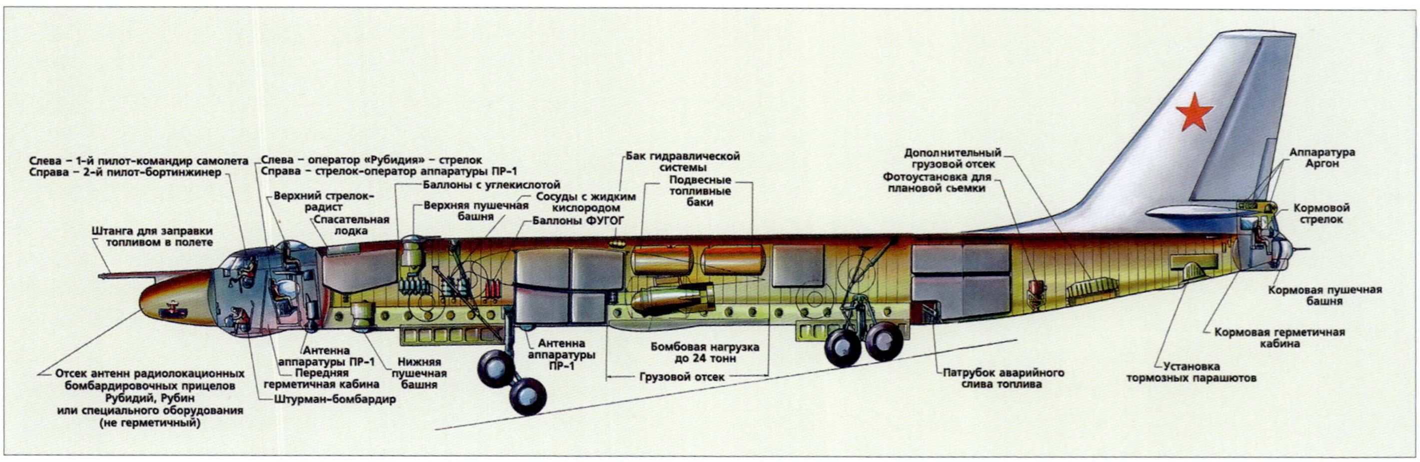

Left: The still-uncoded third prototype 3M (c/n 6320401), pictured during manufacturer's flight tests

Below left: A cutaway drawing of the 3M from the project documents

Right: 3M c/n 7300701 undergoing repeat static tests at SibNIA

to verify the crew rescue system. Eight dummies and three live parachutists were ejected during the tests, which showed that safe ejection was possible at up to 650 km/h (403 mph).

The 3M was not only ordered into production (by CoM directive No. 1249-700, of 1 July 1955) but was actually included into the Soviet air force inventory *before the state acceptance trials had even started*, so dire was the need for the new bomber. Despite the huge amount of test and development work aimed at ironing out the bomber's initial problems, the factory test pilots pointed out that the 3M was quite a handful during takeoff and landing. Doubts were voiced as to whether the Air Force could master the aircraft. A special panel was formed to look into the matter – and came to the conclusion that an average-skilled crew could not cope with the 3M; the airmen needed to be real professionals having practical experience. The air force acknowledged this; however, since the new bomber was not produced in large quantities, finding an adequate number of highly professional pilots was no great problem.

Yet, new problems surfaced soon – and nearly wrecked the 3M's service career before it had even started. Shortly after the completion of the state acceptance trials, two brand-new bombers crashed fatally during predelivery test flights. The accidents were identical: as the nose gear bogie tilted on takeoff, the aircraft pitched up sharply, stalling and crashing immediately after becoming airborne.

Again an investigative panel convened. VVS and TsAGI representatives charged to the attack, demanding removal of the nose gear bogie-tilting mechanism, which, they claimed, was responsible for the crashes. The designers of OKB-23 had a hard time defending their brainchild. Eventually, however, the aircraft was cleared completely. Unbelievably, it was the same story as with the M-4 – old habits had gotten the better of the pilots. Normally, as wing lift increased during the takeoff run, the nose gear was offloaded and the hydraulic damper tilted the bogie, lifting its front wheels off the ground. Later there was a standing joke among *Bison* pilots that all the captain had to do at this moment was sit back, keep the rudder pedals neutral, and wait for payday. However, unable to overcome ingrained habits, some pilots would haul back on the control column to initiate rotation, just as they would on a tricycle-gear aircraft. This was the fatal error; the bogie would tilt faster than it should, kicking the nose up, and the inertia of the heavy forward fuselage would pitch the aircraft up into a super-stalled position.

OKB-23 invested a lot of effort in improving the crew's working conditions. Workstation ergonomics and the air-conditioning system were constantly refined; a special test programme was held to find ways of reducing the psychological and physical strain during night flying. As already noted, a yaw damper was introduced into the rudder control circuit at TsAGI's suggestion, to facilitate flying the aircraft; later, the rudder itself was reshaped on late-production bombers. Nevertheless, some shortcomings 'seen from the cockpit' persisted until the end of production. Notably, the 3M had no crew rest area, galley, or toilet – no small thing on long missions.

The production-standard 3M was inferior in range to the prototype – with an identical 5-ton (11,020 lb.) bombload, the aircraft had an unrefuelled range of only 9,440 km (5,860 miles). On 8 February 1957, a test crew captained by Nikolay I. Goryainov successfully flew the 3M's first long-range mission, which involved refuelling from an M-4-2 captained by B. M. Stepanov; the bomber in question was the second preproduction machine (c/n 6320401).

On 19 July 1957 the 3M (again c/n 6320401) flew for the first time with drop tanks suspended under the wing roots; the takeoff weight was 202 tons (445,330 lbs.). The bomber covered a distance of 12,050 km (7,484 miles) without 'hitting the tanker,' staying airborne for 15 hours, 15 minutes.

At 0020 hrs. Moscow time on 23 August 1957, a 3M bomber took off from Zhukovskiy, starting on a 20-hour ultra-long-range test mission. The seven-man crew comprised captain Nikolay I. Goryainov, copilot Aleksandr S. Rozanov (who was chief of the

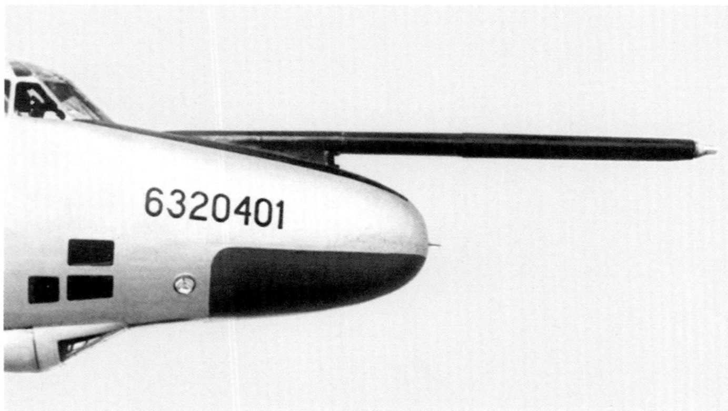

Above: 3M c/n 6320401, seen during state acceptance trials with the tactical code '60 Red'

Left: Close-up of the same aircraft's nose section

Below: A three-view drawing of the 3M with a provisional arrangement of four drop tanks (not implemented in practice)

Right and above right: 3M c/n 6320401 takes off and cruises with two 6,500-litre drop tanks (often misidentified as JATO boosters).

Below right: The same aircraft seen from a tanker during IFR tests (again with drop tanks fitted)

Myasishchev OKB's flight test facility), navigator V. I. Milyutin, systems operator N. Faïzi, project test engineer I. G. Tsar'kov, radio operator L. N. Goosev, and tail gunner S. A. Sokolov. The aircraft was fully fuelled and equipped with two 6,500-litre (1,430 imp. gal.) underwing drop tanks; a 5-ton bomb was suspended in the bomb bay. As on the previous occasion, the takeoff weight was 202 tons. Shortly afterward, an M-4-2 tanker with a 194-ton (427,690 lb.) TOW took off and followed the bomber.

The intended route, which was between 15,000 and 16,000 km (9,320–9,940 miles) long, took the aircraft north to Vologda, where the bomber was to jettison the drop tanks and turn onto an easterly heading toward Irkutsk in eastern Siberia and Nikolayevsk-on-Amur in the Soviet Far East. Thence, the 3M was to double north and west to the Severnaya Zemlya ('Northern Land') archipelago in the Arctic Ocean, proceed to Nar'yan-Mar on the Barents Sea coast, then pass over Moscow before making a big loop that took it over Minsk in Belorussia and Kiev in the Ukraine, and then head back home to Moscow. On the eastbound leg the bomber was to rendezvous with the tanker between Kranoyarsk and Nizhneudinsk in western Siberia, receiving 40 tons (88,180 lbs.) of fuel, and drop the bomb at the designated target range.

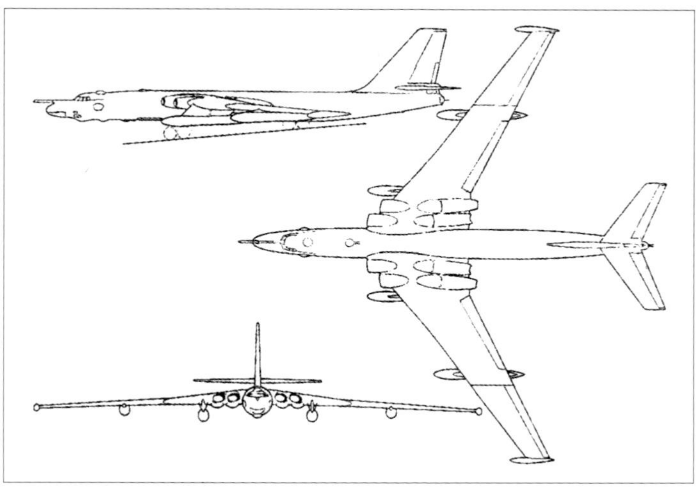

At first, everything seemed to go nicely. At 8,700 m (28,540 ft.), the bomber settled into cruise mode, travelling at 800 km/h (496 mph) and climbing gently as fuel was burned off. Soon, however, things started happening – the aircraft encountered a storm front. This is how Goryainov described it in his postmission report:

'As we approached the drop tank jettison point, we found ourselves over an area of heavy storm activity 300 m [990 ft.] below us. Severe static electrisation of the entire aircraft began, with long, luminous stripes snaking all over the aircraft skin; a large and bright greenish-blue electrical "lantern" appeared at the tip of the refuelling probe, which we tried to put out by scavenging the probe with nitrogen. The entire aircraft was aglow with powerful electrical discharges, and at first it appeared to the crew that fires had broken out in several places at once.

At the same time the aircraft was hurled sharply downward, with bank angles approaching 50°–55°; despite all efforts, my copilot and I made to hold the aircraft steady.

Seconds later, the aircraft was in the heart of the thundercloud. Powerful updrafts and downdrafts hurled the aircraft every which way, which was accompanied by sharp rolls and strong electrical discharges. At this point the No. 3 (starboard inner) engine flamed out, followed a few seconds later by the nos. 1 and 4 (port and starboard outer) engines. In these extraordinary circumstances the entire crew never stopped fighting the elements, trying to bring the machine back to normal flight. Copilot Engineer-Colonel Rozanov kept relighting the engines, but they flamed out again after the next gust and sharp roll. The tail gunner reported seeing long sheets of flame coming out of the nozzles with each relight – so long that they reached the tail of the aircraft. The probable reason for this is that

This uncoded 3M (c/n 6320601) was used for ejection system tests. The top two photos show the navigator / radar operator and the ECM officer ejecting; in the other two, the tail gunner can be seen ejecting. Note the photo calibration markings.

substantial amounts of fuel accumulated in the combustion chambers due to disruptions in the electrical igniters' operation.

The powerful electrical discharges also affected the electrical system: the lighting went out intermittently and the operation of the radio and the intercom was disrupted. At the same time, the aircraft began to ice up – first and foremost the pitot heads, which caused an airspeed indicator failure; therefore we activated the pitot and engine deicing systems.

While the aircraft was passing through the thunderclouds, we had to rely solely on our instruments, since there was no sight of the ground or celestial objects; all we could see through the glazing was the electrical discharges, which dazzled us. The Rubidiy radar sight helped us avoid the areas where the electrical discharges were at their strongest, even though the Rubidiy itself was disabled three times by electrical discharges, which "fried" the fuses.

At a point when three engines were inoperative and the aircraft was losing altitude sharply, the drop tanks and the bomb were dumped to shed excess weight. Since we had lost radio contact with the ground, we radioed the M-4 [tanker] aircraft, reporting that we had hit a storm and suffered multiple engine flameouts. Powerful downdrafts had dropped us from 9,300 m [30,510 ft.] to 4,800 m [15,750 ft.]; only then did we break clear of the storm front. The aircraft was inside the thundercloud for 10 or 12 minutes.

Analysis of the barograph [flight data recorder – *Auth.*] *chart revealed that the aircraft had lost 4,800 m in 40 seconds; that is, the sink rate had been more than 100 m/sec. [19,680 ft./min.]. When we were clear of the storm front, all engines were restarted and ran normally.*

Considering that the airframe had been subjected to severe stresses and the mission schedule had been wrecked anyway, I chose to abort the mission and return to base. To make a normal landing, it was necessary to reduce the all-up weight by burning off fuel; therefore we remained airborne for about eight hours after leaving the storm front.

We landed at Lookhovitsy airfield because LII's airfield was closed down by fog. Despite being thrown about by strong gusts, with large bank angles, the aircraft responded well to the controls and turned out to be tough enough, which allowed it to survive this disastrous situation.'

On 20–21 May 1958 the same 3M '60 Red' (c/n 6320401) flew nonstop from Zhukovskiy to Nikolayevsk-on-Amur via Krasnoyarsk and Irkutsk and back again, supported by M-4-2 '03 Red' (c/n 6302528). The 14,660 km (9,105-mile) route was flown in 18 hours, 8 minutes.

The introduction of IFR capability required the bombers to be equipped with RSBN-2S Svod and RSBN-7S ***Vstrecha*** (Rendezvous) SHORAN.

When the production 3M was discovered by the West, the new 'solid-nosed' version received the reporting name *Bison-B*.

The 3M also had its share of world records. On 29 October 1959 the first prototype set a series of payload-to-height records, lifting payloads of 5, 10, 15, 20, 25, 30, 35, 40, 45, 50, and 55 tons (11,020, 22,045, 33,070, 44,090, 55,115, 66,140, 77,160, 88,180, 99,200, 110,230, and 121,250 lbs.) to an altitude of 13,121 m (43,047 ft.). The aircraft was piloted by B. M. Stepanov (captain) and Boris I. Yumashev (copilot); the crew of seven also included

Right: General designer Vladimir M. Myasishchev with crew of '201M' after the record-breaking flight. *Left to right*: V. S. Koozovlyov, A. S. Lipko, V. M. Myasishchev, N. I. Goryainov, L. N. Goosev, V. V. Kolosov, an unidentified person, K. P. Ikonnikov, and S. A. Sokolov.

Below right: B. M. Stepanov, V. S. Koozovlyov, and I. I. Rykhlov give an interview at the Shabolovka TV studio in Moscow on 29 October 1959 after the record-breaking flight.

V. I. Sevost'yanov, I. I. Rykhlov, V. S. Koozovlyov, A. P. Okhtyrkov, and A. S. Iksanov. In the documents submitted to the FAI, the aircraft was entered under the designation 201M (which was derived from its c/n 5320101, or 201 for short), while the VD-7 engines were listed under the equally bogus designation D-15, derived from the product code *izdeliye* 15.

3MN-1 (3MN-I, 3M-N) Production Long-Range Bomber

A version of the VD-7 turbofan derated to 9,500 kgp (20,940 lbst) to improve reliability was developed as the VD-7B; this was done by simply limiting the engine speed at takeoff power. While having the same takeoff thrust as the RD-3M-500A turbojet, the new engine boasted a much-lower SFC – 0.7–0.8 kg/kgp·hr. versus 1.0 kg/kgp·hr. – thanks to an overall engine pressure ratio (EPR) of 11.2 versus 6.4 for the RD-3M-500A and a mass flow of 176 kg/sec. (388 lbs./sec.) versus 162 kg/sec. (357 lbs./sec.). Yet, the overall reduction in thrust with VD-7Bs rendered the 3M underpowered,

A high-spirited discussion of the successful record flight of the '201M.' Seated in the first row, I. M. Kargin is second from left, followed by L. N. Goosev, flight test department chief G. N. Inyakin, V. I. Milyutin, A. A. Monin, and V. N. Glushkov; standing behind them are Yu. I. Shashev, B. I. Yumashev, B. M. Stepanov (*fourth from left*), V. S. Koozovlyov, A. P. Okhrytkov, and N. I. Goryainov (*eighth*).

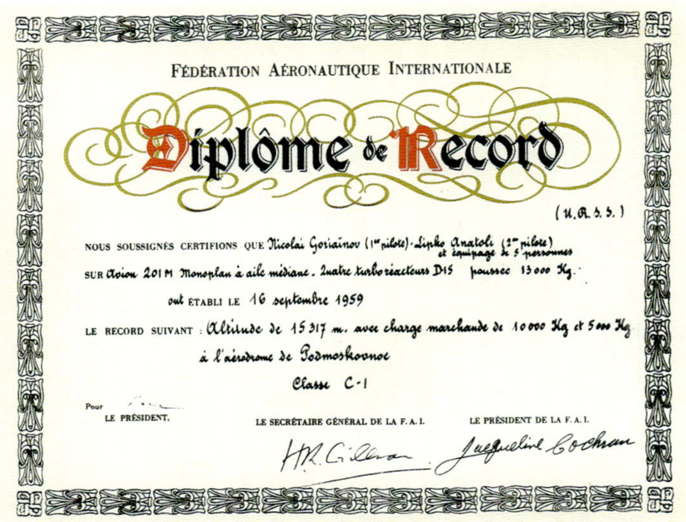

FÉDÉRATION AÉRONAUTIQUE INTERNATIONALE

Diplôme de Record

(U.R.S.S.)

NOUS SOUSSIGNÉS CERTIFIONS QUE Nicolai Goriainov (1er pilote) - Lipko Anatoli (2me pilote) et équipage de 5 personnes

SUR Avion 201 M Monoplan à aile médiane. Quatre turboréacteurs D15 poussee 13 000 Kg.

ont ÉTABLI LE 16 septembre 1959

LE RECORD SUIVANT : Altitude de 15 317 m. avec charge marchande de 10 000 Kg et 5 000 Kg à l'aérodrome de Podmoskovnoe

Classe C-1

Pour LE PRÉSIDENT,

LE SECRÉTAIRE GÉNÉRAL DE LA F.A.I.

LE PRÉSIDENT DE LA F.A.I.

H.R. Gillman

Jacqueline Cochran

FÉDÉRATION AÉRONAUTIQUE INTERNATIONALE

Diplôme de Record

NOUS SOUSSIGNÉS CERTIFIONS QUE Boris Stepanov. Boris Iumachev (U.R.S.S.)

SUR Avion 201 M Monoplan à aile médiane. 4 turboréacteurs D-15 d'une poussée de 13 000 Kg

ont ÉTABLI LE 29 octobre 1959

LE RECORD SUIVANT : Altitude de 13 121 mètres avec charge de 55 000 Kg.. 25 000 Kg. 20 000 Kg et 15 000 Kg. à l'aérodrome de Podmoskovnoe

Classe C-1

Pour LE PRÉSIDENT,

LE SECRÉTAIRE GÉNÉRAL DE LA F.A.I.

LE PRÉSIDENT DE LA F.A.I.

H.R. Gillman

Jacqueline Cochran

Above: The official diploma signed by FAI president Jacqueline Cochran to certify that the 201M aircraft with D-15 engines piloted by captain Nikolay Goryainov and copilot Anatoliy Lipko set a class C-1 record on 16 September 1959 by climbing to 15,317 m with 10,000 and 5,000 kg payloads from Podmoskovnoye airfield [*sic* – that is, Zhukovskiy].
Above right: Another FAI diploma certifying that on 29 October 1959, the 201M aircraft piloted by captain Boris Stepanov and copilot Boris Yumashev set a class C-1 record by reaching an altitude of 13,121 m with payloads of 55, 25, 20, and 15 tons.

Below: A cutaway drawing of the 3MN-2 tanker from the ADP documents. The legend reads (*left to right*): IFR probe, starboard beam gunner / refuelling-system operator, extra fuel tank, HDU, fire protection shroud, defensive fire commander.

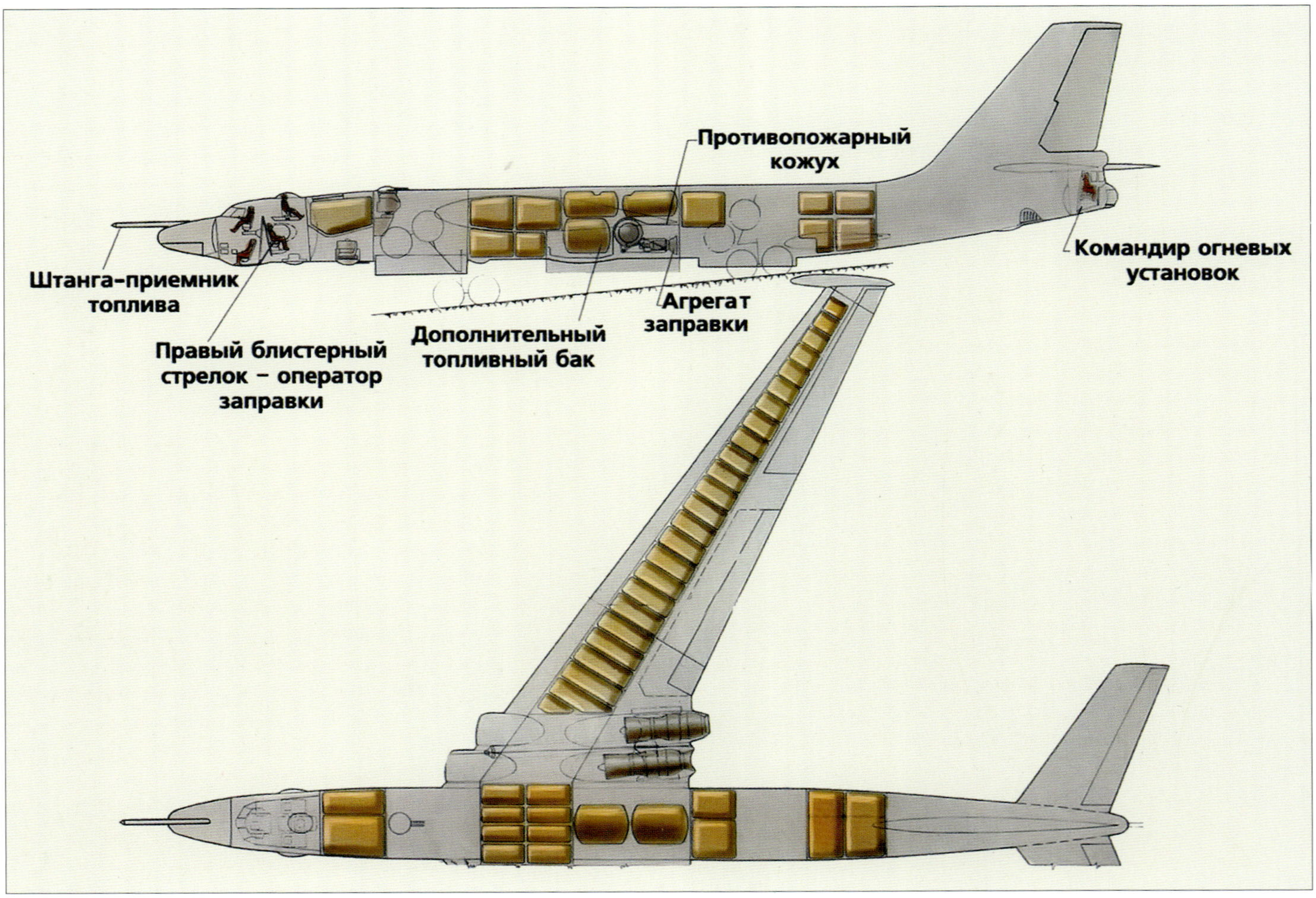

This uncoded 3M (c/n 7300701) underwent state acceptance trials as the first 'phase II nuke carrier' with enhanced protection from the effects of the nuclear explosion.

necessitating a reduction of the fuel capacity to reduce the takeoff weight suitably.

Taking due note of the compressor blade failures encountered on the VD-7 *sans suffixe*, on 18 December 1958 the Council of Ministers issued a directive supporting the joint VVS/MAP proposal to temporarily derate the VD-7 from 11,000 kgp (24,250 lbst) to 9,500 kgp and authorising plant No. 23 to manufacture 12 3M bombers powered by such engines. The version powered by VD-7Bs was designated 3MN-1 (also rendered as 3MN-I or 3M-N) or simply 3MN, the N standing for ***novyye dvigateli*** – new engines. The same directive authorised installation of VD-7Bs on seven in-service aircraft.

When the engine had worked up an acceptable reliability level, plant No. 23 started equipping production aircraft with VD-7B

Top and above: Another uncoded 3M (c/n 7300705) involved in the trials. Note the photo calibration crosses and the almost complete lack of white paint on the fuselage underside.

turbojets. Yet, in January 1961 the engine's TBO was only 200 hours, which was 6.5 times less than the RD-3M-500's.

Production 3Ms from batch 3 onward had a new and simpler c/n system. Thus, c/n 6300301 means year of manufacture 1956, plant No. 23, batch 003, first aircraft in the batch. With this system, production reverted to five aircraft per batch.

On the 10 prototype/preproduction aircraft of batch 2, the inboard pair of wing fences had been omitted. This turned out to be a bad idea, leading to airflow departure, and the inboard wing fences were reinstated on production 3Ms.

The 3MN-1 had a maximum TOW of 203 tons (447,530 lbs.); this configuration involved carrying two 6,500-litre (1,430 imp. gal.) drop tanks under the wing roots on Der5-48 shackles. Even in 'clean' configuration (without drop tanks), the 3MN's range was 15% greater than that of the 3MS thanks to the more fuel-efficient engines. Since the VD-7 engines required structural changes to the *Bison*'s wings, the 'old' and 'new' engines were noninterchangeable; hence the 3MS served on alongside the 3MN.

According to some sources, the VVS took delivery of 47 Dobrynin-engined *Bison-B*s, which made up approximately half of the production run. The 3MN fleet was based at Engels-2 AB, Siauliai, and Ookraïnka AB.

3MN-2 (3MN-II) Refuelling Tanker

In the same way as the 3MS-1, the 3MN-1 could be converted into a refuelling tanker by installing an HDU and auxiliary tanks in the bomb bay. By analogy with the M-4-2, the resulting tanker was designated 3MN-2 (or 3MN-II) and could be reconverted to bomber configuration just as easily by removing the HDU.

The 3MN-2 incorporated revisions to the IFR system. In particular, the pumping station was supplied with AC power from SGS-30B alternators driven by the same engine, instead of the M-4-2's SGS-30TB alternators driven by different engines. The fuel quantity/usage was monitored by the SETS-150E system (*soommeeruyushchiy elektricheskiy toplivomer s signalizahtsiyey* – summing electrical fuel meter with a [low fuel] warning feature). Four ETsN-10 booster pumps (*elektricheskiy tsentrobezhnyy nasos* – electrical centrifugal pump) were provided in the tanks with transferable fuel (the so-called Group 16 tanks), allowing the tanker's engines to run on this fuel in case of need. An updated guillotine cutting the hose in an emergency was fitted.

The 3MN-2 prototype was converted from the second flying 3M built (c/n 6320301) pursuant to Council of Ministers instruction No. 2901RS, dated 20 June 1960; by then the aircraft had been refitted with the derated VD-7Bs. Manufacturer's flight tests (combined with stage A of the state acceptance trials) took place at Zhukovskiy from June to October 1962, the aircraft logging 25 hrs., 45 min., in 12 flights. These included three solo flights, three dry contacts with 3M c/n 6320401, two flights in which 10 tons (22,045 lbs.) of fuel were transferred, one flight in which 20 tons (44,090 lbs.) of fuel were transferred, one flight in which 30 tons (66,140 lbs.) of fuel were transferred, and one flight in which 43 tons (94,800 lbs.) of fuel were transferred. Two night contacts with the transfer of 10 tons of fuel were also made; the total amount transferred was 132 tons (291,000 lbs.).

The trials were not without incident. During the dry run the hose broke because of a poorly adjusted HDU brake, leaving the drogue on the bomber's probe, and was cut by the guillotine. On 3 August the hose broke again because it had run dry, causing a whiplash. Difficulties were also experienced with the drogue's locking mechanism.

The trials protocol was endorsed on 30 January 1963. It noted that refuelling was possible at 470–510 km/h (292–316 mph) IAS and 3,000–9,000 m (9,840–29,530 ft.), with a maximum delivery rate of 2,300 litres (506 imp. gal.) per minute; fuel transfer time was 23 minutes for a quantity of 10 tons and 39–40 minutes for the maximum quantity of 43 tons. The aircraft was recommended for stage B of the trials, where it was to refuel the Tu-95KD *Bear-C* missile strike aircraft, with the proviso that the SPS-2 active jammer, which had been removed, was reinstated, and some revisions were made.

3MS-1 (3MS-I, 3M-S) Production Long-Range Bomber

Even though some of the 3M bomber's prototypes enjoyed the luxury of a complete ship set of VD-7 turbojets, not enough of these engines could be manufactured for mass production of the bomber,

Top left: Head-on view of a production 3M bomber. Note the drogue illumination light buried in the nose cone.

Left and above left: A production 3M coded '54 Red' with the bomb bay doors open. The bomber variant of the *Bison-B* was built with a refuelling probe, unlike the tanker variant, which lacked IFR capability.

Right: A 3MN-2 or 3MS-2 refuelling a Tu-95MS during trials; note the visual reference crosses in the camera's field of view

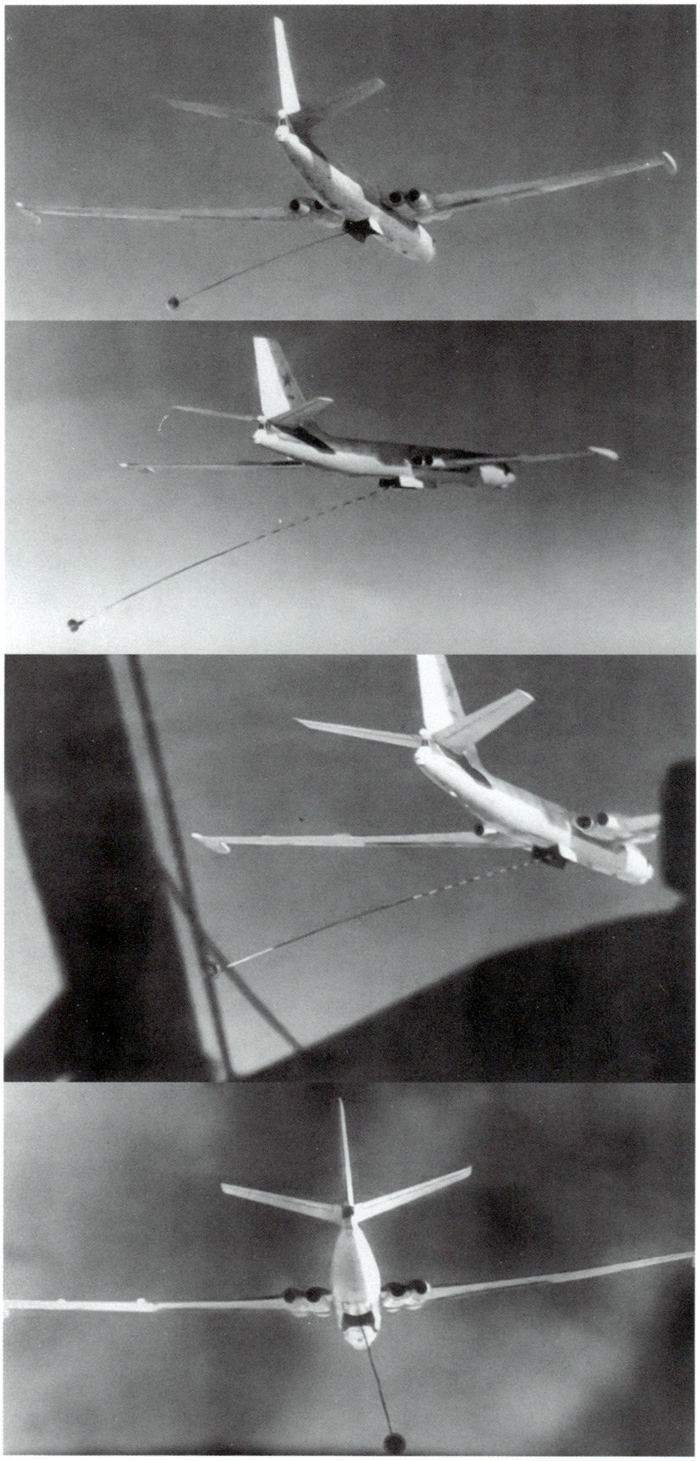

The sequence above shows an in-service 3MS-2 tanker refuelling another aircraft; note the stripes on the hose showing how much has been paid out.

Right and top right: Two 3MS-2 tankers ('23 Red,' c/n 6300304; '24 Red,' c/n 6320601). Apart from the lack of the IFR probe, the *Bison-B* tankers could be identified by the additional strake aerial on the rear fuselage.

Above right: 3MS-2 '27 Red' (c/n 7300604) taxiing at Engels

because it took awhile to overcome the VD-7's teething troubles. The derated VD-7B had no contingency rating; also, in spite of all measures taken, its reliability left a lot to be desired, which affected flight safety. Also, as mentioned earlier, the VD-7B's 200-hour TBO was insufficient. Still, plant No. 23 had to keep turning out *Bison-B*s, which it could not do if no engines were forthcoming.

The situation was resolved when the Myasishchev OKB, the plant, and the air force compromised, agreeing that as a temporary measure until the VD-7 had been fully mastered in production, the early-production 3Ms would be powered by Mikulin turbojets – RD-3M-500As, RD-3Ms, or even AM-3As (of which a considerable stockpile existed). Such aircraft were designated 3MS-1 (sometimes rendered as 3MS-I) or simply 3MS, the *S* standing for ***star**yye **dvig**ateli* – old engines!

The prototypes of the 3MS were two batch 2 aircraft (c/ns 6320501 and 6320801); they underwent state acceptance trials, which revealed that the thirstier engines gave the 3MS about a 2,500 kg (1,550 mile) shorter range. The two versions were interspersed on the production line. In the course of operation, some 3MNs were refitted with RD-3M-500As and converted to 3MSs.

In January–February 1957 the abovementioned 3MN-2 tanker prototype (c/n 6320301) was reconverted to bomber configuration and brought up to *nositel'* I nuke carrier standard, passing tests in this configuration in March. Starting in April 1957, all 3Ms in batches 2–8 (up to and including c/n 7300805) were upgraded to this standard. Later, 3MS c/n 7300701 was further modified to a nuclear-capable configuration called *nositel'* II ('phase II nuke carrier'), undergoing state acceptance trials of the enhanced nuclear protection featured from June to December 1959; all 3MSs and 3MNs from batch 9 onward were built to this standard.

3MS-2 (3MS-II) Refuelling Tanker

The 3MS-1 bomber could be easily reconfigured for the tanker role by installing an HDU and the associated auxiliary tanks in the bomb bay. The Mikulin-engined tanker version was designated 3MS-2 (or 3MS-II); like the 3MN-2, it was immediately reconvertible to bomber configuration. In the 1970s and 1980s nearly all 3MS bombers in service with the VVS were converted to 3MS-2 tankers in accordance with a specially developed schedule.

In the 1980s, to comply with the SALT II treaty, several 3MS-2 and 3MN-2 tankers were rebuilt as dedicated tankers with no provision for reconversion to bombers. The standard bomb bay doors were deleted, leaving only short clamshell doors that were just long enough to deploy the drogue and provide maintenance access to the HDU.

3MSR-2 (3MSR-II) and 3MNR-2 (3MNR-II) Upgraded Tankers

In October 1960, OKB-23 was transferred from the MAP framework to the Ministry of General Machinery (MOM – *Mini**ster**stvo **obsch**chevo ma**shin**ostro**yen**iya*), the agency responsible for the Soviet space and missile programmes, becoming branch office No. 1 of OKB-52, headed by Vladimir N. Chelomey, one of the Soviet Union's top authorities in missile and space systems design (incidentally, Sergey N. Khrushchov, the Soviet leader's son, worked at OKB-52 – though this was hardly a factor that influ-

23
23

27

24

enced the decision). The project materials and know-how amassed by the former Myasishchev OKB were transferred to the Tupolev OKB and Pavel O. Sukhoi's OKB-51, while Vladimir M. Myasishchev was appointed director of TsAGI – that is, 'kicked upstairs.'

The small design office created at the former OKB-23's flight test facility and known as OKB-90 was severely limited in its ability to undertake further improvement and even product support of the M-4/3M. Yet, a number of 3MS-2 and 3MN-2 tankers received an updated avionics fit in the 1960s, featuring a DISS-1 Doppler speed/drift sensor, an additional ARK-42 ADF, an RSBN-2S SHORAN), and an R-802V command radio. Such aircraft were unofficially designated 3MSR-2 (3MSR-II) and 3MNR-2 (3MNR-II), by analogy with the M-4R-2. In fact, the prototype of this upgrade was 3MS-2 c/n 7300501, which thus became the first 3MSR-2 and was tested in April 1963; the crew included project test pilot Neverov and navigator Mezentsev. Such tankers could work with Tu-95KM *Bear-C* missile carriers.

3MSR-1 and 3MNR-1 Upgraded Long-Range Bombers

Being still convertible in 'unSALTed' form, the 3MSR-2 and 3MNR-2 tankers could be refitted back to bombers in case of need. In order to discern them from standard 3MS-1s and 3MN-1s, these aircraft were unofficially designated 3MSR-1 and 3MNR-1, respectively.

Additionally, *Bison-B*s c/ns 8301101, 8301203, 9301301, and 9301302 were equipped with refuelling probes of improved design. The new probe became a standard fit on 3MS-1s from c/n 9301501 onwards.

Test Bed Versions of the 3M

Several production *Bison-B*s were modified under various test and development programmes. In 1968 Myasishchev decided to build two aerodynamics research aircraft based on the 3M; one was to feature a modified wing leading edge for boundary layer control (BLC) as a method of reducing drag, while the other aircraft was

to have laminar-flow outer wings. After the results of calculations had been checked in a series of wind tunnel tests, the OKB manufactured experimental skin panels featuring narrow slits and minute perforations for boundary layer suction. This and other methods of airflow laminarisation were tested on a modified 3M (identity unknown) and in TsAGI laboratories; they included everything from the use of specially developed laminar-flow airfoils and polishing the wing surface to cooling the wing skin.

When the Ilyushin Il-62 *Classic* four-turbofan, long-haul airliner entered production, the need arose to extend its range to permit transoceanic flights. In 1977 Vladimir M. Myasishchev suggested using the 3M as a research vehicle under the Il-62 range increase programme. The work proceeded in close cooperation with TsAGI, LII, SibNIA, and the Siberian branch of the Soviet Union's Academy of Sciences – to be precise, the Theoretical and Applied Mechanics Institute. The latter was then headed by academician Vladimir V. Stroominskiy, a former TsAGI researcher who had worked with Myasishchev in the 1950s, aiding in the creation of OKB-23's swept-wing strategic bombers.

3MA (3M-A) Nuclear Propulsion Test Bed (Project)

In the mid-1950s, both the Soviet Union and the USA were seriously considering the possibility of developing nuclear-powered aircraft – an effort prompted by the encouraging experience with nuclear power plants on ships. In the USSR, such power plants were developed by the Nuclear Energy Institute, headed by academician Igor' V. Kurchatov (IAE – *Institoot* ***ah****tomnoy* ***energ****hii*; now known as the Kurchatov Institute); when the scope of work was widened to include nuclear power plants for aircraft, this effort was supervised by academician Aleksandr P. Aleksandrov. Nuclear ramjets, turbojets, and turbofans were contemplated; reactors with air cooling and intermediate liquid metal cooling were designed. Much attention was paid to protecting humans from radiation; this required development of radiation shields that were effective, yet compact and lightweight enough for aircraft.

On 12 August 1955 the Council of Ministers issued directive No. 1561-868, ordering development of nuclear-powered aircraft. The Tupolev and Myasishchev design bureaus took on the job, since such power plants were obviously suitable only for heavy aircraft. The Myasishchev OKB started work on the M-60 nuclear-powered strategic bomber. While this project lies outside the scope of this book, its nuclear power plant had to be verified in flight before the M-60 could be built. Then the idea came up of converting the 3M into a test bed for this power plant; the aircraft was designated 3MA or 3M-A (***ah****tomnyy* – atomic).

The conversion involved installing a gas-cooled nuclear reactor with appropriate radiation shielding in the former bomb bay; the flight deck was heavily shielded, with reduced glazing (all three observation blisters and most of the side windows were deleted), and the tail gunner's station was unoccupied. Initially the two inboard VD-7B engines were to be replaced with experimental nuclear turbojets in much-longer nacelles, the coolant lines from the reactor passing through the fuselage sides to the engines' heat exchangers, replacing the combustion chambers. The other two engines would be standard VD-7Bs fed by jet fuel from tanks in the wing centre section, inner wings, and forward fuselage (the outer wing and rear fuselage tanks were deleted). In this configuration the 3MA was expected to enter flight test in the fourth quarter of 1959. Later on, it was planned to replace the outboard engines with nuclear turbojets as well, commencing a new round of tests in the second quarter of 1960; such an all-nuclear 3M could form the basis for a long-endurance reconnaissance/surveillance aircraft. The 3MA was never built, because on careful consideration the idea of nuclear power plants for aircraft was abandoned.

3MR Long-Range Reconnaissance Aircraft (Project)

One source mentions that in the 1950s, the Myasishchev OKB was working on a long-range reconnaissance version of the 3M designated 3MR (the *R* stood for *[samo****lyot****-] raz****ved****chik* – reconnaissance aircraft). The aircraft was to have a takeoff weight of 202 tons (445,330 lbs.).

Left: A drawing of the 3MA nuclear propulsion test bed from the ADP documents, showing the reactor and the nuclear turbojets (two different versions) in much-longer nacelles. Note the reinforcement plates on the fuselage next to the inboard engines, the greatly reduced fuel tankage, the reduced flight deck glazing and missing observation blisters, the missing dorsal barbette, and the gap between the flaps and the ailerons.

Right: The systems integration mockup of the 3ME bomber's nose section – an actual 3M pressure cabin mated to a wooden mockup of the unpressurised nose section with an IFR probe and a cutaway radome showing the new radar antenna inside.

The captain's instrument panel on the 3ME

The copilot's instrument panel and engine instruments on the 3ME

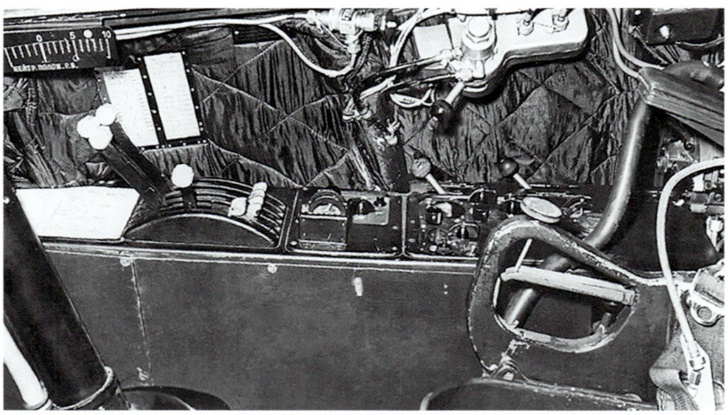

The 3ME copilot's side console with throttles. Note the actuating handle of the ejection seat.

3ME (3M-E) Long-Range Bomber Prototype

By 1958, some of the avionics fitted to the production-standard 3M had become obsolete, no longer meeting the growing requirements of the air force. On the other hand, new advanced radars, LORAN systems, radios, etc. had become available, albeit in prototype form only. Considering this, on 7 August 1958 the Council of Ministers issued directive No. 908-425, ordering OKB-23 to equip the 3M bomber fleet with up-to-date avionics.

Pursuant to this document, the OKB was to issue a set of drawings and specifications for outfitting two *Bison-B*s with the stated avionics before the end of August (that is, within three weeks), and the aircraft were to be transferred to GK NII VVS for checkout trials in October and November 1958. Proceeding from the results of these trials, a decision whether to manufacture new *Bisons* to this standard or not would be taken in 1959. The same directive also tasked OKB-23 with looking into the possibility of arming the 3M with guided missiles while still retaining the capability to carry conventional and nuclear free-fall bombs, and the possibility of upgrading the existing 3M fleet in this fashion.

The new bomber was designated 3ME; the suffix letter did not appear to have any particular meaning in this case. The upgraded prototype was to become the *eta**lon*** ('standard-setter'; that is, pattern aircraft for production) for 1959.

The 3ME was to feature a Rubin-1 (Ruby, pronounced *roo**bin***) navigation/bomb-aiming radar with an FARM-2 photo recorder (equivalent to a gun camera) instead of the RBP-4 radar, an OPB-112 optical bombsight instead of the standard OPB-11R, an NBA automatic navigation / bomb release module (*navigatsi**on**nyy bombardi**rov**ochnyy avto**maht***) instead of the NI-50BM navigation display (*navigatsi**on**nyy indi**kah**tor*), an RV-25 high-range radio altimeter instead of the standard RV-17, a PRS-3Ts Argon-2 gun-ranging radar instead of the standard PRS-1 Argon-1, an SPI-3M LORAN system instead of the standard SPI-1, and a ***Gheliy*** (Helium) transmitter instead of the standard 1-RSB-70. New equipment also included a GIK-1 gyro-flux gate compass (***ghee**ro-indooktsi**on**nyy **kom**pas*) and FRS-200 landing/taxi lights.

The changes in the avionics fit brought about a redesign of the bomber's forward and rear fuselage extremities. The 3ME had a characteristically pointed nose, with the IFR probe installed at the tip rather than on top and flanked by two small windows for the probe illumination lights; the radome was reprofiled, featuring characteristic 'cheekbone' bulges at the sides. This 'nose job' was meant to reduce drag while providing enough room for the bulky RN-1E antenna of the Rubin-1 radar, measuring 1.65 × 0.51 m (5 ft., $4^{61}/_{64}$ in. × 1 ft., $8^{5}/_{64}$ in.).

The radome of the gun-ranging radar was considerably enlarged to accommodate the larger antenna of the Argon-2; this even required the rudder to be cut away at the base, reducing its area by 0.3 m^2 (3.2 sq. ft.). While being more effective, the new radar created an inconvenience for the tail gunner, who could now eject only after sliding his seat all the way back. New wing airfoils with an extended and drooped leading edge were also envisaged.

The Rubin-1 radar was largely automatic, obviating the need for a radar operator; therefore the position near the port beam blister where he used to sit was now the radio operator's workstation. However, a second navigator was introduced into the crew due to the fact that missions could last up to 18 or 20 hours and the workload would be too much for one navigator; the conavigator sat beneath the dorsal blister.

Since OKB-23 was already working on a missile carrier version of the *Bison* designated 3MD (*see below*), all redesign work and crew responsibility redistribution proceeded on the assumption that the features developed for the 3ME would be incorporated in the 3MD without any changes. Indeed, the intention was that 3ME bombers would later be converted to 3MD missile strike aircraft.

The sole prototype 3ME (c/n 8301101) with flaps set for takeoff. Though the redesigned sharper nose is not very obvious here, this three-quarters rear view shows the enlarged radome of the gun-ranging radar and the consequently cutaway rudder, as well as the VD-7B engines.

The experimental shop of OKB-23 built full-size systems integration mockups of the 3ME's new nose and tail sections. Upon inspecting them, the mockup review commission made a few critical comments mostly concerning the choice of equipment. Among other things, it was suggested that the bomber should be fitted with ELINT and ECM equipment, plus a radar homing and warning system (RHAWS).

Two 3MN-1s (c/ns 8301101 and 8301203) were earmarked for conversion as the 3ME prototypes. Eventually, however, only the first aircraft was converted, because deliveries of the intended new systems were running seriously late.

The abovementioned CoM directive of 7 August 1958 obliged the Rybinsk engine factory No. 36 (seat of Dobrynin's OKB-36) to increase the VD-7's TBO to 200 hours in the first quarter of 1959 and commence mass production in the second quarter of the year. Moreover, the modified engines were to have a further reduced SFC of 0.69–0.7 kg/kgp·hr., which would give the aircraft a 300 km (186 mile) greater range.

To enhance the 3ME's fire protection, the aircraft was provided with new flame sensors, an inert gas pressurisation system, and a modified fire-extinguishing system serving the fuel cell bays in the fuselage. An extra RSIU-5 command link radio and a *Pritok* (Tributary) system were installed, ensuring secure rendezvous with M-4-2 tankers; an SOD-57M decimetre-waveband air traffic control / secondary IFF (ATC/SIF) transponder (*samolyotnyy otvetchik detsimetrovyy*) was provided to increase interaction range with friendly ground control radars. No changes were made to the armament; contrary to the original plans, the wing leading edges likewise had to be left unaltered because the programme was seriously behind schedule.

Conversion of 3MN-1 c/n 8301101 to the 3ME prototype was completed in March 1959. Flown by a factory test crew, the uncoded aircraft made seven predelivery test flights totalling 12 hrs., 21 min., in April–May 1959 before it was handed over to OKB-23. Myasishchev OKB test pilots started flying the 3ME on 18 June 1959 after the discovered defects had been eliminated; Nikolay I. Goryainov, Aleksandr S. Rozanov, I. I. Rykhlov, and other pilots made six flights totalling 12 hrs., 55 min. On 18 August 1959 they were superseded by GK NII VVS crews (pilots O. Belyayev, Yuriy V. Sookhov, and Ivan I. Markov; navigators Semovskikh, Mezentsev, Golovin, and Yadryshnikov; gunners Bogdanov, Naoomov, Spirin, Shichkov, and Smolin), which made 13 flights by 22 December, logging 70 hrs., 9 min. The state acceptance trials included simulated and actual bombs (using inert bombs in the latter case), photo gunnery against an attacking fighter, flights to check the effects of gunnery on the aircraft's equipment, and refuelling flights.

In order to assess the 3ME's combat capabilities, a closed-circuit flight was performed on 17 October 1959, the bomber covering a distance of 9,310 km (5,780 miles) in 12 hrs., 8 min., with an average speed of 800 km/h (496 mph). In so doing, it was repeatedly intercepted by Yakovlev Yak-25 *Flashlight-A* and MiG-19PM *Farmer-D* all-weather interceptors, disrupting their attacks by setting up passive ECM (chaff). The planned bomb drop was cancelled because the target range was obscured by bad weather.

The trials were completed on 9 March 1960 with a total time of 106 hrs., 1 min., in 21 flights; the trials report was endorsed on 5 April. As regards performance and handling, the 3ME was no different from a standard 3MN-1 but was a rather more potent combat aircraft thanks to the new radar. However, production entry of the RN-1E radar antenna was being delayed, and the Rubin-1 had to be refitted with a smaller RN-1 antenna measuring 1.15 × 0.3 m (3 ft., 9⁹⁄₃₂ in. × 11¹³⁄₁₆ in.); even so, it had much-greater detection range than the RBP-4. The 3ME made four test flights with this modified radar (which was later fitted to the production 3MD).

The career of the 3ME came to an abrupt end on 12 May 1960. While doing engine runs at the OKB's flight test facility in Zhu-

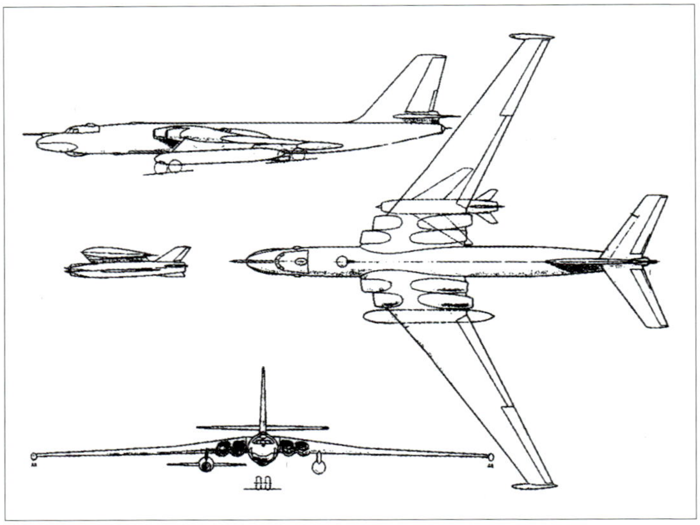

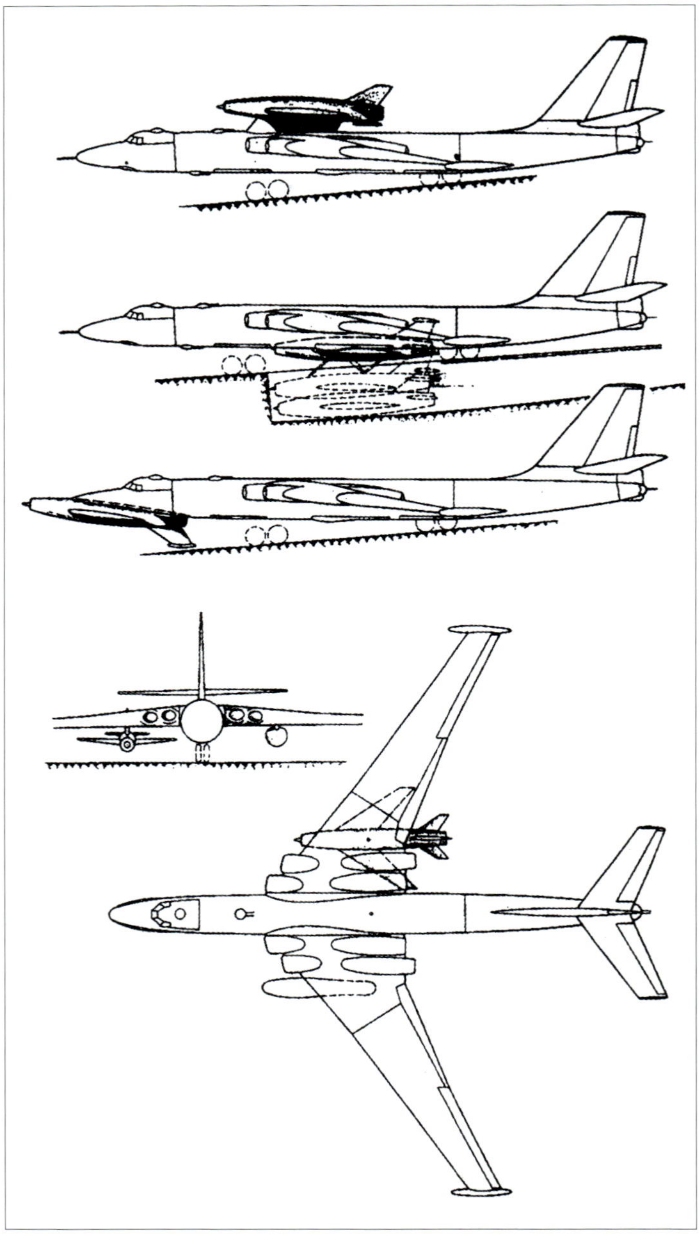

kovskiy, the prototype of the M-50 supersonic bomber ('012 Blue') jumped chocks and crashed into the 3ME parked across the apron from it. The *Bison* suffered serious damage to the nose and starboard wingtip; one man on the 3ME was killed and two more suffered minor injuries. Since the plans to launch production of the 3ME had been dropped by then, the prototype was struck off charge and scrapped. The M-50 was more fortunate, returning to flight status after repairs had been made.

3MK (3M-K, 3M-K-20) Strategic Missile Strike Aircraft (Project)

In the postwar years the efficiency of the Western nations' air defence systems increased dramatically. Penetrating these air defences would be close to a Mission Impossible for bombers carrying ordinary free-fall bombs (especially if they were subsonic aircraft). The answer was to turn the classic bomber into a launch platform for standoff weapons – that is, guided missiles that would be launched from outside the heavily defended areas. By virtue of its high speed and flight altitude, the missile could penetrate the air defences with impunity, then dive onto the target and obliterate it. Importantly, the radius of action of a missile-toting aircraft was bigger than an ordinary bomber's. This logically led to the decision to develop such weapons systems for the DA.

Development of a missile strike version of the production M-4 or 3M was triggered by Council of Ministers directive No. 412-182, dated 11 March 1954. Pursuant to this document, the Myasishchev OKB was to develop the 'aircraft MK' missile platform – one of the main components of the 3M-K-20 weapons system (*K* stood for ***kompleks** [vo'oruzheniya]* – weapons system). The same directive tasked Artyom I. Mikoyan's OKB-155 with creating the Kh-20 long-range cruise missile (NATO code name AS-3 *Kangaroo*), while the KB-1 design bureau, led by Vitaliy M. Shabanov within the Ministry of Defence Industry (MOP – *Ministerstvo oboronnoy promyshlennosti*), was to develop the missile's guidance system. The Tupolev OKB was similarly ordered to evolve a missile strike variant of the Tu-95 armed with the Kh-20. TsAGI was also to participate in the programme as a systems integrator.

The missile was developed by the Mikoyan OKB's Section K (OKB-2-155), specialising in missiles; this was headed by Artyom I. Mikoyan's closest aide Mikhail I. Gurevich, while Aleksandr Ya. Bereznyak led the actual design effort. (In 1967, OKB-2-155 became an organisationally separate enterprise named GMKB ***Raduga*** (*Gosudarstvennoye mashinostroitel'noye konstrooktorskoye **byuro*** – 'Rainbow' State-Owned Machinery Design Bureau.)

The Kh-20 had midset wings swept back 55° at quarter chord, sharply swept conventional tail surfaces with all-movable stabilisers (stabilators), and a sharp-lipped axisymmetrical supersonic air intake in the nose featuring a pointed centre body (shock cone). The power plant was a 9,810 kgp (21,630 lbst) Lyul'ka AL-7FK axial-flow afterburning turbojet (*K* = *korotkoresoorsnyy* – short life) featuring a fixed-area supersonic Laval nozzle; this was to give the Kh-20 a speed of Mach 2. The missile carried a nuclear warhead with a yield variously reported as 350 or 800 kilotons. The Kh-20 had an overall length of 14.603 m (47 ft., $10\frac{59}{64}$ in.) less undernose pitot boom – or 15.415 m (50 ft., $6\frac{57}{64}$ in.) if you include

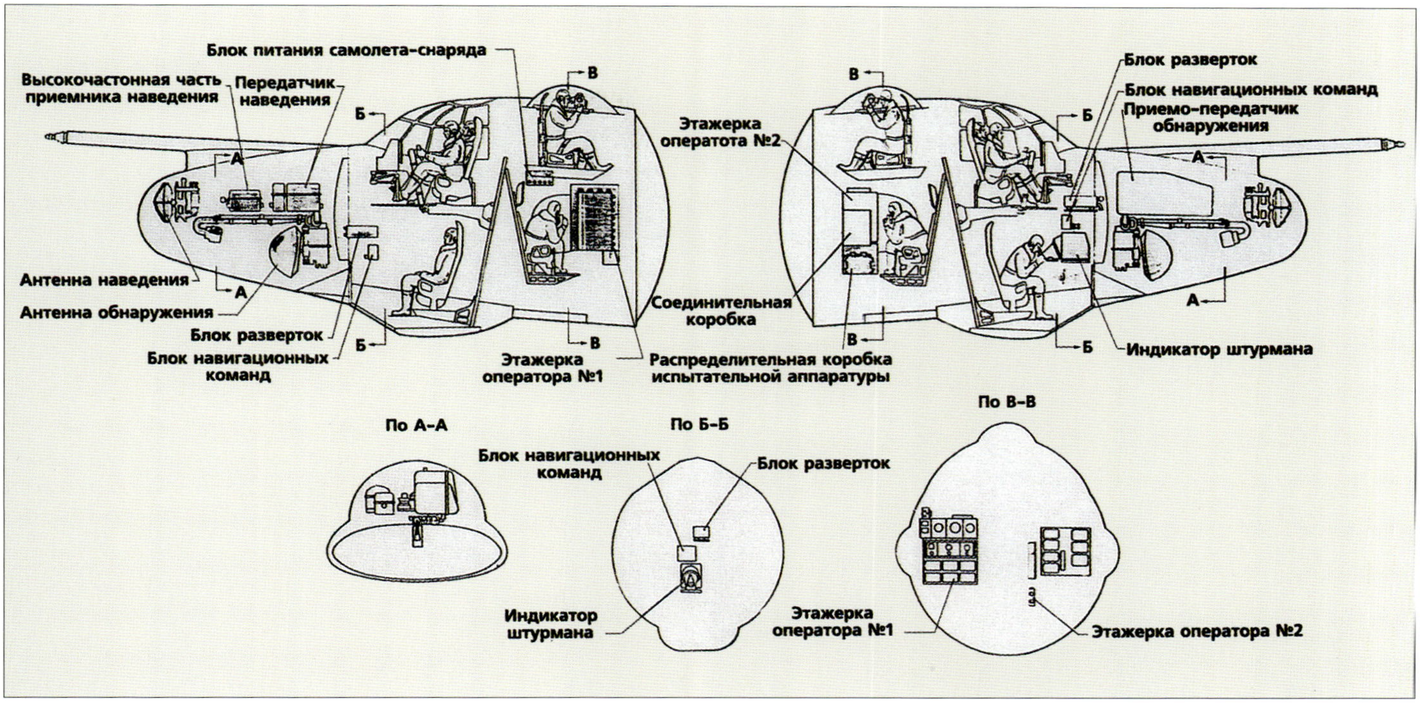

Above left: A provisional three-view drawing of the 3MK carrying a Kh-20 missile to starboard and a drop tank to port

Left: The alternative ways of carrying the Kh-20 considered for the 3MK. The first three were discarded as impractical. Note that the drop tank in the fourth version is depicted differently here.

Above: Port and starboard views of the 3MK's nose section from the ADP docs, showing the YaD radar and other mission equipment

Right: A display model of the 3MK

Below: A drawing showing the 3MK's missile guidance and launch system

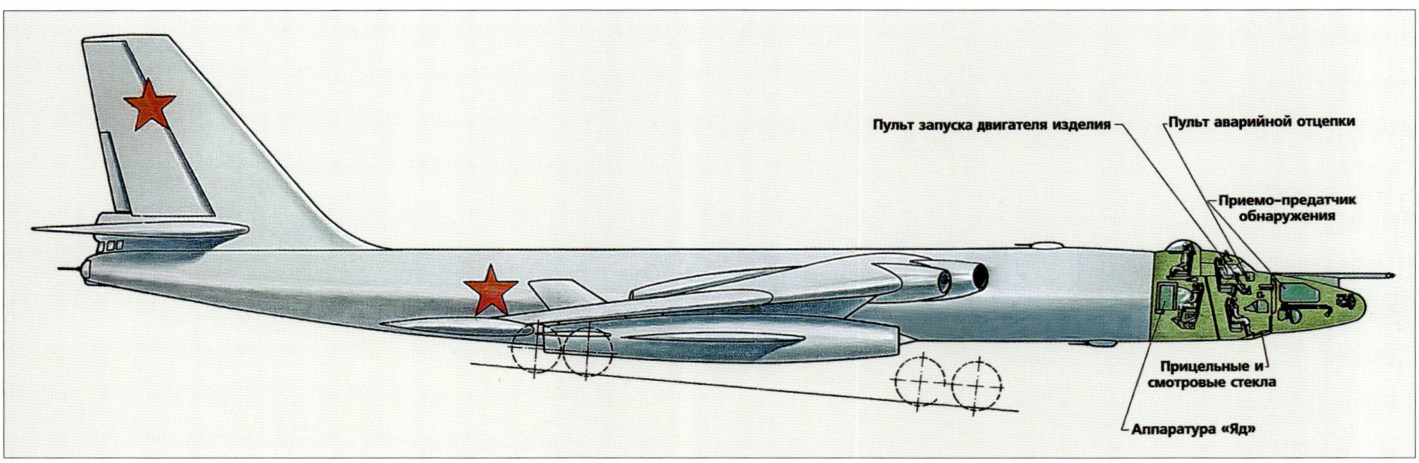

A full-size wooden mockup of the 3MK's forward fuselage, showing the large 'duck bill' radome of the YaD radar's search/tracking antenna, with the small quarter-spherical radome of the missile guidance antenna above it (cut away to reveal the antenna dish). An access panel on the starboard side aft of the main radome is missing. Note that the drogue illumination light is of necessity relocated higher to the base of the IFR probe, just like on the M-4A.

Here the mockup is shown with the main radome and the forward avionics bay access panels above it removed to reveal the bulky elliptical search/tracking antenna and the radar set.

the pitot, a wingspan of 9.03 m (29 ft., 7½ in.), and a launch weight of 11,600–11,800 kg (25,570–26,010 lbs.).

The guidance system developed by KB-1 comprised two subsystems. The aircraft was equipped with a YaD navigation/attack radar (NATO code name *Crown Drum*), which also functioned as a weather radar; it had a search / target illumination channel and a missile guidance channel served by separate antennas. (Curiously, the radar's designation is just like the Russian word *yad*, meaning 'poison,' and some publications misguidedly rendered it just like that.) The missile was fitted with a YaR active radar seeker for terminal guidance (its antenna was housed in the air intake centre body), a midcourse guidance module to pick up signals transmitted by the aircraft's radar, and a YaK autopilot.

Two launch modes were possible. In the so-called navigation mode, when there was no radar contact with the target, the navigator / radar operator either received the target coordinates from a target designator aircraft or calculated them and entered them into the range detection module of the missile's radar seeker, launching the missile from beyond the aircraft's own radar acquisition range (up to 600 km / 372 miles). Once the aircraft was close enough to track the target with its radar, it provided midcourse guidance to the missile. Alternatively, the Nav/Op launched the missile after the radar had acquired the target (in the latter case, the minimum distance to the target was shorter – up to 450 km or 279 miles).

Depending on the flight level, the YaD radar had a detection range of 390–450 km (240–280 miles). Maximum missile launch range was 450–600 km (280–370 miles) in navigation mode and 370–430 km (230–270 miles) in radar-tracking mode, again depending on the altitude. The Kh-20 was launched at 10,500–11,500 m (34,450–37,730 ft.) – or, according to some sources, at

A mockup of the 3MK's starboard inner-wing section, under which a mockup of the Kh-20 missile is suspended on a very provisional pylon (compare this with the picture on page 107).

A mockup of the massive drop tank that is supposed to be suspended under the 3MK's other wing to extend range and counterbalance the missile. Apparently no separate mockup of the port inner-wing section was built, to save time and shop floor space.

any of four prescribed altitudes: 9, 10, 11, or 12 km (29,530, 32,810, 36,090, or 39,370 ft.). The minimum launch distance was 150 km (93 miles); in this case the aircraft was to turn back immediately after the launch, and there was no midcourse guidance.

At 46 seconds after release, when it was a safe distance ahead of the aircraft, the missile pulled up into a climb, levelling off at its cruise altitude of 15,000 m (49,210 ft.). After the first 270 seconds of flight, the missile caught the aircraft's radar beam and the midcourse guidance subsystem took over. The YaD radar determined the positions of the target and the missile relative to the aircraft, with the target illumination channel working on a 10 cm wavelength and the missile guidance channel working on a 3 cm wavelength. On receiving the radar's signal, the YaR system generated a reply on a 2 cm wavelength with a certain delay; when this signal was picked up, the Nav/Op could see the target and missile blips on his radarscope. If the missile strayed from the desired track, he superimposed the missile marker on the target; the radar then computed the missile's deviation angle and generated guidance commands so that the missile made a beeline toward the target. When this had been accomplished, the missile accelerated to maximum speed in level flight, its altitude maintained by a barometric-altitude corrector linked to the autopilot. When the Kh-20 came within 75 km (46.5 miles) of the target, the guidance system switched to autonomous mode, the aircraft made a U-turn and headed for home. When the target was 16.5 km (10.25 miles) away, the missile's engine was shut down and the guidance system initiated a 70° dive onto the target. The warhead was detonated at 3,000–4,000 m (9,840–13,120 ft.).

Preliminary design work at OKB-23 began straightaway after the CoM directive. On 24 December 1954 the Council of Ministers issued instruction No. 13577, specifying that the missile and associated guidance system were to be identical for 'aircraft MK' and the Tu-95's missile-toting version, followed a week later by MAP order No. 763 to the same effect. Plant No. 23 was required to modify a single 3M bomber, delivering to the air force in the second quarter of 1956.

Pursuant to instruction No. 13577, 'aircraft MK' was to have a combat radius of 7,500–8,000 km (4,660–4,970 miles), with one fuel top-off en route to the target. The single missile was to be launched at 11,500–12,000 m (37,730–39,370 ft.), with the aircraft climbing to 12,000–12,500 m (39,370–41,010 ft.) immediately afterward; cruising speed with the missile attached was specified as 770–800 km/h (478–496 mph). The Ministry of Medium Machinery, which was responsible for the guidance system, did not specify any requirements which the carrier aircraft had to meet.

One of the major issues the engineers had to resolve at the ADP stage was how the Kh-20 would be carried by the 3M. Finding the best location wasn't at all easy because of the *Bison*'s bicycle undercarriage. Four options were considered. Version 1 involved carrying the Kh-20 on a pylon under the starboard inner wing, with a large drop tank symmetrically to port to counterbalance it. This configuration required minimum changes to the airframe; aside from the installation of the missile guidance system, all that was necessary was some local reinforcement of the wings and the provision of two hardpoints, one of which was 'wet.'

Version 2 was rather unconventional: the missile was carried on a pylon above the wing centre section, standing proud of the fuselage! Again, the structural changes were minor in this case; however, this configuration required research into the dynamics of the aircraft and missile at the moment of separation, the effect of the missile's jet exhaust on the bomber's airframe, and ways of jettisoning the missile in an emergency.

In Version 3 the missile was semirecessed in the bomber's belly and lowered clear of the fuselage on a trapeze before engine starting and release. This looked straightforward – until you remembered about the bicycle landing gear and small ground clearance, which opened a big can of worms. First, the missile could not be just wheeled in on a dolly and hoisted up into the belly, since there was not enough room for the tall fin; a trench needed to be dug at the loading station, and how do you place a bicycle-gear aircraft over a trench? Second, this arrangement rendered carriage of free-fall bombs (which was a requirement stipulated by the instruction) impossible. Also, it seems that the 3M's bomb bay placed between the wheel wells was just too short for the Kh-20.

The fourth version was the most bizarre of all. The missile was carried inverted under the forward fuselage so that its rear fuselage was semirecessed in the bomber's extreme nose, directly under the flight deck, creating the eerie impression that the bomber was chewing the missile! Apparently the recess was to be closed by inward-opening doors after the launch. Even though the structural changes were relatively minor, the missile obstructed the radar's field of view and precluded crew ejection unless it was jettisoned first.

That said, the choice was quickly narrowed to versions 1 and 2, which were tested in TsAGI's T-106 wind tunnel. Calculations and wind tunnel tests revealed that in both versions, the aircraft had identical range and an identical lift/drag ratio. However, the underwing option was clearly preferable, since it did not require research into how the missile's exhaust affected the bomber's tail, and the emergency jettison procedure was straightforward.

Upon completion of the ADP documents, OKB-23 built a partial full-size mockup of 'aircraft MK' with the Kh-20 missile and drop tank installed. Also known as the 3MK, the missile strike version differed outwardly from the standard *Bison-B* in having a reshaped nose. Just like on the competing Tu-95K-20 *Bear-B*, the search/tracking antenna of the YaD radar was wider than the fuselage, necessitating a distinctive 'duck bill' radome, the radome of the missile guidance antenna being faired into the nose above it. The preliminary inspection by the mockup review commission took place on 15 August 1955; after changes based on the commission's findings had been incorporated, the mockup was presented for final review, which took place between 12 and 20 January 1956. Having examined the arrangement of the YaD radar's modules, the pylons for carrying the Kh-20 missile, the Mikoyan/Gurevich SM-20 'parasite fighter' (a missile guidance system test bed version of the MiG-19 *sans suffixe / Farmer-A* fighter), and the drop tank, the commission gave a thumbs-up.

Manufacturer's flight tests of the 3MK were slated to begin in 1957, proceeding in two stages; stage A would involve flights with the SM-20, while inert and live Kh-20 missiles would be carried during stage B. Alas, the Tupolev OKB was first past the post: on 1 January 1956 the Tu-95K-20 prototype made its first flight. This was the crack of doom: from that day on, the VVS and MAP started favouring the Tupolev OKB. Besides, the basic 3M prototype had only just entered flight test on 27 March 1956, and this programme taxed the Myasishchev OKB's resources heavily. Hence the 3MK project began winding down. Also, some OKB-23 engineers believed the Kh-20 was outdated and could not meet current requirements, and the idea of arming the 3M with the more capable *izdeliye* 44 cruise missile was already under consideration. Finally, the semirecessed carriage of the Kh-20 on the Tu-95K was admittedly a better solution from an aerodynamic standpoint. The *Bear*'s tricycle undercarriage and large ground clearance (a side effect of the turboprop power plant) posed no problems for this.

3M Strategic Missile Strike Aircraft with *izdeliye* 44 (Project)

In 1956 the Myasishchev OKB began development of its own supersonic air-launched cruise missile – the M-44 (*izdeliye* 44). Actually it would be more accurate to say 'continued development,' because the M-44 was apparently based on the RSS (*re'aktivnyy samolyot-snaryad* – jet-powered cruise missile) developed under the guidance of Pavel V. Tsybin, whose OKB-256 design bureau had been pooled with OKB-23 in October 1959. The M-44 had a circular-section fuselage with a pointed nose, sharply swept midset trapezoidal wings of low aspect ratio, and short trapezoidal tail surfaces, the vertical tail pointing downward. Two Sorokin KR-5-26 turbojets or 5,650 kgp (12,455 lbst) MKB Granit R3-45F afterburning turbojets were mounted at the wingtips. The missile was 14 m (45 ft., 11³⁄₁₆ in.) long, with a wingspan of 5.725 m (18 ft., 9²⁵⁄₆₄ in. and a body diameter of 1.38 m (4 ft., 6²¹⁄₆₄ in.).

The M-44's design performance was far superior to that of the Kh-20. The missile was expected to attain a speed of 3,000–3,200 km/h (1,863–1,988 mph) at a cruise altitude of up to 21,000 m (68,900 ft.); it had a launch weight of 11 tons (24,250 lbs.)

A production 3MD bomber coded '10 Red'

and was to carry a 2,700 kg (5,950 lb.) warhead over a range of 2,000–2,300 km (1,242–1,430 miles). The missile was to be carried semirecessed in the 3M's bomb bay, with appropriately modified doors and fitted with solid-fuel rocket boosters; the vertical tail's ground clearance when attached to the aircraft was extremely small. The M-44 was destined to remain a paper project.

3MD Long-Range Bomber / Missile Strike Aircraft (3M-D, M-6K; Project)

While the Tu-16K-10 *Badger-C*, Tu-95K-20, and some other missile strike aircraft had entered service with the DA and Naval Aviation (AVMF – *Aviahtsiya Voyenno-morskovo flota*) by the late 1950s, they were seriously flawed in certain respects. This prompted OKB-23 to create a more advanced weapons system based on standoff missiles. The Myasishchev OKB already had some experience in the field – the 3MK project described above.

In May 1958 the OKB completed the advanced development project of an aircraft designated M-6K (again, *K* stood for ***kompleks** [vo'oruzheniya]* – weapons system). In missile-armed configuration it was intended for destroying surface ships, primarily aircraft carrier groups, yet it retained the capability to carry free-fall bombs. With the not-too-distant future in mind, the Myasishchev OKB contemplated development of a weapons system comprising an aircraft designated 3MD and an air-launched version of the P-6 antishipping cruise missile (alias P-35; NATO code name SS-N-6 *Shaddock*). Developed by OKB-52, the missile featured radio command guidance, a range of 800 km (496 miles), and a top speed of 2,000 km/h (1,240 mph). If the P-6 were selected, the aircraft would have to be equipped with a missile guidance system developed by OKB-283; an alternative guidance system offered by the NII-10 research institute required modifications in order to be integrated with the 3M's systems and equipment.

On careful consideration, OKB-23 found it more advisable to use the KSR-2 (AS-5A *Kelt*) and K-10S (AS-2 *Kipper*) cruise missiles designed by OKB-2-155, with minor changes to their systems. The intention was to modify the Rubin-1 radar as well, turning it into the core of the K-4-2 guidance system (working with KSR-2 missiles) or the K-14 guidance system (working with K-10S missiles). The system's range would be 160 km (99 miles) or 250–300 km (155–186 miles), respectively, increasing to 800–1,000 km (496–620 miles) if the *Yantar'* (Amber) radio command guidance unit was used. To work with the K-4-2 and K-14 guidance systems,

'30 Red' (c/n 0301804), another production 3MD, at Zhukovskiy on 11 August 1971, with the fin deicer section detached. This aircraft was the first to be upgraded to 3MD-K standard with an improved navigation system.

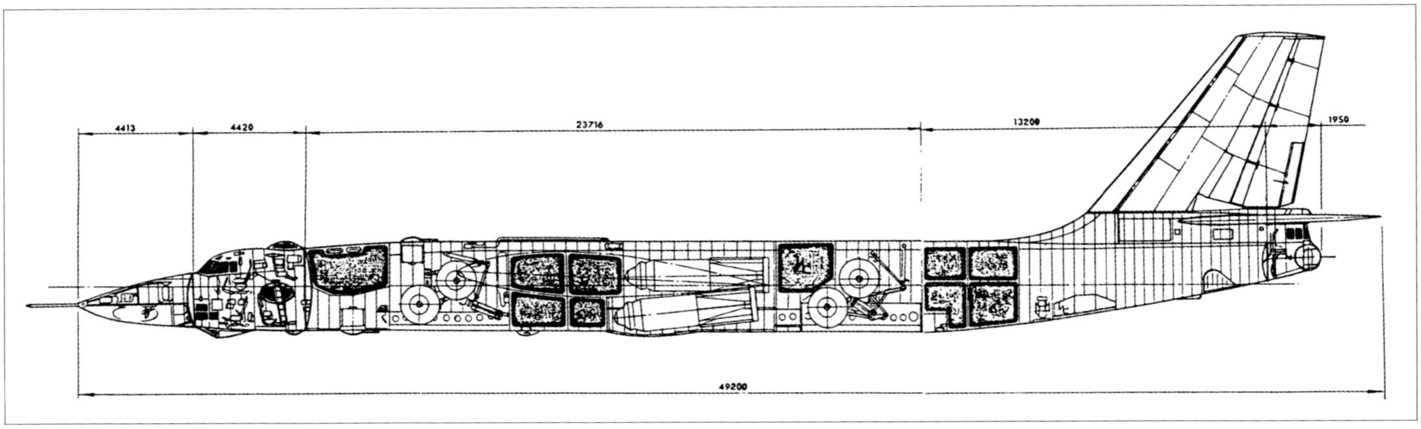

The internal layout of the 3MD

the KSR-2 and K-10S needed to have their standard guidance modules replaced with homing seeker heads developed by OKB-283.

The KSR-2 developed for the Tu-16KSR-2 *Badger-G* had mid-set swept wings swept back 55° at quarter chord and conventional tail surfaces with low-set stabilisers; vertical and horizontal tail sweep at quarter chord was 62° and 55°, respectively. The cigar-shaped fuselage had a diameter of 1.0 m (3 ft., $3\frac{3}{8}$ in.), and the fuselage nose was occupied by a large parabolic radome housing a KS-PM radar seeker head. The missile was designed around an Isayev S5.6.0000 single-chamber liquid-propellant rocket motor, with an initial thrust of 1,200 kgp (2,650 lbst) in launch mode and 600 kgp (1,320 lbst) in cruise mode. The missile was 8.647 m (28 ft., $4\frac{7}{16}$ in.) long, with a wingspan of 4.522 m (14 ft., 10 in.) and a launch weight of 4,077–4,100 kg (8,988–9,040 lbs.), including a warhead weight of 840–850 kg (1,850–1,870 lbs.). As standard, it had an FK-2 conventional warhead (*foogahsno-kumulyativnaya [boyevaya chast']* – high-explosive/shaped-charge, armour-piercing warhead) for use against ships, or an FK-2N HE/SCAP/fragmentation warhead with a so-called active liner for use against ground targets (N = *nazemnaya tsel'* – ground target), the two versions differing in explosive power. Provision was made for fitting a nuclear warhead with a yield of 1 megaton. The missile could be launched at an altitude of 4,000–10,000 m (13,120–32,810 ft.), having a maximum speed of 1,250 km/h (776 mph) and a range of up to 160 km (99 miles), though some sources state 120–140 km (74.5–87 miles).

In contrast, the K-10S developed for the Tu-16K-10 *Badger-C* was turbojet powered. The missile had a midwing layout with a slender circular-section fuselage, whose forward section, tipped with a large ogival radome, incorporated a pressurised heat-insulated bay housing the guidance system. Aft of it was the warhead, followed by the fuel tank, which was the main structural element absorbing all the loads and was made of steel. The wings had 55° sweep at quarter chord; the conventional tail surfaces consisted of stabilators with 55° leading-edge sweep and a fin-and-rudder assembly swept back 55°30' at quarter chord. The 3,250 kgp (7,160 lbst) Mikulin M-9FK axial-flow afterburning turbojet was installed in a conformal nacelle under the rear fuselage, breathing the afterburner section protruding beyond the tip of the fuselage, which mounted a midcourse guidance antenna. The missile came with a choice of conventional or nuclear warheads. The conventional ones were the FK-10 or FK-1M, the latter model being designed for striking large vessels below the waterline; a TK34 nuclear warhead could also be used. The missile was 9.75 m (31 ft., $11\frac{7}{8}$ in.) long, with a wingspan of 4.18 m (13 ft., $8\frac{9}{16}$ in.), a height of 2.27 m (7 ft., $4\frac{35}{64}$ in.), and a body diameter of 1.0 m (3 ft., $3\frac{3}{8}$ in.). The launch weight was 4,418–4,555 kg (9,740–10,040 lbs.), including 850–940 kg (1,870–2,070 lbs.) for the warhead.

Concurrently, changes would be made to the aircraft itself. The nose was reshaped to improve the navigator / bomb aimer's field of view; the radio operator's workstation was moved to the port beam blister. An inert gas pressurisation system was introduced, along with a more effective fire-extinguishing system serving the fuel cell bays in the fuselage. The avionics fit was upgraded with new items, including the Gheliy transmitter, RV-25 radio altimeter, and ARK-54B ADF. Finally, rudder area was reduced.

As already mentioned, the CoM directive of 7 August 1958 concerning the 3ME bomber tasked OKB-23 with looking into the possibility of integrating guided missiles on the 3M while still retaining normal bomber capability and upgrading the existing 3M fleet to this standard. Hence the Myasishchev OKB evolved the 3M-K-14 strategic weapons system comprising the 3MD bomber armed with K-10S missiles. With one fuel top-off, the 3MD would be able to attack targets up to 6,850–7,000 km (4,250–4,350 miles) away from base; the missiles would be fired 250 km (155 miles) from the target. The aircraft had full conventional/nuclear bomber capability; this versatility was an advantage over the Tu-16K-10 and Tu-95K-20, which were dedicated missile carriers.

The 3M-K-14 weapons system allowed various missile types to be used, provided they were made compatible with the Rubicon guidance system, which was based on a modified Rubin-1 radar. The conversion of a 3M bomber to 3MD standard involved mainly installation of the Rubicon system and missile pylons. In overall performance the 3M-K-14 was similar to the data set forth in the M-6K's ADP.

In normal configuration the 3MD was to carry two K-10S missiles on pylons located 6.66 m (21 ft., $10\frac{13}{64}$ in.) from the centreline – just outboard of the engines, with two SU-50 JATO boosters (***star****tovyy ooskor****itel****'*) mounted under the wing roots.

The ground clearance under the missiles' engines was a mere 27.5 cm ($10\frac{53}{64}$ in.). In high gross weight configuration, two more missiles were carried under the engine nacelles, slightly ahead of the outer ones, while the JATO boosters were relocated to the rear fuselage sides. An extra fuel tank was installed in the bomb bay.

Until mid-1959, all work on the 3MD, which began in the previous year, proceeded at the OKB's own risk. The OKB managed quite a lot during this time frame, performing wind tunnel tests and scaled-strength model tests with simulated external stores in order to verify the results of calculations and obtaining TsAGI's expert opinion on these tests. The general operating principles of the guidance system were confirmed in a series of tests using the SDK-13 avionics test bed (a modified MiG-17) and a modified Tu-4; a ground rig was built to test the missile release system etc.

On 22 August 1959 the Communist Party Central Committee and the Council of Ministers issued joint directive No. 998-435, concerning the development of the M6K-14 strategic air-launched missile system, including two K-14S cruise missiles (this was a projected derivative of the K-10S optimised for the 3M) and a modified production 3M (M-6) bomber. This was a signal for OKB-23 to step up the design and development work on the project, which until then had proceeded at a leisurely pace. The work called for close cooperation with the OKBs responsible for the missiles and avionics, as well as with MAP's top-notch R&D establishments.

Three views of the sole surviving 3MD, which is on display at the Central Russian Air Force Museum in Monino. This was the next-to-last aircraft of the *Bison* family built.

Left: An artist's impression of the 3MD as originally conceived – a missile strike aircraft armed with four K-10S missiles in high gross weight configuration

Below: A drawing from the ADP documents showing the 3MD with four K-10S missiles and SU-50 JATO boosters on the fuselage sides

Below right: A project drawing showing the 3MD in normal TOW configuration with two K-10S missiles and underwing JATO boosters

In 1959 the Myasishchev OKB released manufacturing drawings for the modified 3MD bomber / missile carrier; the design changes were kept to a minimum so as to ensure that a prototype could be completed before the year was out. Taking versatility still further, at the insistence of the air force Myasishchev had to include provisions for using the aircraft as a refuelling tanker designated 3MT (*see below*) in addition to its missile strike and bomber roles.

Eventually, however, the idea of using air-to-surface missiles on the 3MD was dropped, and the aircraft materialised as a pure bomber (*see next item*).

3MD Long-Range Bomber (3M-D, '1960-Standard 3M')

After doing aerodynamic calculations for the missile carrier configuration, the OKB built a full-size mockup that was ready for

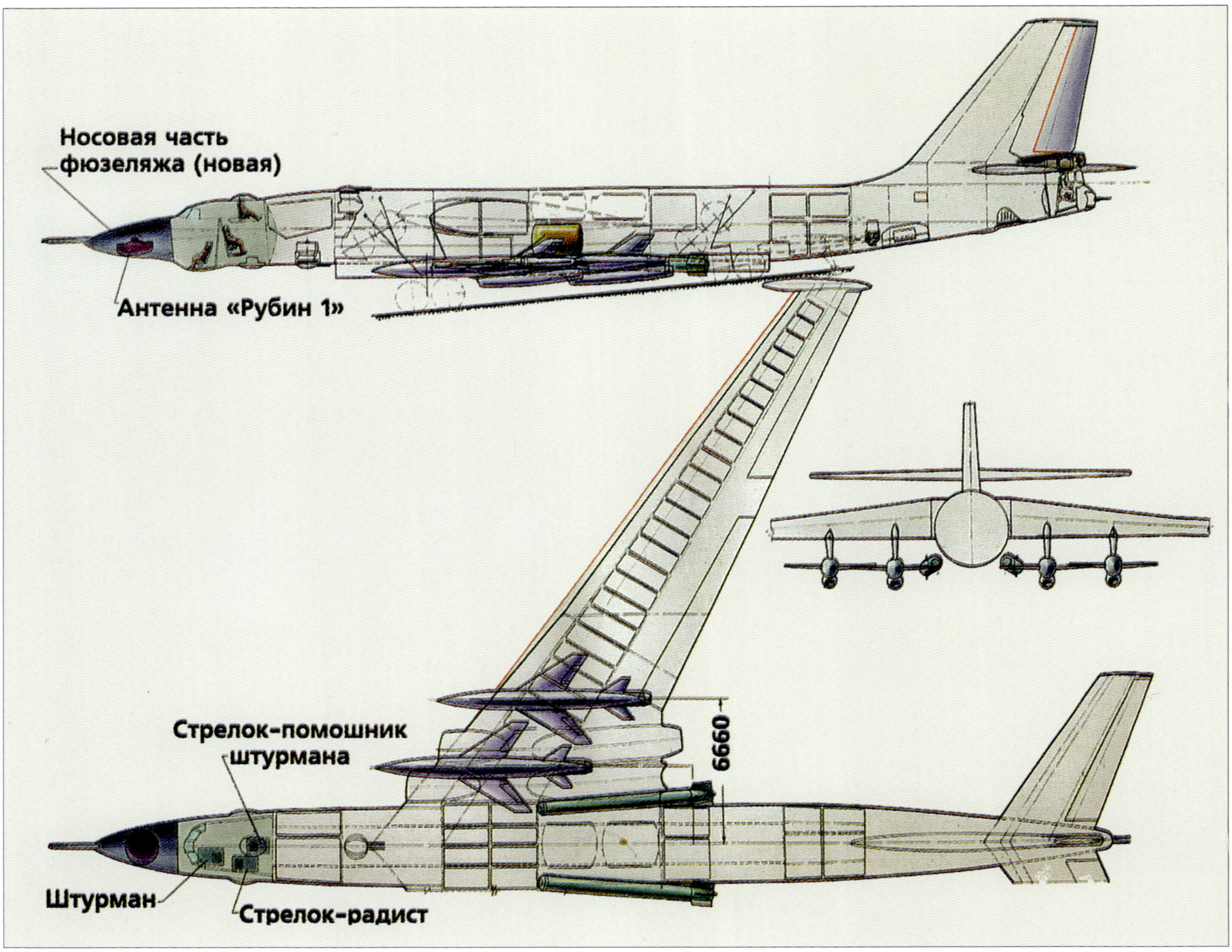

presentation in December 1959. Meanwhile, plant No. 23 was putting the finishing touches to the first prototype 3MD (c/n 9301301), and the second aircraft (c/n 9301302) was substantially complete.

In addition to the changes incorporated into the ill-fated 3ME, the 3MD prototypes introduced other new features, including reprofiled wing leading edges enhancing roll stability at high angles of attack, and a 40% smaller rudder, which considerably eased rudder pedal forces in emergency manual-control mode. The nose gear bogie was equipped with a self-contained steering system, and the IFR probe featured electrical deicing. The inert gas pressurisation and fire-extinguishing systems were different from hitherto; the engine starting panel was revised and the ejection seats were improved. The avionics included a ***Veter**-2* (Wind) Doppler speed and drift sensor, an RSBN-7S Vstrecha SHORAN, an RSIU-5 command link radio, an ARK-54B ADF, FRS-200 landing/taxi lights, etc.

As on the 3ME, the new equipment had brought about a rearrangement of the forward pressure cabin. The radio operator's workstation was relocated from the dorsal blister to the port beam blister, and the dorsal workstation now seated the co-navigator (Nav/Op). It featured a SHORAN switch panel, an AK-53P celestial compass (***astrokom**pas*), an SP-1 panoramic sextant, and a panel for starting and monitoring the missiles' M-9FK turbojets.

Later on, the long-range communications radio was to feature a new Rool' (steering wheel) receiver. The new ADF provided nine-channel, searchless, automatic reception and was more resistant to jamming than the old ARK-5. The RSBN-2S/RSBN-7S SHORAN permitted on-the-route navigation by means of ground ATC beacons, landing approach, and rendezvous/formation keeping.

The first prototype was completed in November 1959, making its maiden flight on 25 November at the hands of B. M. Stepanov and Aleksandr S. Rozanov. A special joint flight test programme held by the OKB and the factory got underway on 1 February 1960; by 31 March the aircraft had made 11 flights totalling 47 hours, 24 minutes, under this programme. An air force test crew flew its first mission the 3MD in April 1960. The trials were concluded in style by a maximum-range flight; taking off with a 190-ton (418,870 lb.) TOW, the aircraft flew nonstop along the Zhukovskiy–Nar'yan-Mar–Amderma–Tiksi–Ambarchik–Tiksi–Nar'yan-Mar–Zhukovskiy route in 14 hours, 13 minutes.

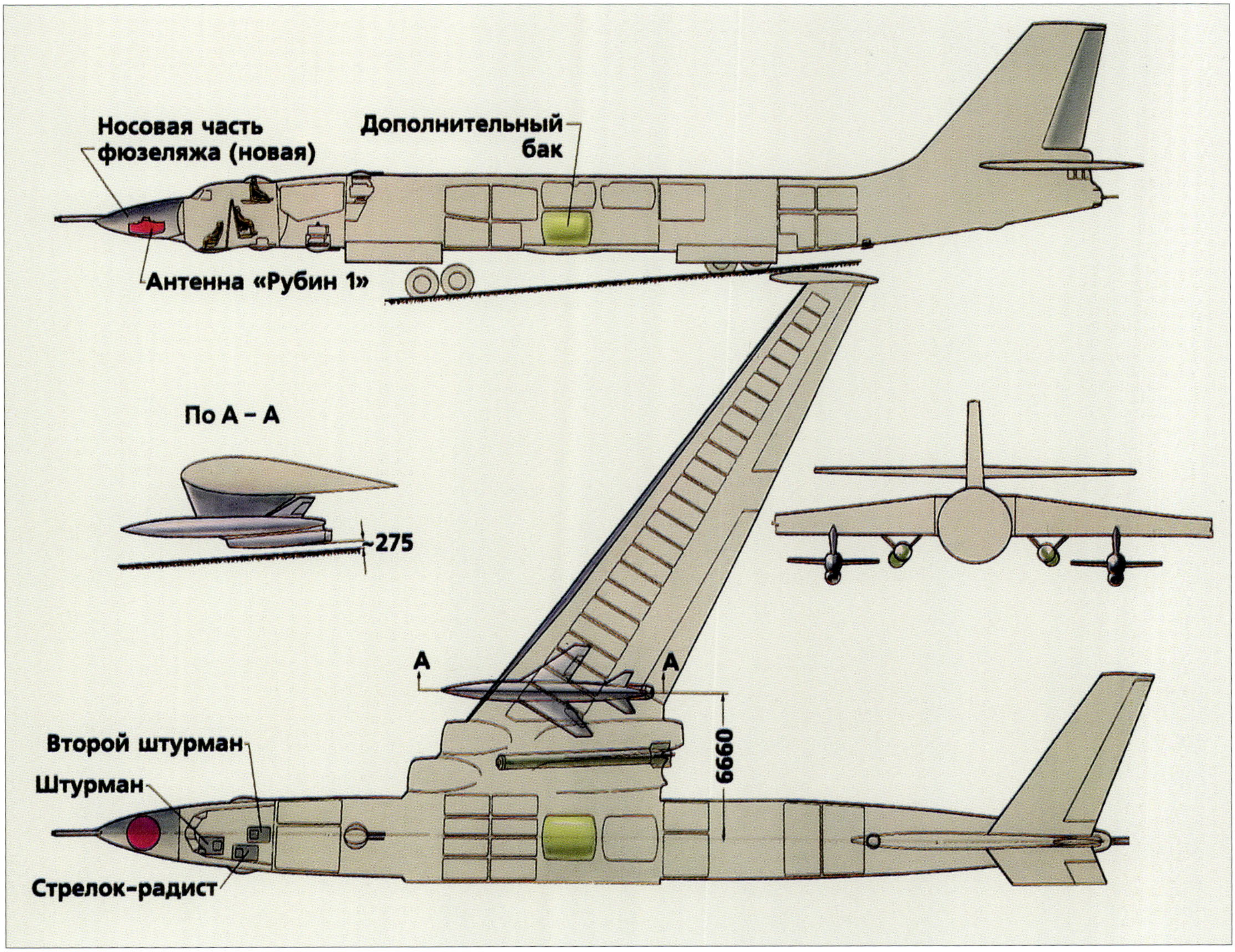

The second prototype 3MD joined the testing in 1960. During state acceptance trials, the aircraft was flown by GK NII VVS pilots Nikolay N. Belyayev and Yuriy V. Sookhov; navigators R. N. Semenovskikh, V. A. Mezentsev, V. S. Golovin, A. N. Yadryshnikov, B. Ye. Kalinovskiy, and M. S. Bogdanov; and gunners V. I. Spirin, A. I. Shichkov, V. A. Naoomov, and V. N. Smolin.

Meanwhile, development and integration of the K-14S missile continued; a full set of drawings were released and a prototype pylon was manufactured. Toward the end of the year, however, work on this system started winding down since the Myasishchev OKB had ceased to exist, being reduced to a division of the OKB-52 missile design bureau. Nevertheless, in late 1960 or early 1961, a K-14S missile (perhaps we should say K-10S after all!) was actually hooked up to 3MD c/n 9301301 at the OKB's former flight test facility in Zhukovskiy. However, when the missile was lifted into position under the prototype's wing, it turned out that the electrical connectors did not match! Since the Myasishchev OKB was defunct by then, the issue was not pursued further.

In 1962 the first prototype 3MD was used in another test programme, with a VD-7P development engine replacing the No. 2 VD-7B. This was an uprated version, with a modified compressor increasing the mass flow from 175 to 181 kg/sec. (from 358 to 399 lbs./sec.) and a turbine temperature raised from 820°C to 880°C (from 1,508°F to 1,616°F), which gave it a takeoff thrust of 11,300 kgp (24,910 lbst) – 300 kgp (660 lbst) more than the original VD-7 *sans suffixe* and 1,980 kgp (4,365 lbst) more than the standard VD-7B. Test pilot Lev V. Vlasov made 20 flights, which yielded good results; still, the VD-7P never entered production. Incidentally, some sources misidentify this aircraft as the 3ME or even as the '3MP with four VD-7Ps' – a version that never existed.

In a flashback of similar occurrences in the history of the Myasishchev OKB, the 3MD entered production when the flight tests had barely begun. A mere 10 production examples were built in 1959–60 (c/ns 9301302, 0301702 through 0301705, and 0301801 through 0301805); hence the type was referred to in official documents by the rather incongruous phrase '3M manufactured in 1960' (or '1960-standard 3M'). They were used as ordinary bombers with no missile armament. In service these aircraft were usually the mounts of squadron, regiment, or division commanders.

The new front-end treatment of the 3MD was obvious enough to warrant a new NATO reporting name, *Bison-C*. In the VVS, however, the aircraft's turned-up nose, giving it a haughty air, earned it the nickname *stilyaga* (this word, with a pejorative connotation, translates loosely as 'fop').

The 3MD was unveiled to the general public on 9 July 1967, when a single example was in the static park at the grand air fest held at Moscow-Domodedovo airport. Thankfully, one of the *Bison-C*s ('30 Red,' c/n 0301804) has been preserved for posterity at the Soviet (now Central Russian) Air Force Museum in Monino, near Moscow.

3MD-K Bomber

Reference has been made in one source to a version of the *Bison-C* with updated flight avionics (including navigation equipment), which was designated 3MD-K (the *K* suffix bears no relation to missiles). The upgrade was developed in 1970; the first aircraft to be thus modified was the abovementioned 3MD c/n 0301804, which arrived at Zhukovskiy for modification on 18 January 1971.

3MN-K Bomber with Enhanced ECM Protection (*izdeliye* 3-22)

During the Vietnam War, the USAF used B-52s equipped with active jammers for penetrating enemy air defences during bombardments of North Vietnam. The ECM equipment proved so effective that on occasion, the North Vietnamese SAM crews, to their frustration, could observe the incoming bombers with the naked eye while the missile guidance radars were completely blinded by ECM.

Taking this into account, the Soviet air force demanded that all strategic bombers be outfitted with ECM equipment for self-protection / mutual protection in a group. Hence, in November 1969, after an MAP conference, the Myasishchev OKB suggested fitting the 3M with available active ECM systems to enable low-level air defence penetration at 300–500 m (980–1,640 ft.). On 24 March 1970 the air force sent the OKB a letter signed by Deputy C-in-C Aleksandr N. Ponomaryov and DA commander Vasiliy V. Reshetnikov. This letter contained recommendations on the models of ECM equipment to be used.

In accordance with this letter, the Myasishchev OKB developed an ADP of a 3M bomber with enhanced ECM and electronic-support-measures (ESM) equipment. The latter included a *Sir**en**a-3M* (Siren) radar-warning receiver, three SPS-153D Siren'-D (lilac, pronounced *see**ren'***) active jammers giving 360° coverage of the front and rear hemispheres, an SPS-5 *Fasol'* (string bean) active jammer covering the lower hemisphere, an SPS-100 *Azaliya* (Azalea) active jammer for group protection, an APP-22 chaff dispenser (*avto**maht** **p**assivnykh **p**omekh* – automatic passive ECM device), and an ASO-2IE7R chaff/infrared countermeasures flare dispenser (*avto**maht** **s**brosa [**dipol'**nykh] otrazhateley* – 'dipole reflector dispenser'). This equipment would replace the obsolete SPS-2K and Siren'-2 jammers and ASO-3M chaff dispenser. It added 203 kg (507 lbs.) of empty weight and required more-powerful generators.

The project was given official status by Council of Ministers directive No. 77-27, of 4 February 1971, and MAP order No. 67, of 10 March 1971. Unfortunately, the work was making slow progress because of bureaucratic red tape; it was a full year before the CoM Presidium's Commission on defence industry matters (VPK – *Voyenno-promyshlennaya komissiya*) issued ruling No. 76 on 29 March 1972 (followed up by MAP order No. 121 on 19 April 1972), which, among other things, allocated an aircraft for prototype conversion and authorised the required funding. Appropriate instructions were issued to MAP and Ministry of Electronics Industry (MRP – *Ministerstvo rahdioelektronnoy promyshlennosti*) enterprise involved in the programme.

At the Myasishchev OKB the programme was code-named *tema* 3-22, while the aircraft itself was designated 3MN-K; again, in this case the *K* suffix had nothing to do with air-to-surface missiles. The project was supervised by V. M. Gusarov, while Yu. A. Vasil'yev and his assistant V. M. Kirillin led the actual design work. The SOR was approved on 24 May 1972; the aircraft was to be ready for testing in the second quarter of 1973.

On 7 July 1972, a production 3MNR-2 tanker (c/n 7300503) arrived in Zhukovskiy for conversion as the 3MN-K, and the work started. Yet, this aircraft was not tested as such, being subjected to more modifications (see 3M-5 section below).

3M-5 Strategic Bomber / Missile Strike Aircraft Prototype (*izdeliye* 3-23)

In the 1960s the 3M's American counterpart, the B-52, was undergoing a constant improvement and upgrading process. In particular, the bomber was modified for low-level air defence penetration to make it less vulnerable to SAMs. Some B-52s were equipped to carry two North American AGM-28 Hound Dog supersonic stand-off air-to-surface missiles, which could be used for attacking targets up to 1,125 km (699 miles) away if launched at high altitude. The B-52H introduced more-fuel-efficient Pratt & Whitney TF33 turbofan engines instead of the earlier Pratt & Whitney J57 turbojets; together with new integral tanks in the wings, they gave the Stratofortress a range of 16,000 km (9,941 miles).

Meanwhile, the 3M's capabilities remained at their original level; the bomber was growing obsolescent. Justifiably alarmed by this, the Soviet MoD started seeking ways and means of boosting the *Bison*'s combat potential. Pursuant to Council of Ministers directive No. 1339, dated 14 June 1966, and MAP order No. 492, dated 21 November 1966, the Myasishchev OKB was reborn again – albeit in 'abbreviated' form. The new entity, based at the flight test facility in Zhukovskiy, was officially called the Experimental Machinery Plant (EMZ – *Eksperimen**tahl'**nyy **ma**s**hino**stro**it**el'nyy **zavod***) but will still be referred to hereinafter as the Myasishchev OKB for the sake of convenience.

In October 1968 the OKB presented a memo on the possibilities of upgrading the 3M. This centred on two aspects – performance and weapons – and on extending the aircraft's service life by eight to ten years. The performance increase was to be achieved by installing more-fuel-efficient engines, increasing the fuel supply, and cutting structural weight. The envisaged power plant was the 11,000 kgp (24,250 lbst) Solov'yov D-30K turbofan, with a bypass ratio of 2.31 as used on the Il-62M airliner; this had a much-lower SFC – 0.498 kg/kgp·hr. at takeoff power and 0.685 kg/kgp·hr. at cruise power – but would necessitate larger-section inlet ducts to cater for the higher mass flow. The inboard wing section fuel cells were to be replaced by integral tanks cutting empty weight by 600 kg (1,320 lbs.) and increasing fuel capacity by 5,000 litres (1,100 imp. gal.); lighter non-self-sealing fuel cells were to be used in the fuselage and outer wings, saving another 1,100–1,300 kg (2,430–2,870 lbs.) and adding 3,000–3,500 litres (660–770 imp. gal.) of fuel. Later, integral tanks could be provided in the outer wings as well, saving another 1,000 kg (2,205 lbs.) and adding a further 8,000 litres (1,760 imp. gal.) of fuel. New wing leading-edge sections patterned on those of the 3MD were to be used, improving the lift/drag ratio by 1.2 points.

The new engines alone could increase range by as much as 25% (3,500 km / 2,174 miles) and increase the top speed and gross weight; the new wing leading edges could add another 1.5%–2%. Overall, the bomber's range with 5 tons (11,022 lbs.) of bombs and 5% fuel reserves could be extended to 14,000–15,000 km (8,695–9,320 miles) – or even further if the TOW was increased to 202–205 tons (445,330–451,940 lbs.).

The combat capabilities would be expanded by integrating KSR-5 (NATO code name AS-6 *Kingfish*) supersonic air-to-sur-

Four views of the 3M-5 development aircraft, with two KSR-5 missiles under the wings; note the new 'droopsnoot' radome

An artist's impression of the 3M-5 in flight, carrying red-painted inert KSR-5 missiles

face missiles, two of which would be carried on pylons under the inner wings, and introducing new bombload options with 100, 250, and 500 kg (220, 551, and 1,102 lb.) bombs carried both internally and, if necessary, externally on multistore racks. The maximum ordnance load was increased to 27 tons (59,520 lbs.). Due attention was also given to low-level operation capability; the VVS even undertook a special R&D programme code-named *Nizina* (low ground) in 1968.

The KSR-5 merits a detailed description. Development had begun at OKB-2-155 in 1960. The missile was fielded by Soviet Naval Aviation in December 1969 as part of the K-26 weapons system, which also included the Tu-16K-26 *Badger-G* missile strike aircraft. Known in house as the D-5 (the *D* probably referred to the town of Doobna, near Moscow, where the design office was based), the KSR-5 resembled a scaled-down version of the Kh-22 (D-2; NATO code name AS-4 *Kitchen*), developed for the Tu-22K *Blinder-B*. It had a conventional midwing layout with delta wings having 75° leading-edge sweep; the cruciform trapezoidal tail surfaces arranged in the vertical and horizontal planes consisted of stabilators, an all-movable dorsal fin, and a fixed ventral fin, which folded to port to facilitate ground handling, deploying after launch. The circular-section fuselage was tipped by an ogival radome of glass fibre honeycomb construction. The power plant was an S5.33 twin-chamber, three-mode, liquid-propellant rocket motor (a.k.a. R201-300), developed by OKB-300 under Sergey K. Tumanskiy. It ran on TG-02 fuel (***to**plivo **g**hipergo**li**cheskoye* – hypergolic or self-igniting fuel), a 50/50 mixture of xylidine and triethylamine that self-combusted when combined with AK-27P oxidiser (AK = ***azot**naya **kislota*** – nitric acid; '27' means 27% nitrogen tetroxide added to give a power boost). With both chambers running, the S5.33 delivered 8,350 kgp (18,410 lbst); with only the smaller chamber running, it was 1,400 kgp (3,090 lbst) at intermediate thrust and 650 kgp (1,430 lbst) at cruise thrust. The airframe was made mostly of aluminium alloys, with stainless steel used for the oxidiser tank.

The KSR-5 was 10.6 m (34 ft., 9²¹⁄₆₄ in.) long, with a wingspan of 2.606 m (8 ft., 6¹⁹⁄₃₂ in.) and a body diameter of 0.92 m (3 ft., ⁷⁄₃₂ in.). The launch weight varied from 3,850 to 3,930 kg (8,490–8,660 lbs.), including a 900 kg (1,980 lb.) warhead; the latter could be a shaped-charge/HE warhead ('version M') for use against a pinpoint target, such as a large surface ship, or a 350–500-kiloton thermonuclear warhead ('version N') for use against a large target, such as a group of ships.

The KSR-5 was a highly accurate 'fire-and-forget' missile for use against ground or maritime targets. It had a Vzlyot (takeoff) guidance system comprising a VS-K active radar seeker head and a BSU-7 autopilot (*borto**va**ya sis**tem**a oopravleniya* – onboard control system), a command link system for midcourse guidance, an altimeter, and a speed sensor. The seeker head acquired the target while the missile was still 'on the wing,' using target information supplied by the aircraft's radar. In an ECM environment, the guidance system made use of the target coordinates stored in its memory; should the ECM become really severe and prolonged, the system switched to another operating frequency, which, it was hoped, was not yet jammed.

The KSR-5 was launched at an altitude of 9,000–11,000 m (29,530–36,090 ft.) after the aircraft's radar had acquired the target and achieved a lock-on. The launch range was 200–240 km (124–149 miles). As the rocket motor fired three seconds after release and went to full thrust, the missile accelerated to 2,500–3,000 km/h (1,552–1,863 mph) and pulled up into a climb, cruising at an altitude of 22,500 m (73,820 ft.) and receiving midcourse guidance. At a certain range from the target, the KSR-5's radar seeker head achieved a lock-on; the missile still received course updates from the aircraft as required. When the target was 60 km (37¼ miles) away, the missile entered a 30° dive; the aircraft was then free to manoeuvre at will or return to base. The radar seeker was turned off 1 km (0.62 miles) from the target. The warhead was detonated by an impact fuse in the conventional version or by a radio altimeter at a preset altitude in the nuclear-tipped version.

Originally, the 3MD was envisaged as the launch platform for the KSR-5, and with good reason. First, it had more-refined aerodynamics than the 3MS/3MN; second, it already had the required Rubin radar and wing hardpoints, which meant it would cost 50% less to convert than the *Bison-B*. Among other things, the crew workstations would be rearranged and the electrical system would be updated, as would the flight/navigation suite. In early 1970 the 'missilisation' of the 3MD was included into the air force's list of modifications to be made to the M-4/3M family. Yet, the work was halted because of funding problems and objections from MRP, which cited pressure of other work at NPO Raduga as the reason and argued that the cost of upgrading the small fleet of 3MDs would be prohibitive. The air force insisted that 50 aircraft (nine *Bison-C*s and 41 *Bison-B*s) be converted to carry KSR-5 missiles; however, in the absence of funding and an appropriate

Council of Ministers directive, the job was eventually struck from the list on 17 November 1970.

Still, the air force would not give up its plans for up-arming the 3M (and the Tu-95). On 24 May 1972 it sent the Myasishchev OKB a draft specific operational requirement for a 'missilised' 3M, which was finalised on 4 April that year. The military's persistence bore fruit – or so it seemed. On 13 February 1973, after a good deal of consultations and discussions between the OKB and the air force, the Council of Ministers issued directive No. 104-36, backed up by MAP order No. 103 on 12 March 1973, which required the 3M and the Tu-95 to be armed with KSR-5 missiles. Together with the new ECM suite (*see above*), this would give a major boost in the bombers' combat potential.

The weapons system based on the 3M was designated K-3M-26; at the Myasishchev OKB the programme was code-named *tema* 3-23, while the aircraft itself was designated 3M-5, the '5' referring to the type of missile carried. The design work on the new armament was led by R. V. Zaïtsev. The work received top priority because it gave the OKB a chance to secure a production plant for itself again. An order signed by chief designer Vladimir M. Myasishchev on 7 March 1973 required design department chief Arvid D. Tokhoonts to complete the preliminary design project within a month and temporarily relieved the OKB's departments involved with the project of all other work. The PD project was duly completed and endorsed in late April.

The 3M-5 was to feature the 3MNK-26 missile armament system (this was because the KSR-5 was part of the K-26 air-to-surface missile system). Apart from the basic version, the KSR-5N missile, optimised for low-level use, and the KSR-5P antiradar missile were also envisaged. The KSR-5N (***niz**kovysotnaya* – low-altitude) had a VS-KN radar seeker and an APR-5N autopilot (***av**topee**lot** raket*y – missile autopilot). The KSR-5P (***pro**tivorah-diolokatsi**on**naya [ra**ke**ta]* – ARM) differed in having a VS-P passive radar seeker and was designed for destroying shore-based and shipboard radars working in search or tracking mode – both stable-frequency units and those whose operating parameters change smoothly or incrementally (frequency-hopping).

The aircraft's defensive suite was to include an SPO-15 *Be**ryo**za-P* (Birch-P) radar homing and warning system (RHAWS; SPO = *sis**te**ma preduprezh**de**niya ob obloo**che**niï* – lit. 'irradiation warning system') and an ECM suite. The latter comprised four active jammers – the SPS-5 Fasol,' SPS-6 Los' (Moose), SPS-100 Azaliya and SPS-153D Siren'-D, and an ASO-2I IRCM flare dispenser. The RHAWS gave full 360° protection, analysing the type of radar and indicating range.

The flight avionics were augmented by an RSDN-3S long-range radio navigation system (***rah**diotekh**nich**eskaya sis**te**ma* ***dahl'**ney navi**gah**tsiï* – LORAN) working with the Tropik-2 and LORAN-S ground-based systems.

Models of the 3M with KSR-5 missiles attached were tested in TsAGI's T-103 and T-106M wind tunnels; the tests showed that there was little change in pitch and roll stability and that the aircraft could be operated safely as long as the CG was within the allowed range. The results of previous wind tunnel tests in 1958–59 with models of the 3MK and 3MD (with K-10S missiles) came in handy here, saving time.

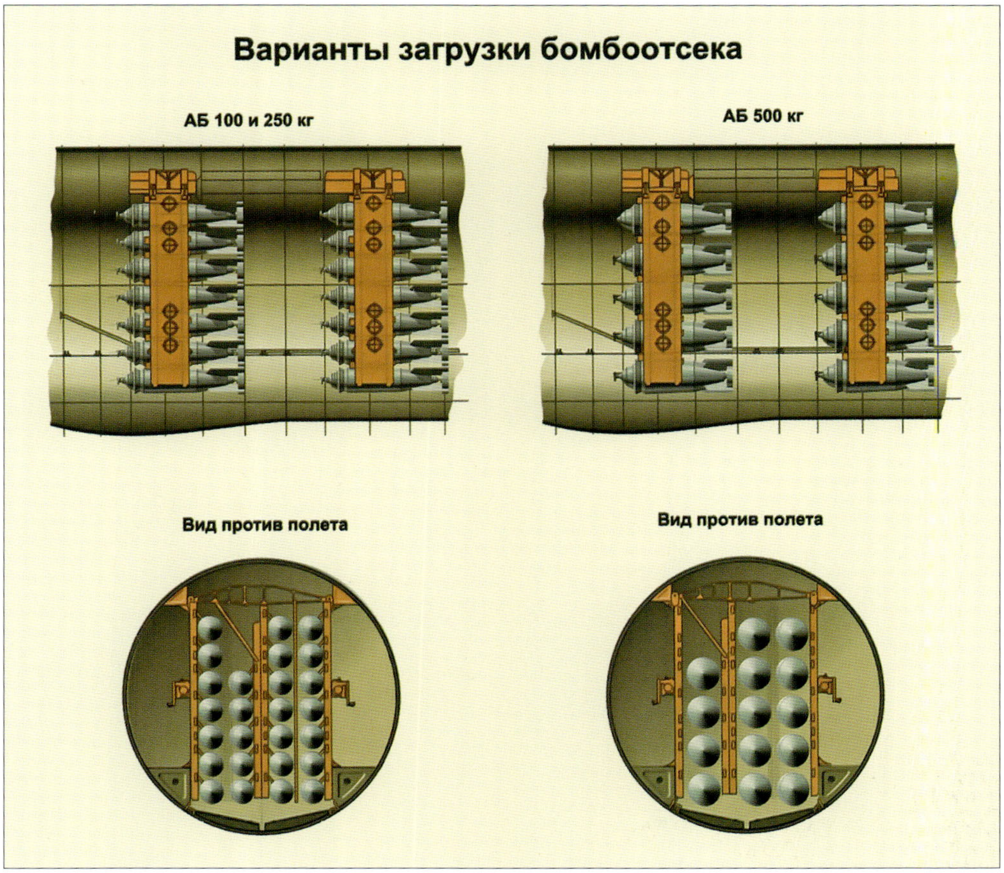

The 3M-5's new internal bombload options. The drawings on the left show 100 and 250 kg (220 and 551 lb.) bombs on eight bomb cassettes; the ones on the right show 500 kg (1,102 lb.) bombs on six bomb cassettes. The lower drawings show a view toward the tail.

Above: 3M '28 Red' (c/n 9301401) was fitted experimentally with a dummy KSR-5 under the starboard wing as part of the 3M-5 work.

Below: The 3M-5 (c/n 7300503) with a joint Myasishchev OKB/GNIKI VVS test crew. *Left to right*: Yuriy I. Shkuratov, I. D. Polyakov, Kir V. Chernobrovkin, S. M. Baranov, A. A. Khalizov, B. Kh. Aizatoollin, A. P. Kucherenko. The aircraft was later used in the 3MT programme.

Various MAP, MRP, and MOP enterprises were involved in the project as subcontractors pursuant to the CoM directive of 13 February 1973. In particular, NPO Mars supplied the Rubin-1ME navigation/target illumination radar, and the ECM equipment was developed by the Central Electronics Research Institute (TsNIRTI – *Tsentrahl'nyy naoochno-issledovatel'skiy rahdiotekhnicheskiy institoot*), the Aircraft Instrument Engineering Research Institute (NIIAP – *Naoochno-issledovatel'skiy institoot aviatsionnovo priborostroyeniya*), and the *Ekran* (Screen) and *Avtomatika* (automatic devices) design bureaus. The AP-15 autopilot was modified with TsAGI's assistance to enable low-level flight, while LII helped choose the optimum location of various aerials to ensure electromagnetic compatibility (EMC).

An updated version of the SOR for the K-3M-26 weapons system appeared on 6 July 1973. The 3M-5 was to be capable of engaging pinpoint and large-area targets with conventional or nuclear munitions, engaging ground or maritime targets having a high radar signature with KSR-5 missiles. The latter were to be used within their entire launch envelope – that is, at altitudes of 500–10,000 m (1,640–32,810 ft.) and speeds of 400–500 km/h (248–310 mph). The aircraft was to be capable of carrying a mix of bombs (of 100–9,000 kg / 220–19,840 lb. calibre) and two missiles; if the latter had nuclear warheads, the environmental control system was to heat and pressurise their warhead bays. The possibility of carrying underwing nuclear bomb pods instead of missiles was also contemplated. The navigation/attack system, based on the Rubin-1ME radar and the flight/navigation suite, were to enable long-range/short-range navigation and formation flying in any weather, day or night; the aircraft was to feature equipment for clear-code and secure communications. The ECM/ESM suite was to provide adequate protection against antiaircraft fire control radars and suppress long-range air defence radars. The modified autopilot was to enable automatic control at 200–4,000 m (660–13,120 ft.) with the pitch control channel disabled and at 2,000–4,000 m (6,560–13,120 ft.) with the pitch control channel enabled.

A standard 3MS ('28 Red,' c/n 9301401) was ferried to Zhukovskiy on 4 May 1973 for use as a 'full-size mockup' on which the placement of the 3M-5's mission equipment would be verified. This work proceeded in accordance with a plan endorsed on 23 May, lasting until July 1973; part of the equipment was real, but mostly wooden dummies were used. A single BD-352-11-5 missile pylon (*bahlochnyy derzhahtel'* – beam-type rack) from the Tu-16K-26 was installed under the starboard wing, and a dummy KSR-5 missile known as *izdeliye* 091 was hooked up; ECM antenna fairings were added to the fuselage. An in-house 'mockup review commission' examined the aircraft between 13 and 15 August 1973, finding that most of the new equipment and relocated existing equipment was convenient to operate; it did voice a few criticisms, which were duly acted on. The commission's report was endorsed on 18 September.

Next, the aforementioned 3MNR-2 (c/n 7300503), which was being outfitted as the 3MN-K prototype, was now further converted as the prototype of the 3M-5 missile strike version. The conversion involved installing a new Rubin-1ME radar, which was enclosed by a new deep radome quite unlike that of the 3MD, and fitting two BD-352-11-5 missile pylons under the wing roots. The Azaliya jammer was installed in an ogival fairing, replacing the tail turret and the bulletproof rear glazing of the tail gunner's station; this fairing, known as the UKhO (*oonifitseerovannyy khvostovoy otsek* – standardised tail compartment), was used on some versions of the Tu-16, Tu-22, and Tu-95. Other ECM antennas were fitted elsewhere on the airframe to give full 360° coverage.

Outfitting and structural modifications began in May 1973, continuing until December; on 25 December the uncoded 3M-5 was presented to the military with modifications as per CoM directive No. 104-36. However, additional equipment had to be installed in accordance with the revised requirement; this work lasted three months and was followed by lengthy ground checks and elimination of the many flaws detected. Thus, it was not until July 1974 that the aircraft was finally ready for the joint flight tests, with a total of some 70 test flights envisaged. Stage A was concerned with stability/handling, field performance, dynamic strength tests with external stores, EMC, and the functioning of the mission equipment; albeit formally manufacturer's flight tests, these were performed jointly with the air force. Stage B – the state acceptance trials – would be held at GNIKI VVS (as GK NII VVS had been restyled by then), centring on flight performance and mission equipment ergonomics. Also, both stages included test launches of missiles.

The OKB assigned pilots Ivan A. Laptev (project test pilot), Kir V. Chernobrovkin and F. Kozyrev; navigators S. M. Baranov and S. N. Sokolov; flight engineer B. A. Aïzatoollin; and radio operator I. D. Polyakov to the 3M-5, and Yu. I. Shkuratov was engineer in charge of the tests. On its part, GNIKI VVS assigned pilots N. M. Bystrimovich, Anatoliy P. Kucherenko, and Boris V. Oleynikov; navigators S. Maksimov, E. Skibin, and A. Khalizov; flight engineer Demidov; and gunner/radio operator A. Illarionov to the programme, with V. Bazilevskiy as engineer in charge. The ground crew was supplied by the OKB.

In late July 1974 the 3M-5 made its belated first flight with Laptev and Kozyrev at the controls. By the end of the year it had made 11 flights under stage A, logging 41 hours with pauses for engine changes and modifications (ground checks of the missile armament were made during this downtime).

Meanwhile, the missile pylon and the new radome underwent static tests at TsAGI. In December the pylon failed at 90% of the design load and had to be reworked; conversely, the radome passed the tests successfully in early 1975. TsAGI also completed several test reports, including one on the results of the autopilot's tests in simulated low-level flight.

The test schedule kept slipping because of ongoing modifications to the aircraft and runway repairs at Zhukovskiy, causing the end of stage A to be postponed until the third quarter of 1975. This led the air force to revise the test programme, leaving out high-alpha tests, IFR tests, test flights with a simulated autopilot failure, and bombing-accuracy checks. Bombing was impossible anyway because the bomb bay was occupied by test equipment recorders. To save time, in the third quarter of 1975 the 3M-5 started making test flights from Zhukovkiy under stage B when stage A was still in progress; it was completed on 26 November, the aircraft logging 229 hrs., 50 min., in 42 flights (including 10 under stage B). Kir V. Chernobrovkin did most of the flying at this stage (33 flights with a total time of 144 hrs., 26 min.) because Ivan A. Laptev was

Myasishchev OKB test pilot Nikolay I. Goryainov

Myasishchev OKB test pilot Ivan A. Laptev

Myasishchev OKB test pilot B. M. Stepanov

Myasishchev OKB test pilot Mark L. Gallay

Myasishchev OKB test pilot Boris I. Yumashev

Myasishchev OKB test pilot Aleksandr S. Rozanov

Myasishchev OKB test pilot Anatoliy S. Lipko

Myasishchev OKB test pilot Kir V. Chernobrovkin

Myasishchev OKB test pilot Lev V. Vlasov

GK NII VVS test pilot Sergey G. Dedukh

GK NII VVS test pilot Semyon M. Antonov

GK NII VVS test pilot Yuriy V. Sookhov

forced to quit flying at the end of 1974. A major contribution to the tests was made by the OKB's flight operations department chief A. I. Nikonov and flight test facility chief Yu. A. Kireyev.

Stage B formally began on 22 December 1975, continuing until 15 March 1976 and involving 23 flights with a total time of 121 hours. All in all, the 3M-5 made 54 test flights, logging 264 hrs., 30 min. It turned out that the aircraft was almost identical in handling and piloting technique to the standard 3M when carrying missiles. The aircraft was flown with takeoff weights up to 185 tons (407,850 lbs.) and landing weights up to 116 tons (255,730 lbs.); the maximum speed attained was 600 km/h (372 mph) IAS at 5,000 m (16,400 ft.) and Mach 0.896 at 10,900 m (35,760 ft.). Both so-called tactical launches (that is, simulated launches) of dummy missiles and actual launches were made; in the latter case the 3M-5 had to redeploy to Engels-2 AB, where the missiles were prepared for hooking up. Three of the four live launches were successful (on the second launch, the missile's rocket motor quit due to a fuel leak, and the missile crashed 3 km (1.86 miles) short of the target). The missile guidance system functioned stably; on one occasion the missile even scored a direct hit on the Luneburg lens, emulating the radar echo of a railway bridge.

The state acceptance trials protocol was endorsed on 20 October 1976; the 3M-5 was recommended for service, with the proviso that its service life be extended. In December 1976 the Myasishchev OKB and GNIKI VVS intended to start a new test programme encompassing the things that had been left out during the trials.

The tests showed promising results, and the OKB proposed upgrading other VD-7B-powered *Bison-B*s to 3M-5 standard in order to increase their punch. However, the OKB still did not have control over a production plant, and the air force's aircraft repair facility at Dyagilevo AB, which handled the M-4 and 3M, would take too long to complete the job of modifying 50 aircraft; besides, in the meantime a new advanced cruise missile – the Raduga Kh-55 – was under development. On 10 October 1977 the Council of Ministers pulled the plug on the programme by issuing directive No. 909-291, which excluded the 3M from the range of aircraft eligible for an upgrade; the issue basically boiled down to money.

As for the 3M-5 prototype, it was transferred to the OKB free of charge, serving for test and development work for more than ten years. On 13 March 1989, this aircraft was struck off charge by MAP order No. 123 and eventually scrapped in Zhukovskiy in September–October 1989.

Incidentally, the air force had also tasked the Tupolev OKB with adapting the Tu-95 for carrying the KSR-5 missile. Accordingly a production Tu-95M *Bear-A* (c/n 8800601) was converted to the Tu-95M-5 prototype in October 1976 by fitting two missile pylons, a modified Rubin-1KV radar, and an SPS-153 jammer in a UKhO tail fairing. The aircraft entered flight test in the autumn of 1976, but the programme was terminated in May 1977. Eventually the VVS settled for the Kh-22 ASM, having both a larger warhead and longer range. The Soviet government decided that only Tu-95 bombers were to be converted into missile carriers, which led to the appearance of the Tu-95K-22 *Bear-G*.

3M Bomber Armed with Cruise Missiles (Project)

In 1974 the USA began developing strategic cruise missiles; in particular, the Boeing Co. started design work on the AGM-86A Air-Launched Cruise Missile (ALCM). A year later, when this alarming news reached the Soviet leaders and the prospect of ALCMs being launched in large numbers by the USAF's Rockwell B-1As and Boeing B-52s loomed large, serious research in this field began in the Soviet Union as well. The Tupolev OKB and MKB Raduga (the latter was by then headed by Igor' S. Seleznyov) submitted a joint proposal for a low-level subsonic cruise missile with a nuclear warhead to be used against ground targets and surface ships with a low radar signature. Designated Kh-55 (NATO code name AS-15A *Kent*), the missile was to have a correlation navigation system and a terrain-following feature helping it to avoid detection and interception.

The missile changed appreciably in the course of the work. The definitive version, known in house as *izdeliye* 120, had a circular-section fuselage of 0.514 m (1 ft., 8$^{15}/_{64}$ in.) diameter with a blunt ogival nose, a midwing layout with unswept tapered wings (which had the same shape of rounded tips folding aft into a slit in the fuselage), double-jointed 'snap-action' tail surfaces (which were completely within the fuselage diameter when folded), and a small turbofan engine stowed inside the rear fuselage in prelaunch configuration and extended immediately before release. The Kh-55 had a length of 5.88 m (19 ft., 3½ in.), a wingspan of 3.1 m (10 ft., 2 in.), and a weight of 1,185 kg (2,610 lbs.), including a 410 kg (904 lb.) warhead; the specially developed compact thermonuclear warhead had a yield of 200 kilotons. After being launched at 200–10,000 m (660–32,810 ft.) and 540–1,050 km/h (335–652 mph), the missile would descend to 40–110 m (130–360 ft.) and follow the preprogrammed route to the target, making use of the terrain-following feature. Cruising speed was 720–830 km/h (447–515 mph) or Mach 0.8; maximum range was 2,500 km (1,552 miles).

There is evidence that in 1976 the Myasishchev OKB considered the possibility of modifying the 3M for carrying cruise missiles similar to the future Kh-55. A drawing from a publication authored by Myasishchev OKB employees shows several alternative options of cruise missile carriage. Option 1 envisaged carrying five missiles on a rotary launcher in the bomb bay. The launcher was installed in a nose-down position (probably to facilitate loading of the missiles on the ground), which required a large bulge to be made on the fuselage underside between fuselage frames 46 and 63. Option 2 likewise envisaged internal carriage, but only of two missiles side by side on ordinary racks.

Option 3 had the aircraft carrying four such missiles externally on two pairs of short pylons flanking the bomb bay, which left the latter free for other weapons. In Option 4, the missiles were to be carried on underwing pylons – either singly or in side-by-side pairs. Confusingly, the drawing is titled 'Accommodation of missiles on the 3M-5 aircraft.'

The project was not implemented, because in 1977 all further work on improving the 3M was terminated.

ASW Version of the 3M (Project)

In 1970 the Myasishchev OKB floated the idea of developing an antisubmarine warfare (ASW) version of the 3M, equipped with appropriate mission systems (search radar and sonobuoys) and antisubmarine weapons. A PD project was drafted in the second quarter of 1970 and presented to MAP; no details are known. The

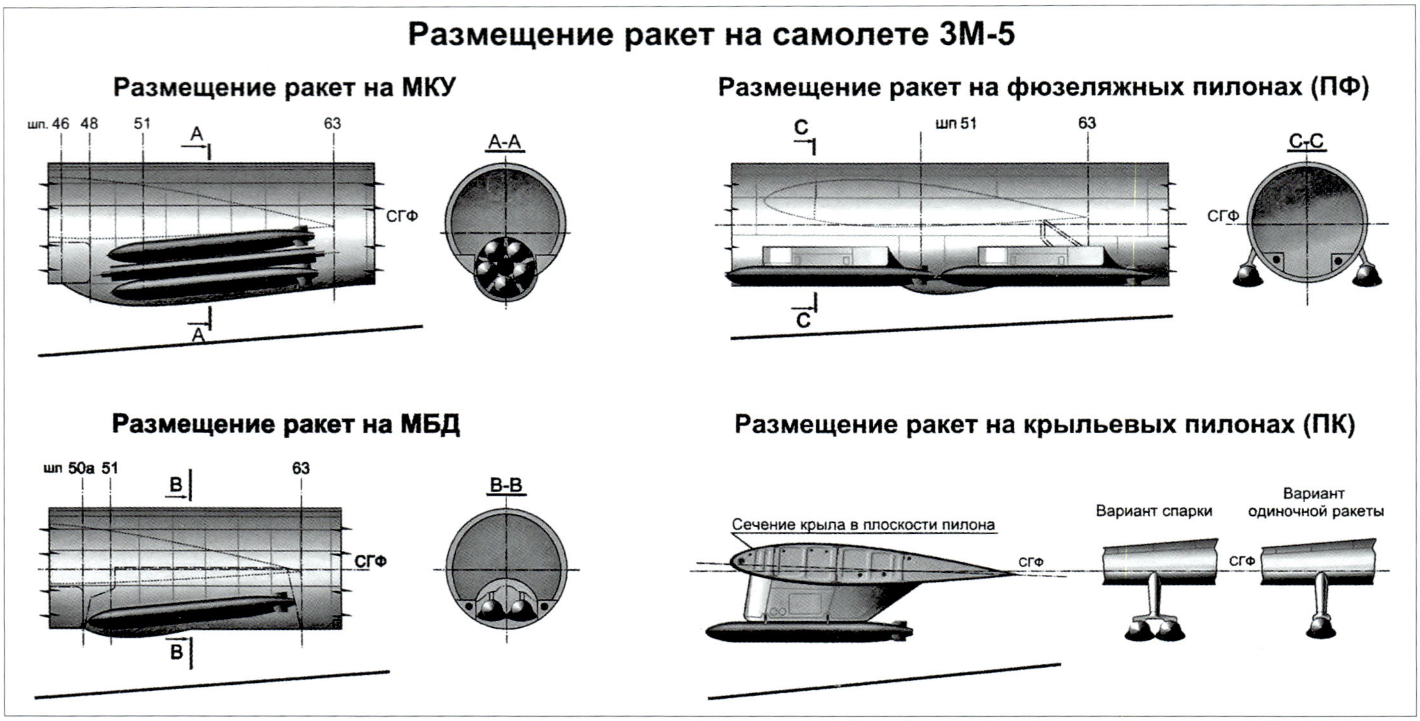

This drawing (titled 'Missile carriage on the 3M-5'!) shows a projected version of the 3M carrying two to five cruise missiles, similar to the Kh-55 internally or externally.

project was not implemented because MAP preferred Tupolev's offer – the Tu-142 *Bear-F* ASW aircraft based on the Tu-95.

3MT (3M-T) Refuelling Tanker (First Use of Designation, Project)

On 31 June 1958 the Council of Ministers issued directive No. 867-408, which, among other things, tasked OKB-23 with developing a refuelling tanker to support the M-52 supersonic missile strike aircraft (the proposed production version of the M-50 that never flew), the 3M and other bombers using the probe-and-drogue system. The aircraft received the provisional designation 3MT, the *T* standing for ***top**livozaprahvshchik* – refuelling tanker.

The 3MT was based on the ultimate production version, the 3MD, and was to outperform the M-4-2 by a substantial margin. For instance, according to calculations it would have given a 3M bomber a range of 3,400 km (2,110 miles) with a single fuel top-off, which was 800 km (496 miles) longer than what the M-4-2 could give. Maximum transferable fuel was 73 tons (160,935 lbs.) versus the *Bison-A*'s 40 tons (88,180 lbs.); the fuel delivery rate was increased from the M-4-2's 2,400 litres (529 imp. gal.) per minute to 4,500–5,000 litres (990–1,100 imp. gal.) per minute.

Importantly, the 3MT was conceived as a convertible tanker – it could also fill the bomber and missile strike aircraft roles if necessary after the HDU and auxiliary tanks had been removed, which could be done rapidly in a hangar. The 3MT's fuel system differed from that of the 3MD mainly in that extra fuel transfer pumps were installed in certain groups of tanks, and several additional pipelines were fitted; otherwise the two were identical.

Being part of the M-6K-14 strategic weapons system, the 3MT was assessed by the same mockup review commission as the 3MD. However, it remained a 'paper aeroplane' due to the closure of OKB-23 in September 1960. Much later, when the Myasishchev OKB had been reestablished for the second time, the designation 3MT was reassigned to an outsized cargo transporter based on the 3M, which was developed for supporting the Soviet space programme (see chapter 6).

3MP (3M-P) Refuelling Tanker (Project)

In the 1960s the air force considered outfitting in-service 3MDs for the tanker role by analogy with the 3MS-2 and 3MN-2. The would-be conversion received the somewhat unusual designation 3MP instead of the logical 3MD-2 (the meaning of the *P* suffix is unknown). The design of the 3MP's refuelling system gave it a quick-change capability – the aircraft could be quickly equipped as a tanker or refitted back to bomber configuration in the field. The conversion would involve nothing more than installation of fittings for the HDU and fuel connectors, plus some additional avionics ensuring the rendezvous with the receiver aircraft.

As on the 3M, the copilot doubled as the flight engineer. In tanker configuration, the refuelling-systems officer's station was fitted out instead of the ECM officer's station; the tail gunner also participated in the refuelling operation, monitoring the contact with the receiver aircraft and activating the HDU servo loop in full auto mode. In the receiver aircraft, the copilot controlled the machine during the refuelling procedure and monitored the fuel status.

The 3MP was to support all types of heavy aircraft featuring the probe-and-drogue IFR system, transferring up to 43.5 tons (95,900 lbs.) of fuel at a rate of up to 2,500 litres (550 imp. gal.) per minute. This would give the receiver aircraft a range extension of up to 3,750 km (2,330 miles).

Since the few *Bison-C*s were used chiefly as unit commanders' 'hacks' after the liquidation of the Myasishchev OKB and the end of 3MD production in 1960, converting them was not worthwhile, and the project was dropped.

'Izdeliye 39' Strategic Bomber (Project)

In 1957 the Myasishchev OKB launched an effort to improve the *Bison*'s range and service ceiling. Known in-house as *izdeliye* 39, the aircraft was to be powered by Dobrynin VD-7V turbojets, offering better high-altitude performance; it was to have new wings of greater span and area, while the vertical tail area was to be reduced. Two versions were contemplated; one had the usual three cannon barbettes and a crew of six. In the other version the defensive armament was limited to a tail turret with a Ksenon gun-laying radar, and the crew was reduced to four, reducing the weight by up to 2 tons (4,410 lbs.). The carriage of the Kh-20 cruise missile was envisaged.

Preliminary performance figures included an AUW of 22 tons (44,090 lbs.), a cruising speed of 850 km/h (527 mph), a top speed of 980–1,000 km/h (608–621 mph), a service ceiling of 15,000–15,500 m (49,210–50,850 ft.), and a range of 16,500–17,000 km (10,248–10,559 miles). The prototype was expected to fly in 1958, with production commencing in 1958–59. *Izdeliye* 39 was never built.

3M-M Strategic Bomber Seaplane (Project)

Perhaps the most unusual spin-off of the *Bison* was developed to meet an early 1960s requirement in accordance with which several design bureaus were tasked with developing a long-range flying boat intended for operations over ocean areas. This was to be primarily a strike aircraft, with potential other uses, such as maritime search and rescue.

Within the framework of this task, in 1960 the Myasishchev OKB studied the feasibility of converting the 3M bomber into a seaplane. The approach chosen by the designers did not envisage a radical revision of the bomber's airframe. Instead, the existing fuselage would be mated with a deep conformal float running almost the full length of the fuselage; its bottom would have a V-shaped planing step amidships and apparently incorporate weapons bay doors. Nonretractable stabilising floats were to be mounted at the wingtips instead of the outrigger strut fairings.

Preliminary calculations suggested that the aircraft, which was designated 3M-M (the second *M* denoted ***morskoy*** – maritime; that is, seaplane), would attain a speed of 800 km/h (497 mph) and have a range within the 8,500–25,000 km (5,283–15,538 mile) bracket. The project was abandoned at an early stage because of the Myasishchev OKB's liquidation; no experimental work was conducted to test the hydrodynamics of the proposed arrangement.

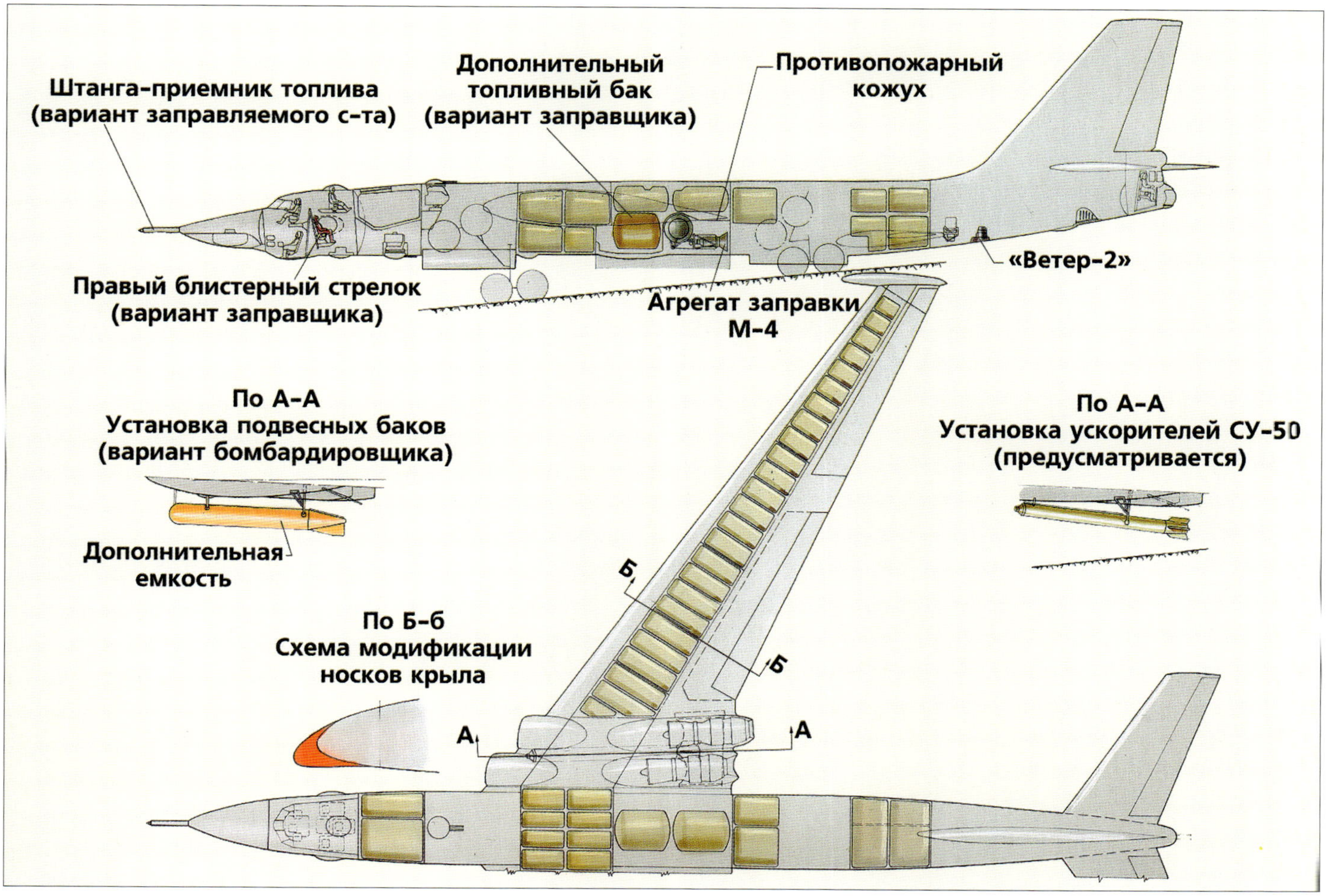

A drawing of the 3MP tanker from the project documents, showing the extra tank in the bomb bay, the optional underwing drop tanks and SU-50 JATO boosters, and the modified wing leading-edge sections

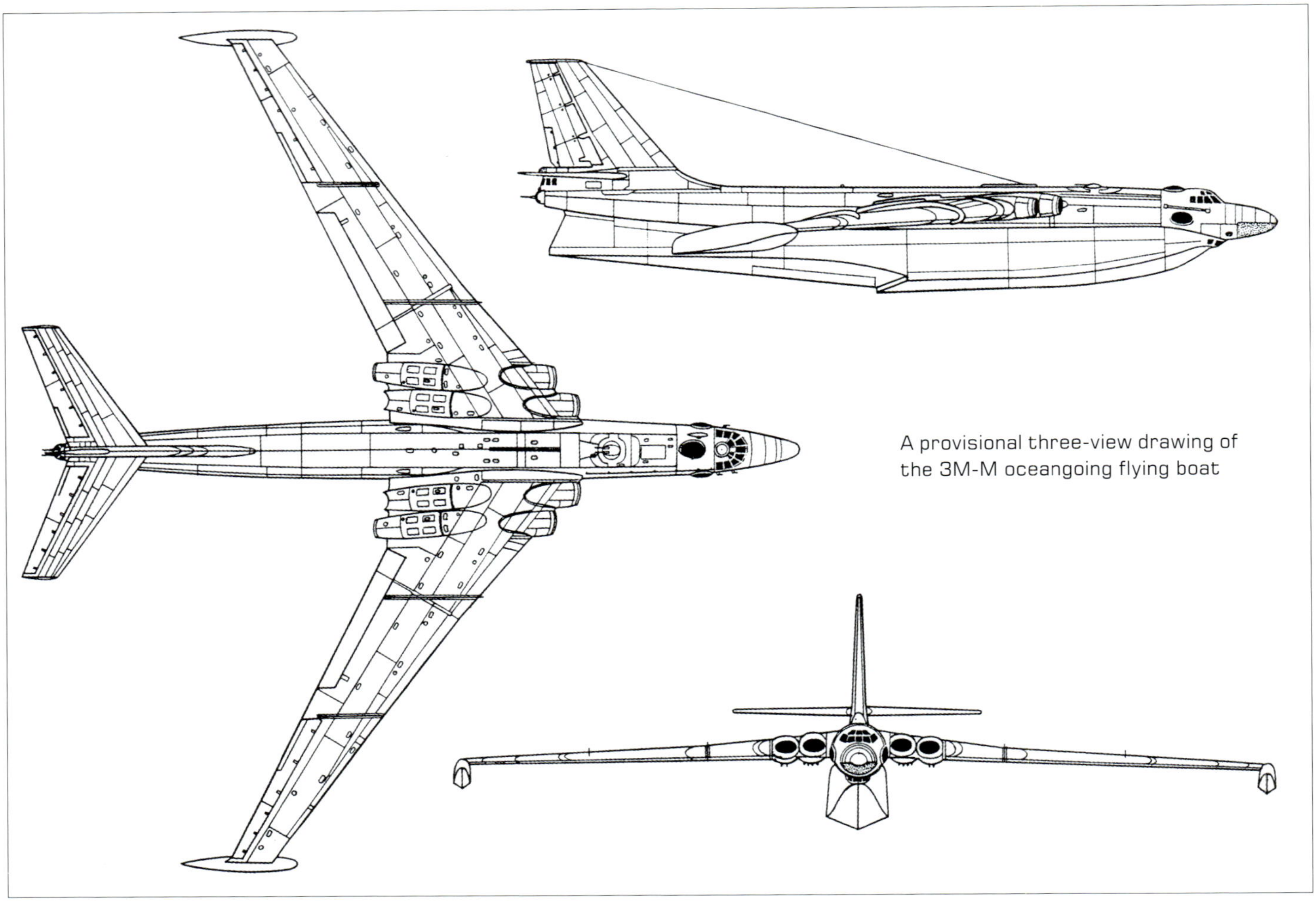

A provisional three-view drawing of the 3M-M oceangoing flying boat

Left: A display model of the projected 3M-M flying boat, based on the 3M bomber

Below left: This view shows how deep the 3M-M's hull would have been, and the considerable draught of the seaplane (note the position of the waterline).

* * *

A total of 125 aircraft of the M-4/3M family had been completed when production ended in 1960. However, the following table showing production by model and by year gives a different figure (116); unfortunately, no version-by-version breakdown of the 3M series is given. Note that the table apparently does not include the M-4 prototypes; the year given is the year when the aircraft was accepted by the air force, not when it left the factory.

M-4 and 3M production by year

	1954	1955	1956	1957	1958	1959	1960	Total
M-4	2	9	20					31
3M			15	26	15	20	9	85
Total:	2	9	35	26	15	20	9	116

Chapter 4

The *Bison* in Detail

The following structural description applies to the baseline production M-4. Details of other versions are given as appropriate.

Type: (M-4) high-speed, long-range bomber designed for operation in visual (VMC) and instrument (IMC) meteorological conditions, in daytime or at night, attacking ground or maritime targets of strategic importance singly or as part of a group; (M-4-2) in-flight refuelling tanker. The aircraft has a crew of eight: captain, copilot, navigator / bomb aimer, flight engineer / gunner, gunner / radio operator, navigator / ECM operator (Nav/Op), dorsal gunner, and tail gunner (defensive stations commander).

The airframe is of all-metal riveted construction, with V95-series aluminium alloy and D16-series duralumin as the primary structural materials. Other structural materials include AK6 and AK8 aluminium alloys (*alyu**mini**y **ko**vochnyy* – aluminium optimised for forging), MA8, AMTsAM, and ML5-T4 magnesium alloys (ML = ***magniy** **l**iteynyy* – magnesium optimised for casting; the T4 is a reference to the thermal treatment mode). Limited use is made of various grades of steel (including 30KhGSA, 30KhGSNA, 40KhNMA, and EI-402) and BrAZh9-14 bronze. Glass fibre composites are used for some nonstressed components.

Fuselage: Semi-monocoque stressed-skin structure with a smooth skin supported by frames/bulkheads, some of which are reinforced mainframes, and formers plus a set of extruded stringers. Some of the frames have duplicate numbers (including 50/50A

An M-4-2 tanker. The M-4's fuselage is of basically cylindrical shape.

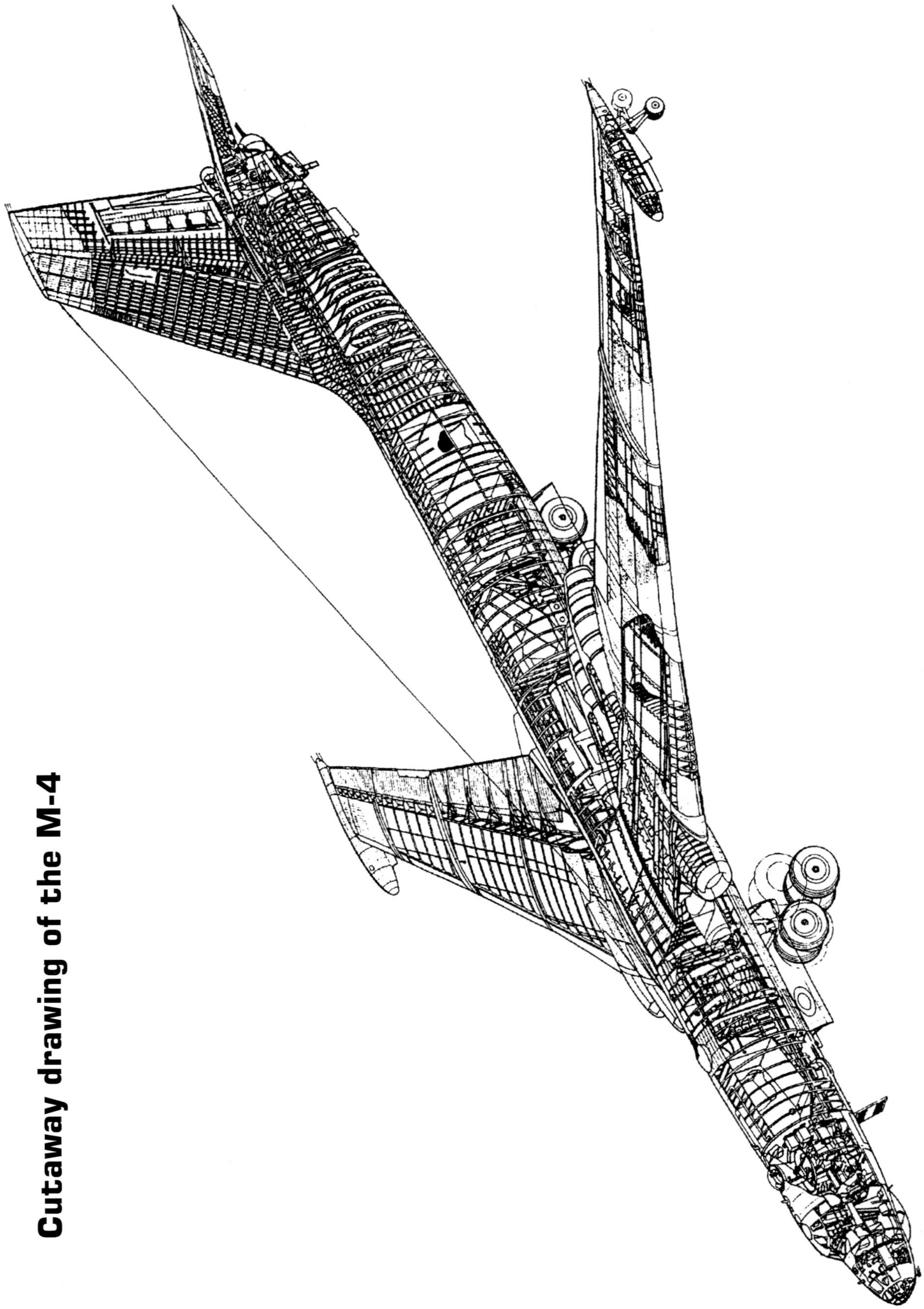

Cutaway drawing of the M-4

Right: This view shows the characteristic 'hump' of the flight deck roof, the dorsal observation/sighting blister, and the star tracker of the SP-1 astrosextant offset to port ahead of the windshield. Two of the four pitot heads can be seen in front of the port beam observation/sighting blister, with the portside Venturi tube below them.

and 78/78A). The skin is made of V95AT and D16ATL sheet; its thickness varies from 1 to 4 mm (≈ 3/64–5/32 in.), and flush riveting is used throughout. The fuselage is of basically circular cross-section, changing to pear-shaped toward the rear extremity, with a maximum diameter of 3.5 m (11 ft., 5 51/64 in.) and a length of 45.6 m (149 ft., 7 9/32 in.). The nose section has a parabolic shape, the centre portion is a perfect cylinder whose axis coincides with the fuselage waterline (FWL), and the rear fuselage is slightly conical; the axes of the nose section and the rear fuselage are inclined down and up, respectively, from the FWL.

For ease of assembly the fuselage is divided by manufacturing breaks into four sections: the forward fuselage, the centre fuselage, the rear fuselage, and the tail section. These are held together by hardened steel bolts and attachment fittings made of AK6 aluminium alloy, which are located at the ends of the stringers around the circumference of the end frames.

Above right: The entry hatch of the forward pressure cabin (used by the dorsal gunner for ejection, *lower left corner*) is visible here, as are the ejection hatches of the flight engineer (below the starboard blister; the radio operator's hatch is symmetrically to port), the navigator/operator (aft of the radome), and the navigator / bomb aimer (ahead of the radome). The radome is jettisoned to enable ejection of the pilots. Unfortunately, the missing circular transparency on the starboard side of the navigator's station has been replaced by a metal patch on this preserved aircraft.

Right: The navigator / bomb aimer's hatch mounting a forked ILS aerial and two of the landing lights is lopsided. Note the antiflash shutters on the nose glazing and the beam blisters.

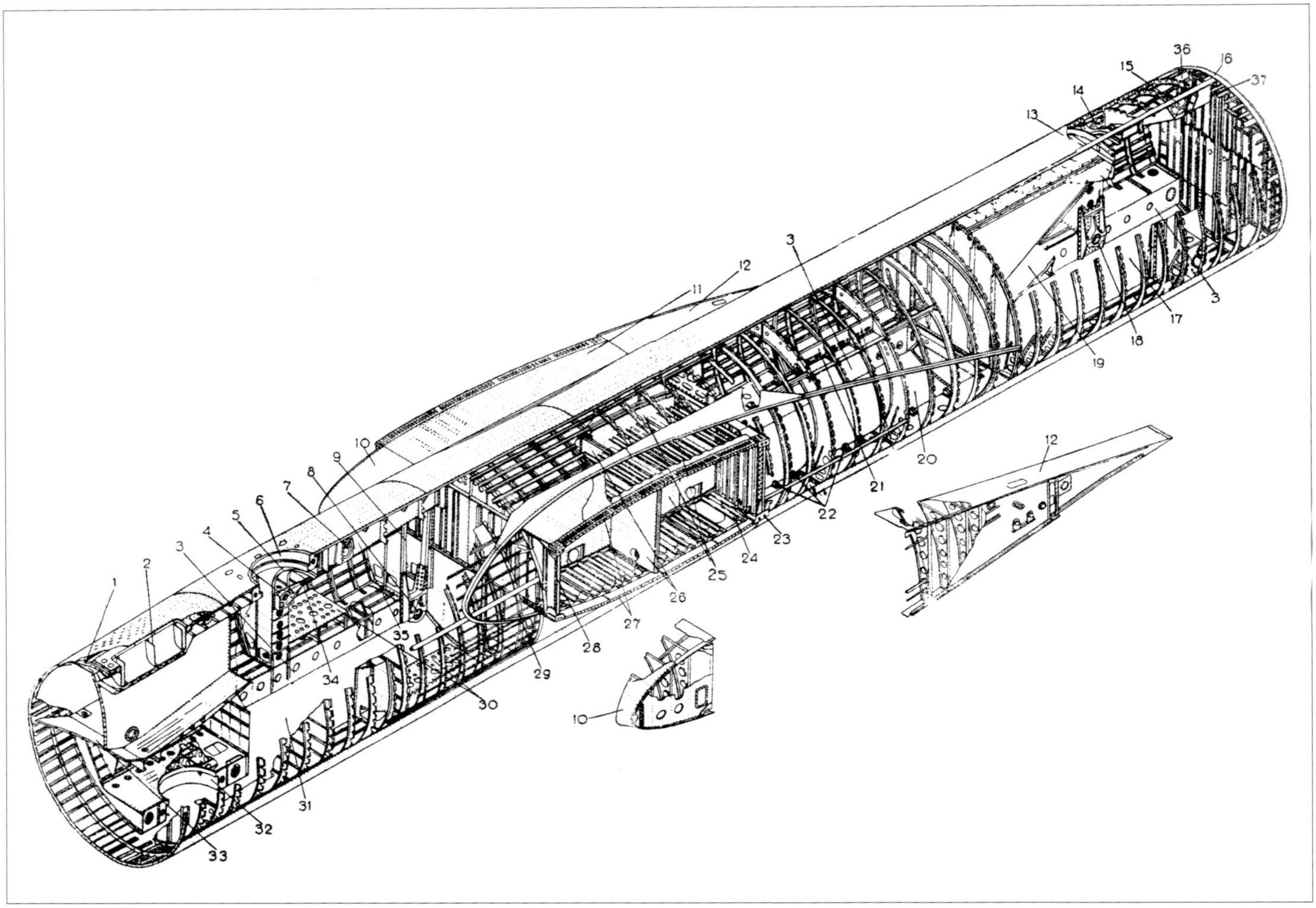

Above: The centre fuselage (section F-4), showing the wheel wells and the bomb bay. The port leading-edge and trailing-edge fairings of the integral wing centre section are shown separately for clarity.

Opposite, above left: The unpressurised rear fuselage (section F-5). *Opposite, above right*: The rear pressure cabin (section F-6), showing the ventral entry hatch (with ladder) and dorsal escape hatch.

Below: The forward fuselage of an M-4. The LPS-53M port beam gunsight can be seen through the blister.

Below: The forward fuselage, showing the dorsal escape hatch and the navigator's direct-vision window to starboard

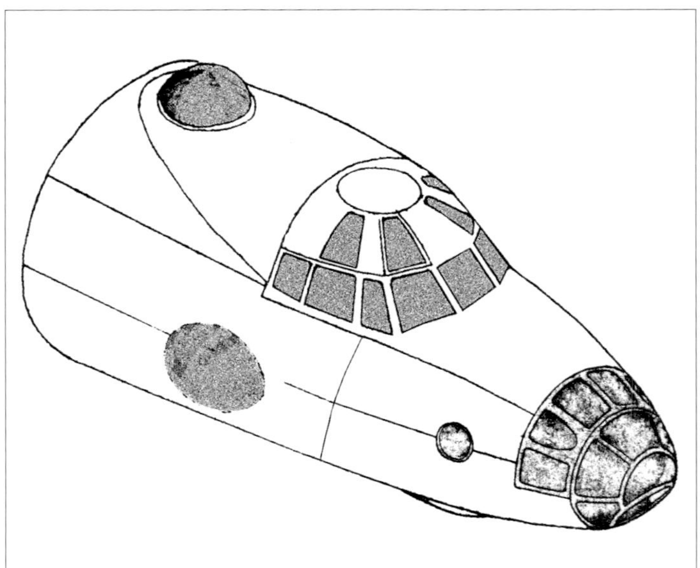

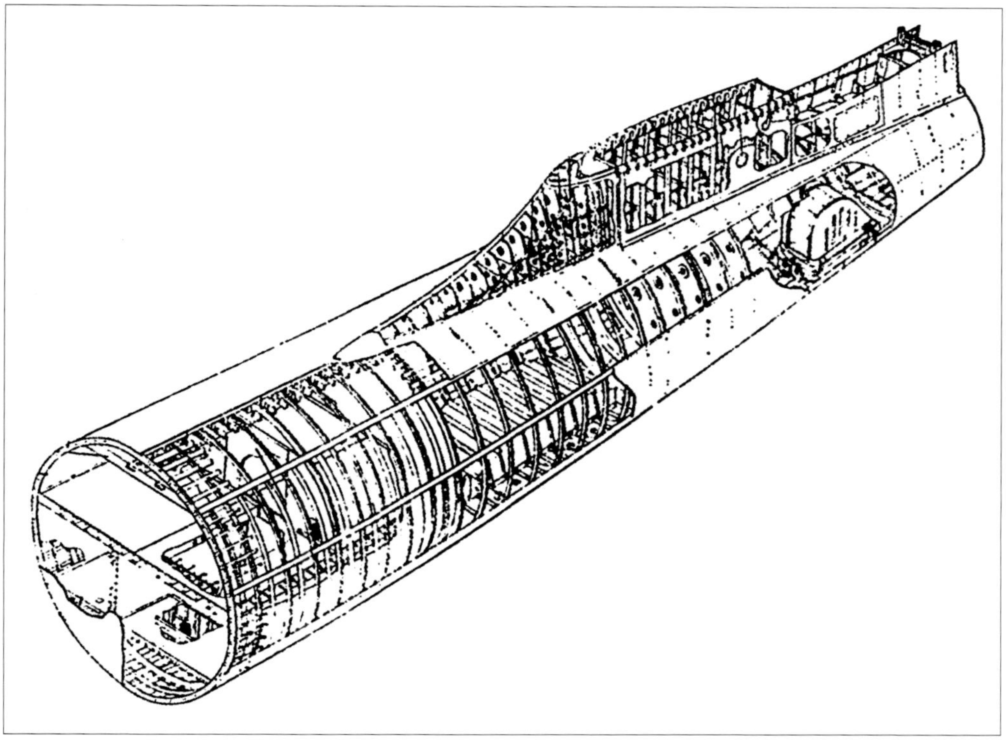

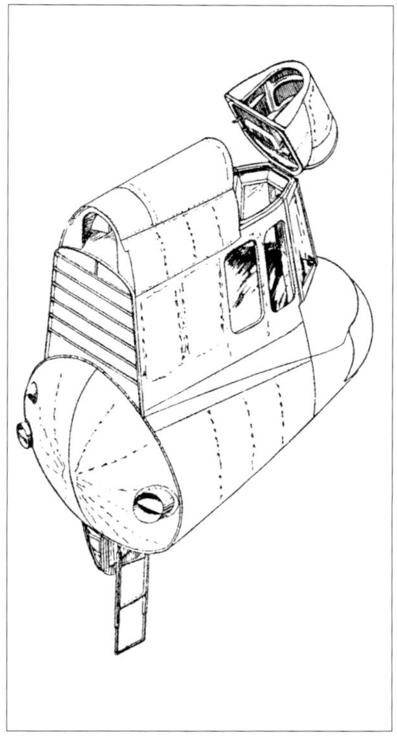

The *forward fuselage* (frames 0–16) is the main crew section, comprising the navigator's station glazing framework (section F-1, frames 0–2) and the forward fuselage proper (section F-2); the latter includes the navigator's station (frames 0–5) and the flight deck, which, together with section F-1, form the forward pressure cabin, with a volume of 18.5 m³ (653.32 cu. ft.). Section F-2 terminates in a gently convex pressure dome. To ensure adequate pressurisation of rivet joints, Thiokol rubber tape is placed between the skin and internal structural members before riveting, and a sealing compound is brushed on over the joint. The pressure cabin walls are lined with quilted heatproofing and soundproofing mats made of ATIM-1 (*aviatsi**on**nyy **tep**lo'izolyatsi**on**nyy materi**ahl*** – aviation-specific thermal insulation material) and ANZM (*aviatsi**on**nyy nevosplame**nya**yemyy **z**voo**ko**'izolyatsi**on**nyy materi**ahl*** – aviation noncombustible soundproofing material).

The forward pressure cabin accommodates seven of the eight crew members, as well as the greater part of the navigation, communications, and targeting/weapons control system equipment. The navigator's station is extensively glazed. The front row has curved windows (five trapezoidal ones, two triangular ones, and one strongly convex, crescent-shaped transparency at the tip) and an optically flat elliptical lower front panel set in a one-piece cast frame made of ML5-T4 magnesium alloy (section F-1); it has a modified parabolic shape and is bolted to frame 1. The rear row has a riveted frame with four rectangular windows and two semielliptical ones, the latter being optically flat and serving for celestial navigation. The three optically flat windows have electrically heated triplex silicate glass panes; the others have moulded birdproof Perspex glazing, the outer panes being 15 mm (19/32 in.) thick and the inner ones 4 mm (5/32 in.) thick, the gap in between being provided with silica gel cartridges to prevent misting up. The transparencies are attached to the framework by duralumin strips (with rubber gaskets) and bolts. The navigator's station features an inward-opening, circular direct-vision window on the starboard side (frames 3–4) for illuminating the portside instrument panel (the navigator / bomb aimer normally sits facing left) and is accessed via the flight deck.

The flight deck accommodates the captain to port, the copilot to starboard, and the aft-facing flight engineer / gunner, gunner / radio operator, Nav/Op, and dorsal gunner behind them. The flight deck glazing framework (frames 5–10) is welded from steel members, featuring three optically flat bird-proof triplex panes at the front (the port and starboard ones are electrically deiced) and dual Perspex panes with silica gel demisters elsewhere (three side windows and two eyebrow windows on each side). The two foremost side windows are aft-sliding direct-vision windows. The flight deck roof

The jettisonable dorsal blister (looking toward the nose), with the VPS-53M gunsight visible inside

features a circular escape hatch for use in the event of a belly landing or ditching, and a circular observation/sighting blister made of 12 mm (15⁄32 in.) Perspex aft of it; the blister is jettisonable in an emergency. Two hemispherical beam blisters made of 12 mm Perspex are located on the forward fuselage sides between frames 9–12. The glazing is sealed against the frames with strips of U20A rubber and UTZ sealant paste.

An unpressurised bay for the radar antenna is located beneath the pilots' seats. It is enclosed by a bulged radome made of glass fibre Textolite composite and attached by screws and anchor nuts.

Provision is made for in-flight refuelling; hence, a U-section channel closed by flush skin panels runs along the port side of the forward fuselage. This channel was to accommodate the fuel transfer line from the IFR probe; however, no M-4s except the two experimental M-4As were fitted with the probe, and on most *Bison-As* this channel remained unused.

The forward fuselage underside incorporates an entry hatch (frames 13–16) whose aft-hinged cover is operated manually and secured by twin locks when closed; in the open position it is held by a strap. There are also four escape hatches whose covers are jettisoned prior to ejection (*see crew rescue system*). The hatch covers have a cast magnesium alloy frame skinned in duralumin, and the hatches have inflatable perimeter seals.

The unpressurised *centre fuselage* (section F-4, frames 17–76) is 23.7 m (77 ft., 9⁵⁄64 in.) long (for some obscure reason, there is no section F-3). Its internal structure comprises 55 frames – mostly spaced 500 mm (≈ 1 ft., 7¹¹⁄16 in.) apart – and 60 stringers numbered 0 at the top, 30 at the bottom, and 1–29 on each side. The skins are made of V95AT alloy, and the stringers of V95-T; the skin is thickest on the centre fuselage sides between frames 36 and 54. The centre fuselage incorporates the nosewheel well (frames 27–41), the wing torsion box carry-through structure (frames 41–51), the bomb bay (frames 50A–63), and the mainwheel well (frames 65–76), ending in a flat bulkhead. Circular cutouts for the ventral and dorsal cannon barbettes are located between frames 22 and 24 and 28 and 30, respectively. Blast plates made of heat-resistant steel 0.3–1.5 mm (0.011–0.059 in.) thick are riveted to the fuselage skin fore and aft of the barbettes to protect the duralumin skin from the flames of the cannons' muzzle flash.

The bomb bay is closed by outward-opening, one-piece clamshell doors of single-spar riveted construction. They are actuated hydraulically, opening through 90° for bomb release and 180° for bomb loading; the operation of the doors is electrically controlled by the navigator / bomb aimer. Two hefty triangular-section hollow beams set 1.8 m (5 ft., 10⁵⁵⁄64 in.) apart flank the bomb bay and the wheel wells, absorbing the bending and torsional loads near the cutouts; they also act as attachment points for the nose and main gear units, the bay doors, and the bomb racks. In these areas the fuselage frames are formers and the number of stringers is reduced to 1–24 on each side.

Two bays housing the forward pair of fuselage fuel cells are located between frames 17 and 27; a bay above the port forward fuel cell, closed by a detachable cover (frames 17–24), accommodates inflatable rescue dinghies with their associated CO_2 bottles, emergency food ration bags, and paddles. More fuel cell bays are located inside the wing centre section torsion box, beneath it (frames 41–46 and 46–49), and above the mainwheel well (frames 72–76). The bays are accessed via the wheel wells and the bomb bay, except for the tanks below the wing centre section, which are accessible via hatches between frames 43 and 46 and 46 and 48. The bay walls are lined with KAST-V glass fibre Textolite composite, and the internal partitions (which are detachable for fuel cell installation/removal) are made of 47 mm (1⁵⁵⁄64 in.) Styrofoam/KAST-V sandwich.

The tapered, unpressurised *rear fuselage* (section F-5, frames 77–110) is 13.2 m (43 ft., 3¹¹⁄16 in.) long; the skin here is made of V95AT sheet 1.2–3.5 mm (³⁄64 to ≈ ⁹⁄64 in.) thick and supported by 60 U- and Z-section stringers at 5° or 6° intervals. The rear fuselage incorporates six bays for the rear fuselage fuel cells and the brake parachute bay (frames 103–107) and houses the tailplane trim jack, the rudder and elevator hydraulic actuators, the tail turret's ammunition boxes, a strike camera (frames 86–88), ECM equipment, and a bay for DYa-SS flare bomb cassettes (*derzhatel' yashchichnyy signahl'nykh sredstv* – 'box-type rack for signal means') between frames 88 and 94, closed by clamshell doors.

The *tail section* (section F-6, frames 111–115) is a pressurised cabin 1.95 m (6 ft., 4⁴⁹⁄64 in.) long and 2.85 m (9 ft., 4¹³⁄64 in.) high, which is the tail gunner's station, with the same heat- and soundproofing as the front cabin. It incorporates the tail turret attached to frame 115, its control system, the gun-ranging radar, and other equipment; the pressure dome at frame 111 incorporates two apertures for the tail turret's ammunition belt sleeves. The tail gunner's station has a three-section, bulletproof-glass rear window set in a cast magnesium alloy frame and two windows with double Perspex panes on each side. Access is via a 0.7 × 0.93 m (2 ft., 3⁹⁄16 in. × 3 ft., ³⁹⁄64 in.) ventral entry/ejection hatch opening to starboard (frames 112–115) with a telescopic ladder; a dorsal escape hatch (frames 112–114) is also provided in case of a belly landing or ditching.

The armour plating of the crew sections offers protection against 20 mm (.78 calibre) cannon fire and antiaircraft shell fragments. It includes homogeneous armour, Krupp cemented armour, and aluminium antifragment armour. The flight deck floor and sidewalls are made of duralumin armour, with additional armour plates on both sides at frame 12 and in the upper part of frame 16. The pilots' and navigator / bomb aimer's seats feature 10 mm (²⁵⁄64 in.) armoured seat pans and seat backs; the forward gunners are protected by armour 10–20 mm (²⁵⁄64–²⁵⁄32 in.) thick, while the tail gunner is shielded from behind by an armour plate forming the rear wall of his cabin and bulletproof glass panes 145 mm (5⁴³⁄64 in.) thick. In order to minimise the weight penalty imposed by the armour, its thickness was chosen with regard to the extra protection provided by noncritical airframe and systems components.

Wings: Cantilever midwing monoplane with wings of basically trapezoidal planform. Sweepback at quarter chord 35°, anhedral 1°50' on outer wing panels, incidence –2°30', no camber, aspect ratio 7.82. Wingspan 50.526 m (165 ft., 9⁷⁄32 in.); wing area 258.0 m² (2,774 sq. ft.).

The wings are two-spar, stressed-skin structures made of aluminium and magnesium alloys, the skin being supported by ribs and stringers. Structurally they are built in five sections joined together by bolts and fittings. The *centre section*, accommodating fuel tanks, is manufactured integrally with the centre fuselage

Right: This drawing from the structural manual shows the wings' manufacturing breaks, the high-lift devices, and the fuel cell bays inside the wing torsion box. Note the strips sealing the inner/outer wing joints. The inboard wing fence is shown only partially.

Below: An exploded view of the starboard wing, with the nos. 3 and 4 engine cowlings shown in place.

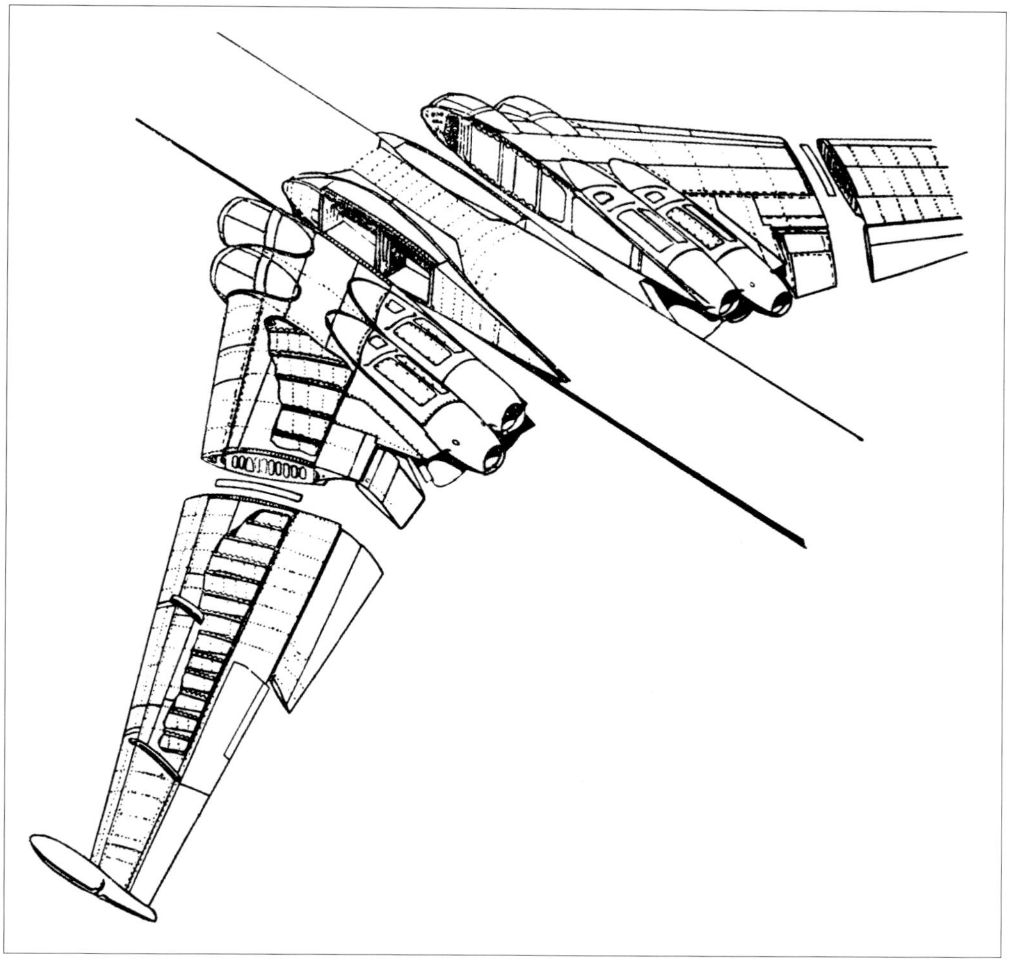

(section F-4), the spars being located at fuselage frames 41 and 51, and is only slightly wider than the fuselage, with leading-edge and trailing-edge fairings fore and aft of the spars. It has three ribs (nos. 1 and 2L/2R) and a false spar at frame 46. The upper skins are made of 3 mm (0.118 in.) V95A alloy sheet, and the lower skins of 5 mm (≈ $^{13}/_{64}$ in.) D16ATV duralumin.

The *inner wing panels* (sections K-1, ribs 1–11; *K* stands for ***krylo*** – wing) incorporate the engine bays and inlet ducts between ribs 1 and 5; ribs 2 and 4 are aligned with the engine bays' axes and therefore were built as vertical pairs of beams. The skin is made of D16ATNV duralumin 5 and 6 mm ($^{13}/_{64}$ and $^{15}/_{64}$ in.) thick.

The *outer wing panels* (sections K-2, ribs 11–30) have greater trailing-edge sweep so that there is a slight kink at the inner/outer wing section joint. The skins are made of D16ATV and D16ATNV duralumin 3 and 4 mm ($^{7}/_{64}$ and $^{5}/_{32}$ in.) thick. The inner/outer wing torsion box is formed by I-section duralumin spars and skins sup-

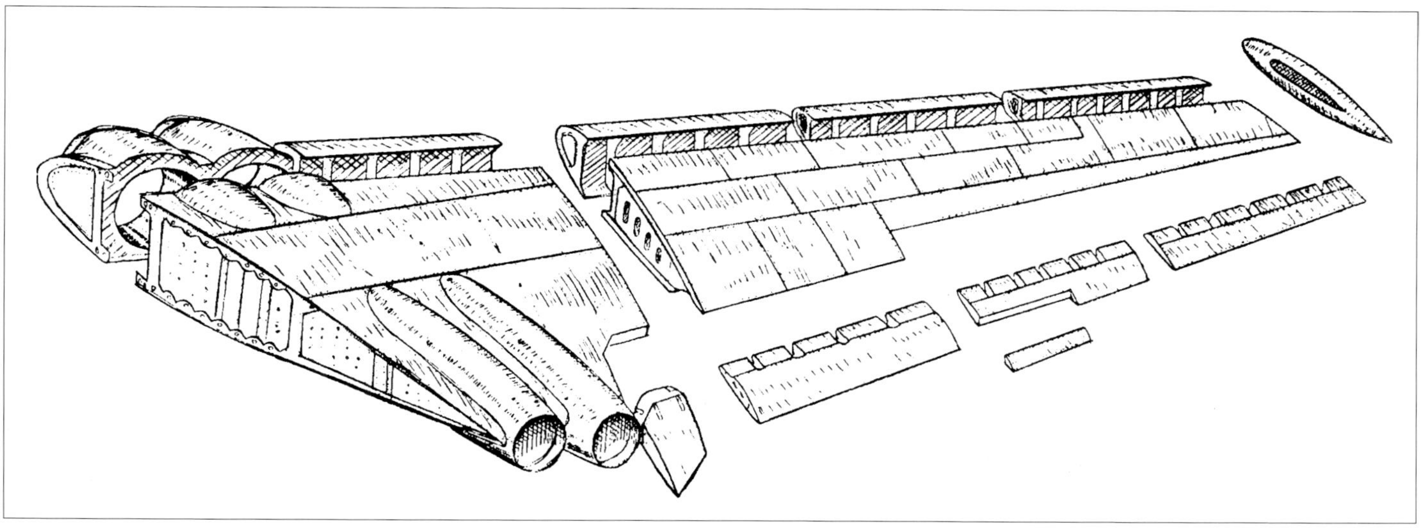

Top: The tail unit of M-4-2 '65 Red' (c/n 4300306), showing the tailplane dihedral. The tail gunner's station is covered by a tarpaulin, which is standard practice in the Soviet/Russian air force.

Above: The horizontal tail and the tail gunner's station glazing

Below: A drawing of the horizontal tail from the structural manual, showing the rib numbers and access hatches (the dorsal ones are marked in black, the ventral ones in white)

ported by ribs and 17 stringers; between ribs 1 and 5, the spars incorporate forged AK6 alloy frames shaped like a horizontal figure eight to accommodate the engine inlet ducts. The ribs are at 90° to the front spar, except ribs 1–5 and 30, which are parallel to the centreline.

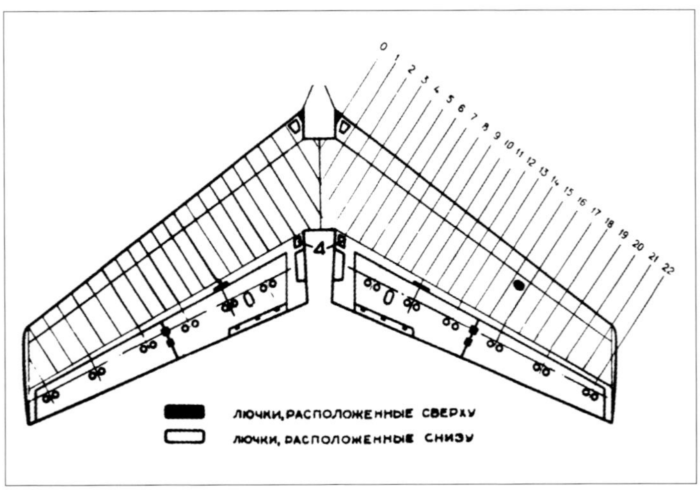

The wing torsion box likewise houses fuel tanks between ribs 6A and 24, the bays being lined with KAST-V composite sheets separated from the ribs by plywood grids, and the lower skin has detachable access panels. The detachable leading-edge sections are attached by screws and anchor nuts.

The wingtips carry large, teardrop-shaped fairings accommodating the landing-gear outrigger struts; these double as antiflutter balances. The fairings are of semimonocoque, stressed-skin construction, with 15 frames and two beams flanking the wheel well.

The wings utilise TsAGI airfoils (P-35-16 at the root and P-36-12 from the inner/outer wing joint to the tip), with a thickness/chord ratio of 15%–15.3%.

There are no leading-edge devices (the leading-edge sections act as deicers). The trailing edge is occupied by simple hinged flaps, which are shaped to conform to the engine jet pipes, and two-section, area-increasing TsAGI flaps (modified Fowler flaps), with ailerons outboard of them. The hinged flaps (ribs 1–8) are carried on three hinges each (at ribs 1, 3, and 5), deflecting to a maximum of 24°. The TsAGI flaps are mounted on both inner and outer wing sections (between ribs 8 and 11 and 11 and 19, respectively), moving on internal curved tracks at ribs 9, 10, 12, 14, 16, and 18 (two for each flap section). They are of riveted single-spar construction with front and rear false spars, stamped ribs, and 0.8 mm (1/32 in.) skins made of D16ATV duralumin. The flaps are electromechanically operated by an MPZ-3M actuator (*mekhanizm privoda zakrylkov* – flap drive mechanism) with twin electric motors and common reduction gear via extension shafts, angle drives, and screw jacks; flap settings are 5° for takeoff and 30° for landing.

The ailerons are located between ribs 19 and 30, being divided into two sections at rib 25. They are of riveted, single-spar construction, with front and rear false spars, stamped ribs, and 0.6 mm (0.0236 in) skins made of D16ATNV duralumin. The inboard sections feature geared tabs. The ailerons are aerodynamically balanced and mass balanced; each aileron is carried on six mounting brackets at wing ribs 20, 22, and 24 (inboard section) and 26, 27/28, and 29 (outboard section).

The wings have two pairs of boundary layer fences on the upper side, located at the flap/aileron joints and the inboard/outboard aileron section joints.

Tail unit: Conventional tail unit of stressed-skin construction made of aluminium and magnesium alloys. The *vertical tail* comprises a fin, whose root section with a small root fillet is manufactured integrally with the rear fuselage, and a one-piece rudder; sweepback at quarter chord 35°, aspect ratio 1.96, taper 2.5. It uses a NACA-00M symmetrical airfoil with a thickness/chord ratio of 10% at the root and 8% at the tip. The fin is a two-spar structure with 23 ribs (nos. 1–23) attached to the fuselage by fittings between frames 92 and 102; the skin, whose thickness varies from 0.8 to 2 mm (1/32 to 5/64 in.), is supported by 14 stringers on each side. The detachable leading edge incorporates a deicer. The rudder is a single-spar structure with front and rear false spars; it features a trim tab at the root and is carried on three mounting brackets at fin ribs 11, 17, and 22 plus a lower support at rib 7.

The cantilever variable-incidence *horizontal tail* is positioned above the rear fuselage, being attached to frames 102 and 109. It

Top: The nose gear unit seen from behind, showing the sway braces. The wheel size is identical on both gear bogies.

Centre and above: These views of the nose gear unit show the massive oleo strut, the bogie hinge, and the torque link at the front.

Above right: The nose gear unit of an early M-4, featuring different wheels with no lightening holes. Note the bulge at the front of the wheel well door to accommodate the rear wheel.

Top: The M-4's main gear unit. Note the recesses in the wheel well doors for the bogie's rear wheels, and the ground power socket (with power cable connected) ahead of the wheel well.

Centre: The starboard outrigger strut and wingtip fairing of M-4-2 '66 Red' (c/n 6302528). Note the extreme rearward incline of the strut in static-load position.

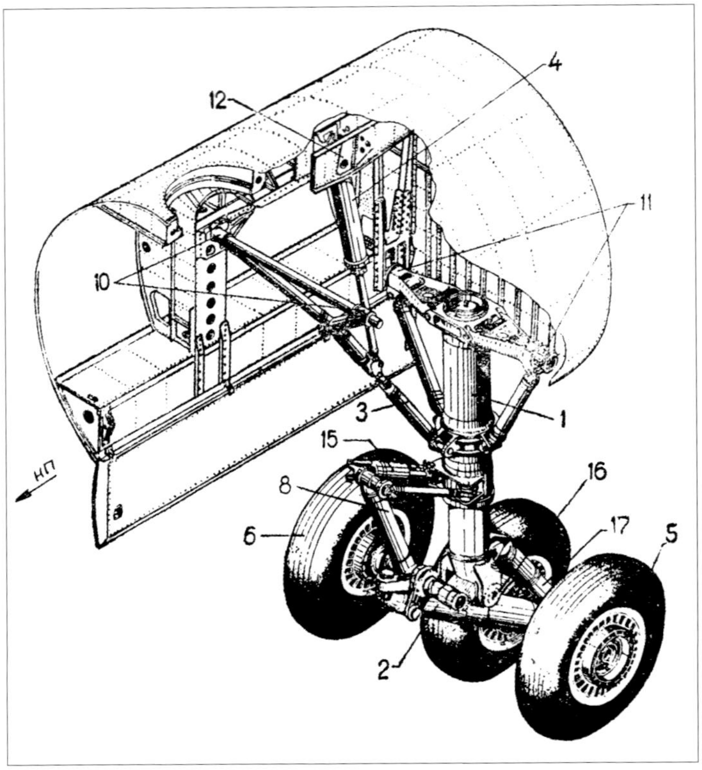

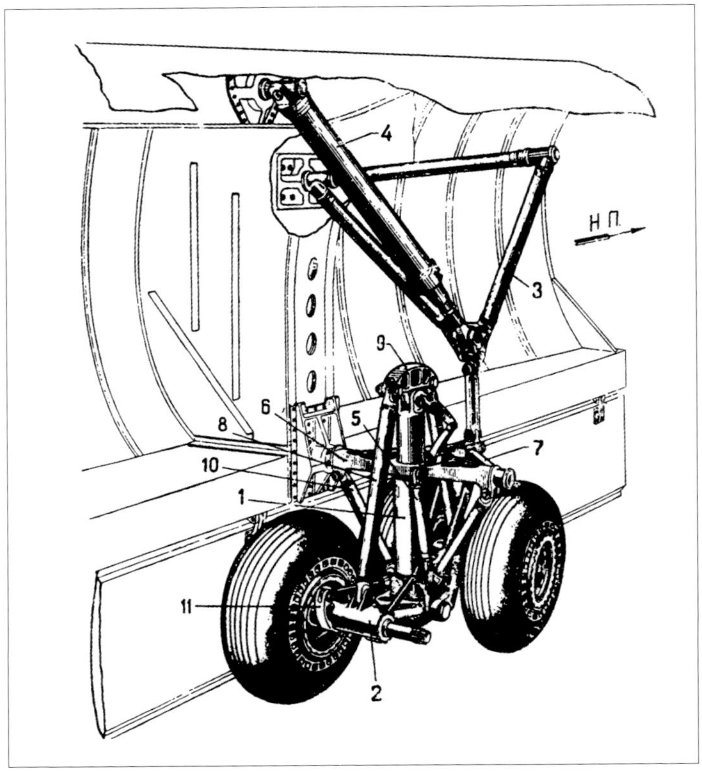

comprises two stabilisers and one-piece elevators; sweepback at quarter chord 33°30', dihedral 10°, aspect ratio 4.0, taper 2.0. It uses a NACA-00M airfoil with a thickness/chord ratio of 11% at the root and 10% at the tip. The stabilisers are joined by bolts on the aircraft's centreline, with 23 ribs each (nos. 0–22) and 12 stringers on both upper and lower skins, whose thickness varies from 0.8 to 3 mm (1/32 to ≈ 1/8 in.). The detachable leading-edge sections incorporate deicers, and the tip fairings feature triple static-discharge wicks. Incidence is adjusted within –0°15'/–2°45' by an electrically operated screw jack housed in the rear fuselage. The elevators are manufactured in two sections that are rigidly fixed together; each section is carried on three mounting brackets at stabiliser ribs 7, 10, 13, 16, 19, and 22. The inboard sections feature trim tabs.

Landing gear: Electrically retractable bicycle type. All four units retract forward by means of screw jacks. Emergency extension is also electric, using the emergency electrical system; in this mode the units are extended one by one, in any order.

The nose unit features a four-wheel bogie with 1,700 × 550 mm (66.92 × 21.65 in.) K2-92/I nonbraking wheels. The front wheels are hydraulically steerable through ±4°, allowing the bogie to turn ±27° left and right for taxiing; nose gear steering is controlled by the rudder pedals. The nose gear bogie also features a special hydraulic

Above left: A drawing showing the nose gear unit with the bogie in no-load position. The arrow is marked NP (*napravleniye polyota* – direction of flight).

Above: A drawing showing the main gear unit with the bogie in no-load position. The rendering of the port wheel well door does not match the doors' position with the wheels down.

Left: This view of an M4 trestled for a gear swing shows how the nose gear bogie tilts on takeoff. Note that the front and rear wheels are of different models on this particular aircraft.

Top right / far right: Drawings from the same manual, showing the design and kinematics of the port outrigger strut

Right: The Mikulin AM-3D turbojet, which powered the M-4. The elongated nose fairing houses an S300-75 turbine starter with lateral air intake slits and the exhaust at the front.

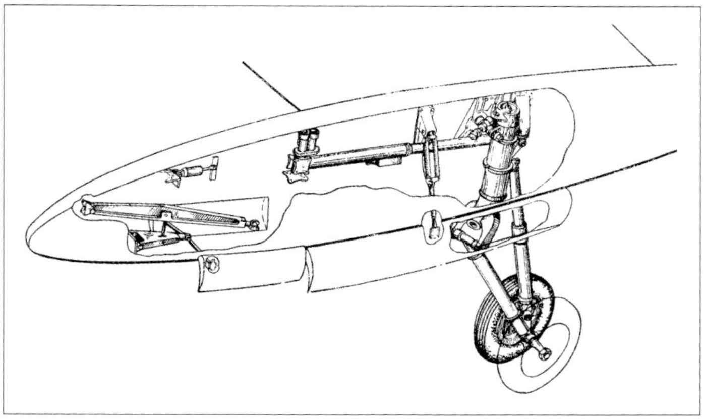

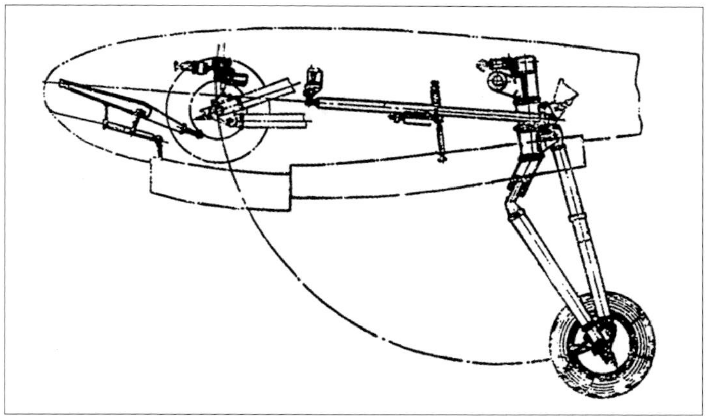

ram aft of the oleo strut, which tilts it when a certain speed is attained during the takeoff run so that the front wheels are lifted clear of the ground; this increases the angle of attack by 3°, facilitating liftoff. The nose gear fulcrum is mounted at frames 37 and 41, while the breaker strut is attached to frames 30–31 and the actuator connected to it is mounted on beams attached to frames 31–37.

The main gear unit also has a four-wheel bogie, which is equipped with 1,700 × 550 mm KT-77M wheels (*koleso tormoznoye* – brake-equipped wheel) and tilts nose-up in no-load condition to facilitate stowage in the wheel well. The main gear fulcrum is at mainframes 71–72, and the actuator is mounted on beams attached to frames 72–76.

The levered-suspension outrigger struts are fully castoring, turning back to front during pushback; they are equipped with twin 660 × 160 mm (26.00 × 6.30 in.) K2-57 nonbraking wheels and shimmy dampers. The outrigger struts are actuated by MPSh-17M drives (*mekhanizm privoda shassee* – landing-gear drive mechanism) with twin electric motors. All four struts are made of 30KhGSNA steel and have oleo-pneumatic shock absorbers.

The nosewheel and mainwheel wells are closed by one-piece clamshell doors of the same design as the bomb bay doors. Each outrigger unit has clamshell main doors at the front and an auxiliary door hinged on the inboard side of the wingtip fairing. All doors remain open when the gear is down. For landing-gear checks and wheel changes, jacking points are provided under the inner wings (at ribs 5L/5R under both wing spars) and the centre fuselage (at frame 76).

To shorten the landing roll, the aircraft is provided with a PT-3237-53 brake parachute system; it has three 63 m² (678 sq. ft.) circular main canopies made of Kapron synthetic fabric, giving a total area of 185–189 m² (1,990–2,034 sq. ft.). The parachutes are stowed in a bay on the rear fuselage underside, closed by clamshell doors, deploying automatically on touchdown at 200–220 km/h (124–136 mph). The parachute cable is secured by a faired lock ahead of the bay; it is released at 40 km/h (25 mph) to avoid damage to the parachutes caused by scuffing against the runway. Parachute deployment and release are operated pyrotechnically; the container is detachable and loaded with the parachutes pre-packed.

Power plant: Early-production M-4s were powered by four Mikulin AM-3D non-afterburning turbojets rated at 8,750 kgp (19,290 lbst) for takeoff and 6,200 kgp (13,670 lbst) for cruise. Later aircraft had Zoobets RD-3M (RD-3M-200) engines – derivatives of the AM-3D with a takeoff thrust increased to 9,500 kgp (20,940 lbst), a cruise thrust increased to 7,650 kgp (16,865 lbst), and a two-minute contingency rating of 10,608 kgp (23,386 lbst);

Top and top right: The M-4's engine air intakes. The top view from the flight deck roof escape hatch also shows the dihedral tailplanes, the dorsal cannon barbette, and the 'towel rail' aerial on the port side of the forward fuselage.
Above and above right: The engine nacelles of an AM-3-powered *Bison*, showing the differing shape of the inboard and outboard nozzles. Some of the cowling panels are already missing on this retired example.

from 1961 onward, these were replaced by identically rated RD-3M-500s, having an increased time between overhauls (500 hours) and higher reliability. Operation time at takeoff power is limited to eight minutes. The engine was manufactured by the Kazan' Engine Factory from 1952 to 1963.

The AM-3 (RD-3M) is a single-spool turbojet having a fixed-area, subsonic air intake with a large, fixed, parabolic centre body supported by six radial struts, an eight-stage axial-flow compressor, an annular combustion chamber with 14 straight-through flame tubes housed in a common shroud, a two-stage axial-flow turbine, and a fixed-area subsonic nozzle with a conical centre body. The spool rotates in three supports, with roller bearings in the first two and dual ball bearings in the third one. The compressor casing is divided into three sections; the front and rear sections are annular, while the centre section is made up of eight segments. The third compressor stage features an automatic antisurge air bleed valve; further bleed valves are provided at the seventh stage (for the air-conditioning system and the engine / air intake deicers) and the eighth stage (for the wing / tail unit deicers). Only about one-third of the air passing through the compressor is mixed with fuel in the combustion chamber; the rest is used to cool the chamber and the turbine. The combustion chamber features throat vortex generators and ribbed flame tubes for better cooling four SPN-4-3 igniters, energised from a KP14-2R1 ignition coil.

Three accessory gearboxes (left, right, and lower) are located on the centre portion of the engine and are driven by a vertical shaft with angle drives in the front part of the compressor casing. The left gearbox mounts two GSR-18000 DC generators, a model 453V hydraulic pump, and a centrifugal breather; the right one mounts an AK-150 air compressor, a PN-28B (or, on the AM-3, PN-28-15) fuel control unit (FCU; PN = ***ploon****zhernyy nasos* – piston-type pump), and a TsD-1 centrifugal engine speed sensor (*tsentro****bezh****nyy* ***daht****chik*) controlling the antisurge valve. The lower one mounts the engine oil pumps (two feed pumps and a scavenging pump) and the TsN-1D fuel feed pump (*tsentro****bezh****nyy na****sos*** – centrifugal pump).

Starting is by means of an S300-75 turbine starter (that is, starter, developed by OKB-300, power output 75 shp) on the AM-3 or the S300M version on the RD-3M et seq. This is a small centrifugal-flow gas turbine engine driving the spool directly via a hydromechanical clutch; the term 'jet fuel starter' is not applicable, since the S300 runs on aviation gasoline. The turbine starter is, in turn, started by a 1.2-kilowatt SA-189B electric starter and has its own PNR10-3M fuel pump (*poosko****voy*** *na****sos****-regoo****lya****tor* – starter FCU); it runs at 31,000–35,000 rpm. The S300-75 (S300M) is housed in the air intake centre body, back to front, breathing through slits in the centre body. The starter air intake is located low on the side of the engine nacelle and is closed by twin doors rotating on longitudinal axles; the dogleg exhaust pipe runs from the tip of the intake centre body to a dorsal outlet. The engine starting procedure is automated. The engine controls (fuel control units and shutoff cocks) are connected to two banks of throttle levers located on the port and starboard consoles in the flight deck.

Engine speed is 4,650 rpm at takeoff power and 4,350 rpm at maximum continuous power. Engine pressure ratio (EPR) is 6.2 at takeoff power and 7.2 at the contingency rating, mass flow at takeoff power is 150 kg/sec. (331 lb/sec.), and turbine temperature at takeoff power is 1,130°K. SFC is 1.0 kg/kgp·hr. (lbs./lbst·hr.) at takeoff power and 0.931 kg/kgp·hr. (lbs./lbst·hr.) at cruise power. Length overall is 5.38 m (17 ft., $7^{13}/_{16}$ in.), casing diameter is 1.4 m (4 ft., $7^{7}/_{64}$ in.), and dry weight is 3,100 kg (6,830 lbs.).

The engines are installed in staggered side-by-side pairs aft of the wings' rear spar between ribs 1 and 3 and 3 and 5; the engine bays are designed to ensure adequate cooling of each engine and are separated from each other and the adjacent structure by firewalls. The engines are installed with a 4° 'toe-in' angle and a 4° nose-up angle relative to the FWL to avoid the exhaust impinging on the rear fuselage sides and horizontal tail, which might cause fatigue problems. Each engine breathes through an individual elliptical air intake, the inlet ducts passing through the wing torsion box. The ducts are made of duralumin sheet and were built in three sections, the front ones being detachable; their cross-section changes from elliptical to circular at the engine inlets. The intakes protrude beyond the wing leading edge and are stepped to preclude harmful interference in all engine operation modes (including start-up and shutdown), the inboard ones being longer. Similarly, the jet pipes protrude beyond the trailing edge in a stepped arrangement, the outboard ones being longer. Each engine has upper and lower hinged cowling panels for maintenance access.

Provisions are made for installing JATO boosters under the inner wings on Der5-48 shackles.

Control system: Fully powered dual controls with manual emergency backup; all control inputs are transmitted by rigid push-pull rods, control cranks, and levers. The control rods exit the forward pressure cabin via pressure seals. An AP-15 autopilot is fitted.

Roll control is provided by two-section ailerons; the inner and outer sections are linked by jointed shafts to make sure they move in concert, while precluding the possibility of jamming as the wings flex. The ailerons have both aerodynamic and mass balancing and are slightly 'nose heavy'; the inner sections feature servo tabs doubling as mechanically controlled trim tabs.

Pitch control is provided by aerodynamically balanced and mass-balanced two-section elevators whose GU-47 actuator (***ghi**drou**silitel'*** – hydraulic booster) is located between fuselage frames 104 and 106. The elevator travel limits are 10° down and 25° up. Electrically operated trim tabs are provided for longitudinal trim.

Directional control is provided by a one-piece rudder likewise featuring aerodynamic and mass balancing and incorporating a servo/trim tab at the root. The rudder travel limits are ±20°.

The pitch and roll control circuits feature irreversible hydraulic actuators, whereas the directional control circuit includes a reversible hydraulic actuator with a reduction ratio of 1:15. To provide feedback, the pitch control circuit included an AU-2 automatic artificial-feel unit (*av**to**maht oo**si**liy*), with simpler spring-loaded mechanisms in the directional and roll control circuits. Each circuit features an MP-100AD electrical trim mechanism (*me**kha**nizm pri**vo**da*) for reducing control forces. Gust locks operated by identical MP-100AD mechanisms are provided.

Fuel system: On the M-4, internal fuel is carried in 70 self-sealing bladder tanks (fuel cells) – 20 in the centre/rear fuselage and 50 in the wing torsion box. The fuel lines from the wing tanks are routed along the front spar. The tanks are accessible for maintenance (removal/installation) via detachable skin panels on the fuselage and wing underside and inside the wheel wells and the bomb bay. Additionally, two elliptical-section metal tanks can be installed in tandem in the upper part of the bomb bay for extra-long-range missions. Theoretically, total fuel capacity is 132,390 litres (29,125 imp. gal.); in reality, however, the aircraft cannot take more than 123,600 litres (27,192 imp. gal.).

The fuel tanks are divided into 18 groups, with several groups being assigned to each engine. In particular, the centre fuselage houses two tanks (group 1 / port and group 2 / starboard) between frames 17 and 27, eight group 11 port/starboard tanks in the wing centre section torsion box (frames 41–50A), with two group 3 tanks (frames 41–46 and 46–49) below them and two tanks (group 13 / port and group 14 / starboard) above the mainwheel well between frames 72 and 76. The rear fuselage accommodates two group 5 tanks (frames 77–80 and 80–85) on the upper tier and four tanks (group 4 / port and group 6 / starboard) below them. The groups are connected by means of a cross-feed valve, enabling each engine to draw fuel from any group of tanks. The fuel system automatically sequences fuel usage to maintain the aircraft's CG within acceptable limits.

The open bomb bay of an M-4-2 tanker (or possibly a 3MS-2 or 3MN-2), showing the KAZ hose drum unit. The swivelling guide frame is tilted to the lower position, from which the massive stabilised drogue is deployed.

A look into the open bomb bay of a convertible 3MS-2, showing the hose drum unit and the yellow-painted extra fuel tank ahead of it

The M-4 has single-point pressure refuelling via a connector in the nosewheel well; refuelling by gravity via individual filler caps for each group of tanks is also possible. Defuelling is via a cock under the fuselage near the wing centre section. Fuel grades used are Soviet/Russian T-1, TS-1, or T-2 aviation kerosene with a specific gravity of 0.83 g/cm³, 0.78 g/cm³, and 0.775 g/cm³, respectively.

An inert gas pressurisation system using carbon dioxide is provided to pressurise the fuel tanks and reduce the hazard of explosion if hit by enemy fire. The system's CO_2 bottles are mounted in the nosewheel well.

The fuel tanks have a common vent system. An emergency fuel jettison system is provided for reducing the all-up weight quickly; the fuel jettison pipes are located on the rear fuselage underside and the wing underside (between ribs 9 and 10, 13 and 14, 15 and 16, and 22 and 23).

The engines' turbostarters have a separate fuel system (the so-called starter fuel system). This uses 70-octane B-70 aviation gasoline with 1% engine oil added; the starters of each pair of engines are fed from a common Avgas tank.

The M-4-2 tanker is equipped with a KAZ hose drum unit and a third metal tank in the bomb bay connected to the fuel system.

Hydraulics: Three separate hydraulic systems (main system plus two control actuator power systems). Hydraulic pressure is supplied by four engine-driven Model 453V pumps.

The *main hydraulic system* powers the wheel brakes, nose gear steering mechanism, bomb bay doors, and windshield wipers. It also maintains pressure in two hydraulic reservoirs for emergency wheel braking and emergency bomb bay door actuation.

The *control actuator hydraulic systems* (port and starboard) cater for the aileron, rudder, and elevator actuators. They can also be used for emergency landing-gear extension and nose gear steering if the main system fails.

The systems use AMG-10 mineral oil–type hydraulic fluid (*aviatsionnoye **mahslo** ghidravlicheskoye* – 'aviation-specific hydraulic oil'); total capacity is 120 litres (26.4 imp. gal.). Nominal pressure in all three systems is 150 kg/cm² (2,140 psi).

Electrics: The electrical system caters for the avionics, armament, engine-starting system, fuel system, lighting equipment, indication and warning systems, and part of the deicing system.

Primary 28–28.5 V DC power is supplied by eight 18 kW GSR-18000 (or GSR-18000M) generators connected in parallel to power a common circuit, each engine driving two generators. Backup DC power is provided by two 12SAM-53 or 12SAM-55 lead-acid batteries (*startyornaya, aviatsionnaya, monoblochnaya [akkumulyatornaya batareya]* – aircraft-type monobloc starter battery) accessible via the nosewheel well.

Ground AC power can be supplied via a RAP (*razyom aerodromnovo pitahniya* – ground power receptacle) or ShRA-400LK connector, located on the starboard side just ahead of the mainwheel well. The operation of the electrical power sources is monitored by means of four type A-3 ammeters, a type A-2 (or A-1) ammeter, and a switchable V-1 voltmeter.

The 115 V/400 Hz single-phase AC power is provided by two 4.5 kVA PO-4500 converters (*preobrazovahtel' odnofahznyy* – single-phase AC converter). One converter is used at any one time, while the other is a backup; they alternate between active and backup modes, with the purpose of equalising the wear on the carbon brushes and using the converters' service life equally.

Exterior lighting includes port (red) and starboard (green) BANO-45 navigation lights (*bortovoy aeronavigatsionnyy ogon'*) on the outrigger strut fairings and a KhS-39 white tail navigation light (*khvostovoy signahl*) under the tail gunner's station. Four retractable LFSV-45 landing/taxi lights (*lampa-fara samolyotnaya vydvizhnaya* – aircraft-type retractable sealed-beam lamp, 1945 model) are mounted side by side on the navigator's station ejection hatch and singly under the wing roots. For night operations in a group, PSSO-45 formation lights (*plafon siniy stroyevykh ogney* – blue formation light, 1945 model) are mounted flush with the upper and lower surfaces in several places on the fuselage and wings.

Pneumatic system: The pneumatic system works the seat actuators (which slide the seats toward the escape hatches prior to ejection), the entry and escape hatch perimeter seals, engine compressor bleed bands, deicing system valves, wing flap uplocks, brake parachute door and release locks, flare bomb bay doors, dorsal blister emergency jettison ram, and fuel jettison valves. It also serves for emergency gear extension, pressurises the hydraulic reservoirs and the radar set, cocks the cannons, and demists the crew section glazing. Compressed air is supplied by two AK-150 engine-driven compressors on the inboard engines; nominal pressure is 150 kg/cm^2.

Deicing system: Hot-air deicing on the wing, fin and stabiliser leading edges, engine air intakes, and auxiliary air intakes for the cabin pressurisation system, fuel tank vents, and the AU-2 artificial-feel unit, using engine bleed air. Electrical deicing on the flight deck / navigator's station windows, pitot heads, and static ports. The icing detector is mounted low on the starboard side of the nose.

Antiflash protection: The airframe is adequately protected against the heat and flash of a nuclear explosion. The fuselage, wing and horizontal tail undersurfaces, and the rudder are coated with special AS-2SP high-gloss, heat-resistant white enamel. All crew section windows are provided with shutters composed of metal segments, and the glazing is sealed from the outside with TG-18 heat-resistant sealing compound (*teplo**stoy**kiy gher**met**ik*). The aerials and dielectric fairings are likewise coated with AS-2SP enamel.

Fire suppression system: The airframe is divided into 11 'fire precincts' (fire protection zones) – seven for the various groups of fuel cells and one for each engine. The stationary fire suppression system comprises eight FUGOG-1 fire extinguisher bottles charged with CO_2, mounted in the nosewheel well. The first shot is triggered automatically by flame sensors in the engine bays; all fire extinguishers can be fired manually by the flight engineer. The AZS-5 and AZS-10 fire suppression system circuit breakers are installed on the pilot's electrics panel. There are also three portable CO_2 fire extinguishers in the crew cabins.

Air-conditioning and pressurisation system: At altitudes up to 2,000 m (6,560 ft.), the crew cabins are pressurised and ventilated by ram air supplied by air scoops. At higher altitude, the system automatically switches to air bled from the seventh compressor stage of both engines; the maximum possible flow of air through each of the two cabins is 500 m^3 (17,567 cu. ft.) per hour. The quantity of air supplied to the two cabins from both engines is 2,000 m^3 (70,629 cu. ft.) per hour, which allows the pressurisation/air-conditioning system to work adequately on one engine.

In so doing, the cabin pressure exceeds the ambient pressure, decreasing slowly as the aircraft climbs. Up to an altitude of 7,140 m (23,430 ft.), the cabin pressure is equal to 2,000 m above sea level; above that altitude, a pressure differential of 0.4 kg/cm^2 (5.7 psi) is maintained by ARD-54 automatic pressure governors (*avtoma**tich**eskiy regoo**lya**tor dav**len**iya*), one in each cabin.

Before entering the pressure cabins, the engine bleed air is cooled by a cooling turbine. The air temperature is maintained automatically by TRTVK-45 or TRTVK-45M cabin air temperature regulators (*termo**staht**-regoo**lya**tor tempera**too**ry **voz**dukha v ka**bine***) within set limits between +15.5° and +26.5°C (59.9°F–79.7°F), or manually between +10° and +20°C (50°F–68°F).

Oxygen system: The aircraft's oxygen system consists of a KP-24M stationary breathing apparatus with a KM-32 oxygen mask (*kislo**rod**naya **mah**ska*) for each crew member, KP-23 parachute oxygen sets used for bailing out, four KPZh-30 liquid oxygen converters (*kislo**rod**nyy pri**bor** **zhid**kosnyy*), and a KAB-16 onboard oxygen fixture (*kislo**rod**naya arma**too**ra borto**va**ya*).

Avionics and equipment: The M-4 is fully equipped for all-weather day/night operation.

a) navigation and piloting equipment: The navigation suite comprises gyroscopic, gyromagnetic, radio, and celestial navigation systems. It includes an RSBN-2S Svod SHORAN system with flush antennas built into the fin, a DISS-1 Doppler speed and drift sensor, an MRP-48A marker beacon receiver with a ventral aerial between frames 41 and 43, a GPK-52 directional gyro (***gheero**-polu**kom**pas*), RV-2 Kristall and RV-17 radio altimeters with dipole aerials under the wing roots, two ARK-5 Amur automatic direction finders (main and backup), and an NI-50BN navigation display (*navigatsi**on**nyy indi**kah**tor*), a.k.a. ANU-1A (***aero**navigatsi**on**noye oo**stroy**stvo* – air navigation device). An SP-1 star tracker (*sek**stant** periskopi**ch**eskiy* – periscopic [astro]sextant) is mounted on top of the navigator's station.

b) communications equipment: A 1-RSB-70 HF communications radio with a US-D receiver, a second 1-RSB-70 radio with a US-8 receiver, two RSIU-4V UHF command radios (main and backup), served by port and starboard 'towel rail' aerials on the forward fuselage. A Kedr (Cedar) emergency radio is provided for sending distress signals. An SPU-10 intercom (*samo**lyot**noye pere**go**vornoye oo**stroy**stvo*) serves for communication between the crew members.

For air-to-ground communication in radio silence mode (for example, in the event of a radio failure), three EKSP-39 electric signal flare launchers (*elek**tri**ch**eskaya kas**se**ta sig**nahl'**nykh pa**tro**nov*) are installed above one another low on the starboard side of the forward fuselage between frames 21 and 22. Each launcher fires four 26 mm (1.02 in.) flares of different colours (red, green, yellow, and white).

c) targeting equipment: Navigation and accurate bomb delivery are enabled by an RBP-4 panoramic navigation/ground-mapping radar in a chin radome. The radar has two operating frequencies, which made for higher jamming resistance. Maximum accurate aiming range is 70 km (43 miles), while a large ground target can be detected from a distance of 150–180 km (93–110 miles); the error margin is about 75–100 m (250–330 ft.) in range and 2°30' in azimuth. The RBP-4 enables bomb delivery throughout the M-4's operational envelope; that is, at speeds up to 1,250 km/h (776 mph) and altitudes between 2,000 and 15,000 m (6,560–49,210 ft.).

A special adapter links the radar with an OPB-11S optical bombsight. This is suitable for night and day operations and increases bombing accuracy considerably.

Four optical gunsights are provided for the gunners – the VPS-53M dorsal sighting station (***verkh**nyaya prit**sel'**naya **stahn**tsiya*), the LPS-53M port beam sighting station (***le**vaya prit**sel'**naya **stahn**tsiya*), the PPS-53M starboard beam sighting station (***pra**vaya prit**sel'**naya **stahn**tsiya*), and the KPS-53M aft sighting station (*kormo**va**ya prit**sel'**naya **stahn**tsiya*). Ranging and aiming the tail guns is done by means of a PRS-1 Argon gun-laying radar installed at the base of the rudder above the gunner's station. The radar has a 70° field of view. Importantly, it works in concert with the IFF system, precluding 'friendly fire' incidents at night. The defensive armament system also includes three PAU-457 gun cameras and a VB-53M ballistic computer.

d) IFF system: SRZO-2 Khrom-Nikel' IFF transponder. The distinctive triple IFF aerials, giving rise to the NATO reporting name *Odd Rods*, are located ahead of the flight deck glazing and under the rear fuselage.

e) photographic equipment: An AFA-42 day camera (*aerofo**toapparaht*** – aerial camera) or NAFA-75 night camera (*noch**noy** aero**fo**toappa**raht*** – aerial camera for night operations) is installed in the rear fuselage for photographing targets of opportunity and performing bomb damage assessment.

Armament:

a) offensive armament: The M-4 is equipped to carry nuclear or conventional free-fall bombs internally. The maximum ordnance load determined by the bomb bay capacity is 24 tons (52,910 lbs.). Possible payloads include four 6,000 kg (13,230 lb.) BRAB-6000 armour-piercing bombs, two 9,000 kg (19,840 lb.) FAB-9000 or 5,000 kg (11,020 lb.) FAB-5000 HE bombs (in each case on two KD6-M bomb cassettes), six 3,000 kg (6,613 lb.) FAB-3000 HE bombs on two KD4-M bomb cassettes, up to 28 500 kg (1,102 lb.) FAB-500 HE bombs, or up to 52 250 kg (551 lb.) FAB-250 or 100 kg (220 lb.) FAB-100 HE bombs on six KD3-MP bomb cassettes. For maritime missions, provision is made for carrying six Lira, Serpey, APM, AMD-2M, or IGDM naval mines or RAT-52 rocket-propelled torpedoes.

In maximum payload configuration, two FAB-9000s or four FAB-6000s are carried in a vertically staggered arrangement. Alternatively, if one FAB-9000 is carried on an MBD6-M gantry-type rack, it is suspended in a slightly nose-down position; to permit this, the bomb bay doors are bulged.

For target illumination purposes, two KD3-MP racks are installed in a DYa-SS cassette in the rear fuselage. These can hold two FotAB-100-80 flash bombs (*fotogra**fi**cheskaya avia**bom**ba* – lit. 'photo bomb') or one FotAB-250-215 each. Two more DYa-SS cassettes, each holding six TsOSAB-10 coloured marker/signal flare bombs (*tsvet**na**ya oriyen**teer**no-sig**nahl'**naya **a**via**bom**ba*) or OMAB-25-12D/OMAB-25-8N day/night maritime marker bombs (*oriyen**teer**naya mor**ska**ya **a**via**bom**ba*), are housed in the same bay.

The DB-33A dorsal barbette with the fairing detached, exposing the twin AM-23A cannons

b) defensive armament: SPV-25 cannon armament system comprising six 23 mm (.90 calibre) Afanas'yev/Makarov AM-23 cannons in three remote-controlled powered turrets – a DB-33A dorsal barbette, a DB-34A ventral barbette, and a DB-35AM tail turret. The AM-23 weighs 43 kg (94.8 lb.). It is belt fed and fires 200-gram (7.05 oz.) projectiles; the muzzle velocity is 690 m/sec. (2,263 ft./sec.), and the rate of fire is 1,300 rounds per minute.

The dorsal barbette is enclosed by a flattened circular fairing and has an ammunition supply of 600 rounds (300 rounds per gun). It is aimed by means of the VPS-53M sighting station; the field of fire is 360° in azimuth and –3° (in certain sectors) / +80°–90° in elevation. The ventral barbette is enclosed by a similar fairing and has an identical ammunition supply. It is aimed by means of the LPS-53M (port) and PPS-53M (starboard) sighting stations; its field of fire is 360° in azimuth and +3° (in certain sectors) / –80°–90° in elevation.

The tail turret has an ammunition supply of 800 rounds (400 rpg) and is enclosed by a four-piece fairing whose lateral parts are detachable for maintenance access. It is aimed by means of the KPS-53M rear sighting station; the field of fire is ±70° in azimuth and +60°/–40° in elevation. Unlike with the other cannons, spent cases and belt links are ejected, not collected in special bays.

The total ammunition supply is 4,400 rounds. The ventral and dorsal barbettes have self-contained ammunition boxes; the ammunition boxes for the tail turret are located in the unpressurised rear fuselage. Their actual capacity is 1,150 rounds each for the dorsal and ventral barbettes and 2,000 rounds for the tail turret.

Crew escape system: All crew members are provided with cartridge-fired ejection seats, which are rotated and propelled into a safe position for ejection by a pneumatic mechanism. All seats fire downward, and ejection is sequenced automatically. The pilots eject consecutively through a circular escape hatch of 1.2 m (3 ft., 11¼ in.) diameter, which mounts the revolving antenna of the RBP-4 radar; the radome and antenna are thus jettisoned before ejection. The navigator ejects through a trapezoidal hatch between frames 2 and 4, the flight engineer and radio operator use port and starboard hatches measuring 0.68 × 1 m (2 ft., 2⁴⁹⁄₆₄ in. × 3 ft., 3⅜ in.) between frames 9 and 12, and the Nav/Op ejects through a centreline escape hatch in the same area. The dorsal gunner and tail gunner eject through the two pressure cabins' respective entry hatches. The seat harness is cut automatically when the seat is clear of the aircraft, allowing the ejectee to use his parachute. All crewmen are provided with KAP-3 parachute timers (***koo**pol'nyy avto**maht** para**shoo**ta* – automatic parachute-canopy-opening device), ensuring that the parachute opens at a safe altitude even if the ejectee is unconscious and unable to open it manually.

In the event of a belly landing or ditching, the crew evacuates via the flight deck roof escape hatch, the jettisonable upper blister, and a hatch beneath the gun-ranging radar mounted above the tail gunner's station. Two LASM-49 inflatable rubber dinghies (***lod**ka ava**reey**no-spa**sah**tel'naya moderni**zee**rovannaya* – rescue dinghy, updated, 1949 model) are stowed in a bay in the upper forward fuselage; the locks of the bay cover are controlled from the flight deck.

The 3M differs from the M-4 in the following respects:

Type: (3MS-1, 3MN-1, and 3MD) high-speed, long-range bomber and (3MS-2 and 3MN-2) in-flight refuelling tanker designed for operation in VMC and IMC. The aircraft has a crew of seven: captain, copilot / flight engineer, navigator / bomb aimer, ECM officer / gunner, gunner / radio operator, dorsal gunner, and tail gunner (defensive stations commander).

Fuselage: Similar to that of the M-4, with a maximum diameter of 3.5 m (11 ft., $5\frac{51}{64}$ in.), except that the forward fuselage is redesigned and the length is increased to 46.7 m (153 ft., $2\frac{19}{32}$ in.). The fuselage is skinned in V95AT and D16AT sheet 1–4 mm (≈ $\frac{3}{64}$–$\frac{5}{32}$ in.) thick. Section F-1 (frames 1A–6A) is an unpressurised structure ahead of the navigator's station; on the 3M it has a parabolic shape and a length of 3.08 m (10 ft., $1\frac{17}{64}$ in.). Its metal upper half has six crescent-shaped frames and a 'deck,' housing an

Upper view of a 3MD bomber, showing its wing planform

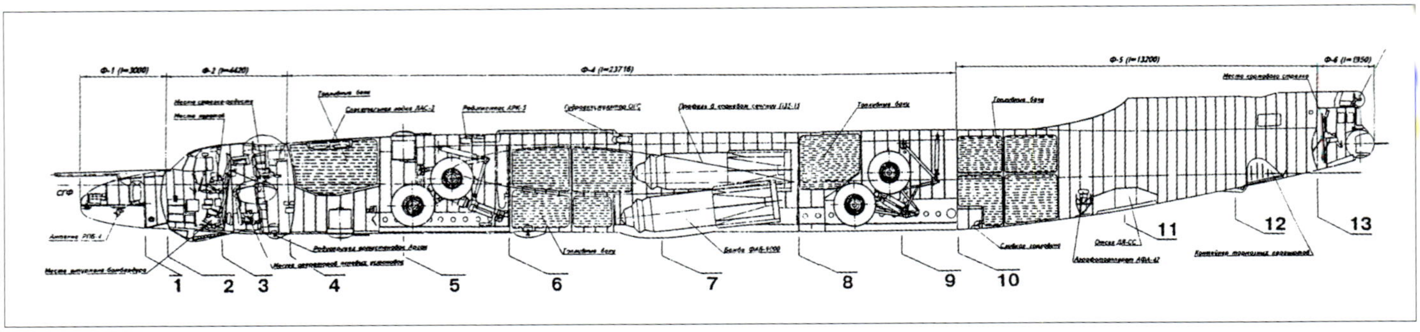

avionics/equipment bay with two hinged access panels and carrying a fixed IFR probe dorsally between frames 2A and 6A; the tip has a small trapezoidal transparency for the drogue illumination light. The lower half is a radar bay enclosed by a large dielectric radome blended into the nose contour (up to frame 4A); the revolving antenna of the RBP-4 radar is mounted on the underside of the 'deck.' A ventral recess closed by a flush Perspex panel is provided aft of the radome to improve the navigator's forward view.

The later 3MD has a different, more pointed nose (section F-1) featuring a radome with a reshaped joint line and lateral bulges to

Top: The fuselage layout of the 3MS-1 bomber with two FAB-9000s in the bomb bay

Above and above left: The nose of a 3MS-2 tanker, lacking the IFR probe. Note the ventral ECM blister aft of the sighting blisters.

Left: The nose of a 3MD bomber

Below: The fuselage layout of the 3MD (shown with the nose gear bogie tilted for takeoff and two long-range tanks plus a FAB-9000 in the bomb bay)

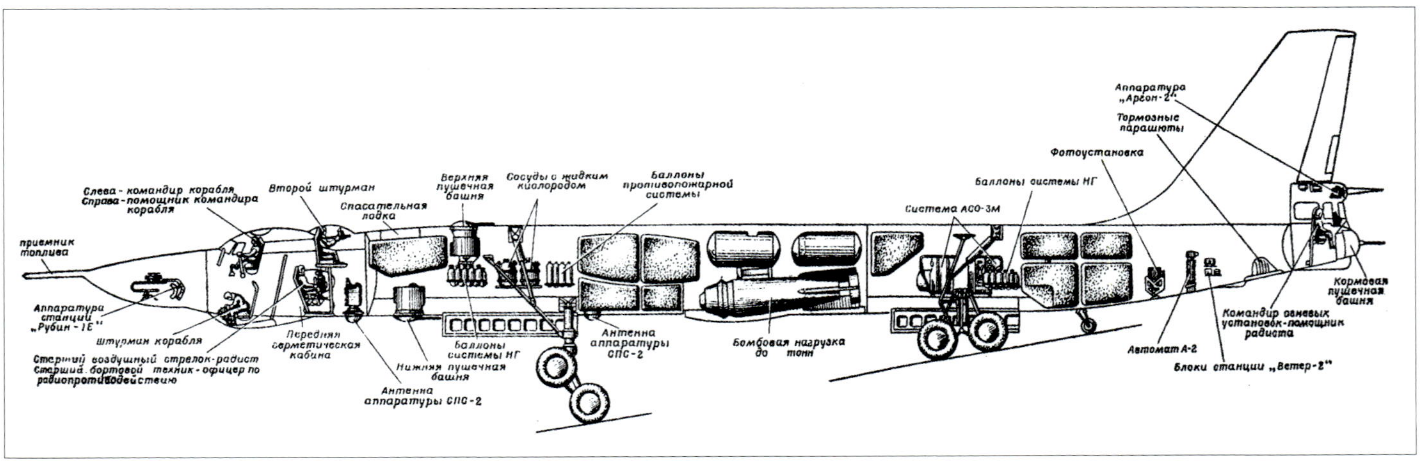

Right: An exploded view of the 3MS

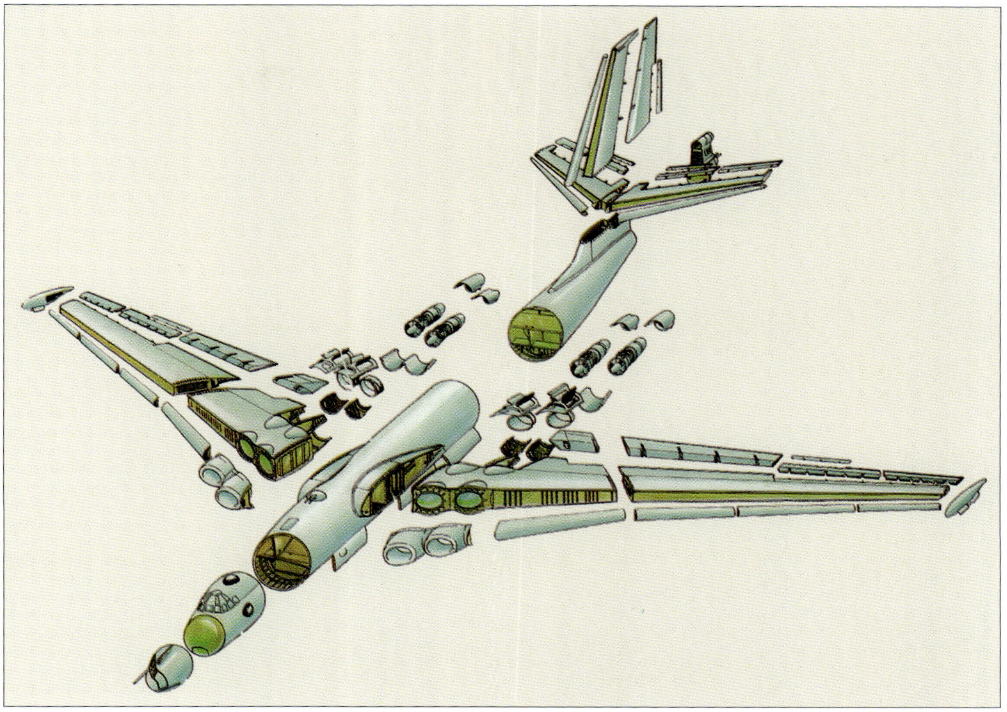

Right: The unpressurised Section F-1 of the 3M's nose, showing the radar mounting deck and the avionics bay hatches

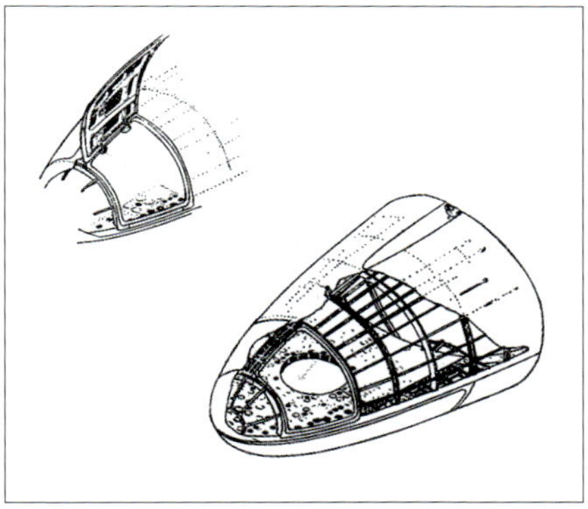

Far right: The forward pressure cabin (section F-2) of the 3M, showing the IFR pipeline channel and the offset platform for the SP-1 star tracker

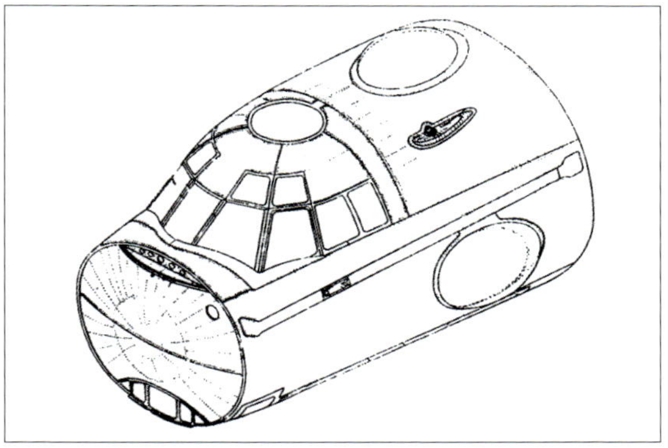

Right: The underside of a 3M's nose, showing the redesigned navigator's station glazing. The three small panes at the front are actually inside the fuselage and are visible through the large horizontal transparency, not reflected in it; the result looks like clenched teeth. Note the inward-opening entry hatch and the lowered built-in boarding steps (though a ground ladder is still required).

Far right, above: The flight deck glazing of the 3M. Note the elliptical beam blister.

Far right, below: The extreme nose of the 3M, showing the radome and the IFR probe

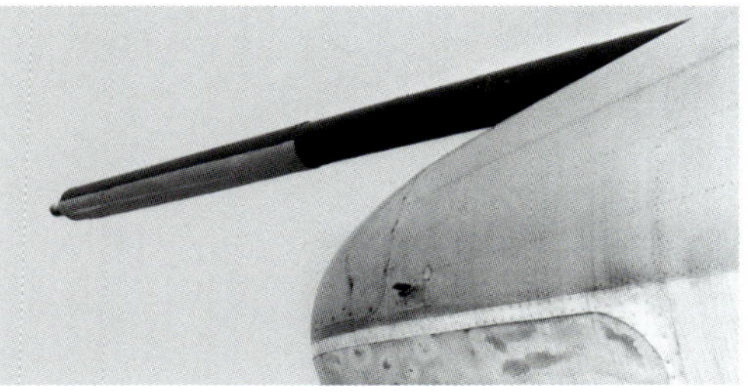

accommodate the larger antenna of the Rubin-1 radar. Also, the IFR probe is repositioned to the tip of the nose and flanked by elliptical transparencies for the drogue illumination lights. Hence, overall fuselage length is even greater at 47.752 m (156 ft., 8 in.).

Section F-2 (frames 1–12) is 7.5 m (24 ft., $7\frac{17}{64}$ in.) long, with a maximum diameter of 3.167 m (11 ft., $4\frac{11}{16}$ in.). It commences with a flat front pressure bulkhead; the navigator's station is moved aft to a position directly below the pilots. Its glazing is reduced to three rectangular windows in an L-shaped arrangement on each side, three ventral windows (the centre one being an optically flat rectangular triplex glass pane for bomb aiming), and three matching windows in the pressure bulkhead. The three ventral windows are built into a teardrop fairing similar in size (but not in shape) to the M-4's radome; it incorporates a 0.99 × 1.18 m (3 ft., $2\frac{63}{64}$ in. × 3 ft., $10\frac{29}{64}$ in.) ejection hatch (frames 4–6) for the navigator and the pilots.

The design of the flight deck glazing and the dorsal blister is basically unchanged, but the beam blisters are elliptical, not circular. Again, all windows are provided with segmented metal shutters to protect the crew from the flash of the nuclear explosion. The 0.64 × 0.7 m (2 ft., $1\frac{13}{64}$ in. × 2 ft., $3\frac{9}{16}$ in.) entry hatch is relocated to frames 7–12 and opens inward. A channel running along the port side of the flight deck accommodates the fuel transfer line from the IFR probe; it is closed flush with the fuselage skin. Oddly, the pressure dome at frame 12 is mated directly to centre fuselage frame 17; thus there are no frames 13–16.

Wings: The wings of the 3MS/3MN are structurally similar to those of the M-4 but have slightly less sweepback (34°45' at quarter chord – though some structural manuals say 35°); anhedral 1°50' on the inner wing sections (K-1) and 0°50' on the outer wings (K-2), incidence –2°30'. While the inner wing sections (ribs 1–11) are unchanged, the outer wing sections (ribs 11–32) are new, increasing the span to 53.14 m (174 ft., $4\frac{1}{8}$ in.) and wing area to 351.78 m^2 (3,786.53 sq. ft.) overall or 324.1 m^2 (3,488.58 sq. ft.) less the centre section; the mean aerodynamic chord (MAC) is 6.84 m (22 ft., $5\frac{19}{64}$ in.). The wing planform is also different, with no trailing-edge kink. The wingtip sections have 5° negative camber.

The area-increasing TsAGI flaps are one piece, not two section. They terminate at rib 21 and the ailerons accordingly at rib 32, with the joint between the aileron sections shifting to rib 27; the outer flap tracks are at ribs 12, 14, 16, 18, and 20, while the aileron mounting brackets are at ribs 22, 24, 26, and 28. The flaps are operated by an MPZ-5 electrical drive; maximum flap deflection is increased to 32° for the inboard (hinged) flaps and 37° for the TsAGI flaps; the speed limit with flaps deployed is 420 km/h (260 mph). Again, there are two pairs of wing fences, but these are located differently (at half span of the flaps and at the flap/aileron joint).

The 3MD has extended and drooped wing leading edges, with a modified airfoil and the MAC increased to 7.154 m (23 ft.,

Top left: This view illustrates the new one-piece TsAGI flaps of the 3M. Note the recesses for the flap tracks and screw jacks.

Left and above left: The rear fuselage and tail unit of a 3MD, showing the zero-dihedral stabilisers and details of the tail turret

The nose gear unit and the starboard outrigger strut of a 3MD

$5^{21}/_{32}$ in.); hence, despite an identical span of 53.14 m, its overall wing area is even greater at 359.82 m² (3,873.07 sq. ft.).

The main gear fairings mounted at ribs 32L/32R are semi-monocoque, stressed-skin construction with 15 frames, eight of which (nos. 4–11) are formers, and two beams. They are attached to the wing spars at frames 7 and 11.

The inner wings incorporate 'wet' hardpoints with Der5-48 shackles for carrying drop tanks.

Tail unit: Conventional tail unit of all-metal, stressed-skin construction. The *vertical tail* is similar to that of the M-4, with 35° sweepback at quarter chord.

The cantilever variable-incidence *horizontal tail* is again structurally similar to that of the M-4 and has the same 33°30' sweepback at quarter chord but has no dihedral; incidence is adjustable from +1°30' to –2°30'. The airfoil and thickness/chord ratio are the same as for the M-4; the elevator travel limits are likewise 10° down and 25° up.

Landing gear: Retractable bicycle type; all four units retract forward. The nose and main gear units are electrohydraulically actuated and have a new design; in an emergency, hydraulic or pneumatic extension is possible. The aircraft's ground angle is 4°40' at the aircraft's empty weight and 5°13' at the maximum takeoff weight.

The nose gear bogie has smaller 1,500 × 550 mm (59.05 × 21.65 in.) wheels; the front pair are KT-58 wheels fitted with light-duty, pneumatically operated brakes for automatic braking during retraction to cancel gyroscopic forces, while the rear pair are KT-43 wheels with heavy-duty hydraulic brakes. The front wheels are hydraulically steerable through ±4° by means of a servo and are pneumatically locked neutral for takeoff and landing; the bogie steering limits are ±23°, giving a turn radius of 41 m (134 ft., 6 in.). The nose gear unit has a bogie tilt actuator to increase the AoA on takeoff.

The main gear bogie has four KT-43 wheels and a UA-24 antiskid unit. The castoring outrigger struts are actuated by MPSh-17M electrical drives and fitted with 660 × 160 mm (26.00 × 6.30 in.) K2-57 nonbraking wheels. Tyre pressure 11 bars (156 psi) for the nosewheels and mainwheels and 7.5 bars (106.5 psi) for the outrigger strut wheels.

All four struts have oleo-pneumatic shock absorbers filled with AMG-10 hydraulic fluid (some structural manuals say a 50/50 alcohol/glycerine mixture!). The nose gear oleo is charged to 40 kg/cm² (571 psi), the main gear oleo to 90 kg/cm² (1,285 psi), and the outrigger strut oleos to 50 kg/cm² (714 psi).

The nosewheels and mainwheels are braked at the push of a button prior to retraction; emergency braking is done by using hydraulic accumulators. Normal braking is performed by pressing brake triggers on the control wheel grips. The nose and main units have hydraulic uplocks; normal gear retraction/extension time is 40 seconds. The speed limit with the gear down is 420 km/h (260 mph); flying with the gear down above 1,000 m (3,280 ft.) is prohibited.

The nosewheel and mainwheel well doors are operated by MSSh-2 electrical drives (*mekha**nizm** **stv**orok **sh**assee*); the outrigger strut doors are mechanically linked. All doors remain open when the gear is down.

From the second production 3M (c/n 6300302) onward, brake parachute deployment/release is operated electropneumatically.

Power plant: The 3MN and 3MD are powered by four Dobrynin (RKBM) VD-7B non-afterburning turbojets rated at 9,580 kgp (21,120 lbst) for takeoff and 2,050 kgp (4,520 lbst) for cruise. The VD-7B is a single-spool turbojet having an inlet assembly with six radial struts and a fixed parabolic spinner, a nine-stage axial-flow compressor with a supersonic first stage and fixed inlet guide vanes, a straight-through can-annular combustion chamber with 14 flame tubes, a two-stage axial-flow turbine, a fixed-area subsonic nozzle, and two accessories gearboxes. Starting is electric, using an STG-12TP starter-generator.

Engine speed is 7,425 rpm at takeoff power, 7,200 rpm at maximum continuous power, and 3,300 rpm at idle power. EPR at takeoff power is 11.2, mass flow at takeoff power is 187 kg/sec. (412 lbs./sec.), turbine temperature at takeoff power is 1,090°K. and SFC is 0.8 kg/kgp·hr. at takeoff power and 0.85 kg/kgp·hr. at cruise power. Length overall is 4.247 m (13 ft., $11^{13}/_{64}$ in.), inlet diameter is 1.216 m (3 ft., $11^{7}/_{8}$ in.), and maximum diameter is 1.288 m (4 ft., $2^{43}/_{64}$ in.). Dry weight is 2,765 kg (96,095 lbs.).

Left: A Dobrynin VD-7B engine – unfortunately a cutaway example. The much-smaller intake centre body is evident; again, it is carried on six radial struts.

Below: This view of a 3MD shows the dorsal slits at the rear of the engine cowlings, which are a distinctive feature of the VD-7B-powered versions. Note also the stainless-steel blast plates fore and aft of the dorsal turret. Unlike the 3MS/3MN, the 3MD has only one dorsal ADF strake aerial instead of two.

Right: The centre fuselage of a 3MS-2 tanker modified to nonconvertible configuration, with short bomb bay doors and no ventral bulge

Below: By comparison, here is a convertible 3MS-2 retaining the bomber's bulged bomb bay doors. Note the starboard dipole aerial of the RV-2 radio altimeter under the No. 3 engine intake; another one is mounted on the opposite side.

Below right: The inside of a 3MS-2's bomb bay, showing the HDU with the drogue in the stowed position, the rear extra tank, and details of the bay door actuation mechanism

Bottom: This view of a retired 3MS-2 tanker shows how the engines could be exposed completely for maintenance or removal. The No. 1 engine is already gone.

The 3MS-1 bomber and 3MS-2 tanker are powered by four Mikulin AM-3D nonafterburning turbojets rated at 8,750 kgp (19,290 lbst) for takeoff and 6,200 kgp (13,670 lbst) for cruise, or Zoobets RD-3M-500 or RD-3M-500A engines with a takeoff rating of 9,500 kgp (20,940 lbst) or 10,500 kgp (23,150 lbst), respectively.

Regardless of engine type, the engines are installed in the same way as on the M-4. The No. 4 engine is started first.

Control system: Fully powered dual control system with irreversible hydraulic actuators and manual backup in an emergency. All control surfaces are both aerodynamically balanced and mass balanced. An AP-15 autopilot is fitted.

Roll control is provided by two-section ailerons operated by GU-79KF actuators, with trim tabs controlled by MP-100M electrical trim mechanisms on the inner sections. The aileron travel limits are 14° down and 22° up.

Pitch control is provided by two-section elevators with a GU-47KF actuator. On the 3MS/3MN, the elevator travel limits are 10° down and 20° up; on the 3MD they are increased to 11° down and 22° up. Electrically operated trim tabs are provided for longitudinal trim by means of an MP-250 trim mechanism.

Directional control is provided by a one-piece rudder with a GU-48KF/2 actuator and a servo tab / trim tab at the root actuated by an MP-100M electrical trim mechanism. The rudder travel limits are ±15°30' for the 3MS/3MN and ±20° for the 3MD. The directional control circuit includes an AP-102M yaw damper with an RA-102M servo (*roolevoy agregaht*) deflecting the rudder ±3°. Gust locks operated by MP-100AD mechanisms are provided.

Fuel system: Internal fuel is carried in bladder tanks located in the centre/rear fuselage and in the wing torsion box; internal fuel capacity is increased to 133,500 litres (29,370 imp. gal.) on the 3MS/3MN and 134,150 litres (29,513 imp. gal.) on the 3MS. Generally, the numbering of the tanks matches the M-4, but the tanks in the wing centre section are renumbered as group 17 (the outer ones) and group 18 (the inner ones). The inner/outer wings house five groups of tanks located between ribs 5 and 8 (group 15), 8 and 11 (group 14), 11 and 15 (group 13), 15 and 21 (group 11), and 21 and 30 (group 10). Some of the tanks (groups 1, 8, 10, 17L, and 18R) are no longer self-sealing, to save weight. Each group of tanks has ETsN-10 and ETsN-M booster pumps (*elektricheskiy tsentrobezhnyy nasos* – electrical centrifugal pump); groups 11 and 13 also have one PTsR-1 transfer pump each. The booster pumps feed the fuel to the engines' DTsN-13D supply pumps, which feed the NR-7B fuel control units (*nasos-regoolyator* – FCU) and ART-10B fuel flow adjusters (*avtomaht regooleerovaniya topliva*).

The 3M (3MS, 3MN) and 3MD have in-flight refuelling capability as standard, using the probe-and-drogue system. On the 3MS/3MN the fixed IFR probe is located on top of the nose ahead of the flight deck windscreen; in the case of the 3MD, it is at the forward extremity of the fuselage. The fuel from the tanker is fed to the group 1–3, 5–7, 9, 10, 14, 17, and 18L tanks; up to 3,830 litres (842.6 imp. gal.) can be transferred at a time.

Again, two elliptical-section metal tanks (the group 12 tanks) can be installed in tandem in the upper part of the bomb bay for extra-long-range missions. Additionally, the 3M has provisions for carrying two cigar-shaped drop tanks under the wing roots.

Emergency fuel jettison valves are provided. Fuel can be dumped from the group 1, 2, 4, 6, 7, 10, 11, 13, 14, 17, and 18 tanks.

3MS-2 and 3MN-2 convertible tankers carry an HDU and a third extra tank in the bomb bay and have the IFR probe removed. Some 3MS-2s were modified to dedicated (nonconvertible) tankers, lacking bombing capability (the greater part of the bomb bay is faired over, with much-shorter doors for the drogue).

Hydraulics: The 3M has simpler hydraulics, with two hydraulic systems instead of three. Hydraulic pressure is supplied by two 419M-150 hydraulic pumps on each inboard engine, giving a pressure of 150 kg/cm² (2,142 psi).

The *main hydraulic system* (a.k.a. port hydraulic system) powers the wheel brakes, nose gear steering mechanism, bomb bay doors, and windshield wipers. It also maintains pressure in two hydraulic reservoirs for emergency wheel braking and emergency bomb bay door actuation.

The *control actuator hydraulic system* (a.k.a. starboard hydraulic system) caters for the aileron, rudder, and elevator actuators. It can also be used for emergency landing-gear extension and nose gear steering if the main system fails. Both systems use AMG-10 fluid; the capacity is 27 litres (5.94 imp. gal.) for the port system and 32 litres (7.04 imp. gal.) for the starboard system.

Electrics: Main DC power is provided by four engine-driven STG-12BP starter-generators, with DC batteries as a backup. 115 V/400 Hz single-phase AC power is provided by two 6 kVA PO-6000 AC converters. The 3MD has four SGO-80 engine-driven alternators.

Exterior lighting includes retractable FRS-200 landing/taxi lights on the sides of the nose aft of the radome and under the wing roots. To facilitate night contacts with the tanker, the 3MS/3MN has a drogue illumination light behind a flush Perspex panel built into the tip of the nose; the 3MD has two drogue illumination lights on both sides of the tip-mounted IFR probe. A green light is built into the fin fillet's leading edge, illuminating when the IFR probe locks into the tanker's drogue to show the refuelling-systems operator that contact has been made and fuel transfer can begin. The 3MS-2 and 3MN-2 tankers have special lights for illuminating the undersides of the rear fuselage and horizontal tail during night contacts, to provide a visual reference for the crew of the receiver aircraft.

Pneumatic system: Air is supplied by an AK-150D engine-driven compressor. The pneumatic system's functions are expanded to include operation of the IFR probe and emergency landing-gear extension if all hydraulics fail.

Fire suppression system: The 3M is equipped with either FUGOG-1 or POG-4 fire extinguisher bottles charged with CO_2. The first shot is triggered automatically by the SSP-2A fire-warning system (*sistema signalizahtsii pozhara*) with DPS-1AG flame sensors (*dahtchik pozharnoy signalizahtsii*) in the engine bays and the wing fuel cell bays. All fire extinguishers can be operated manually by the flight engineer.

Right: The captain's and copilot's instrument panels and control columns of a 3M tanker, with the bank of engine instruments between them

Below and below right: Two of the other crewmen's workstations on the same 3M tanker

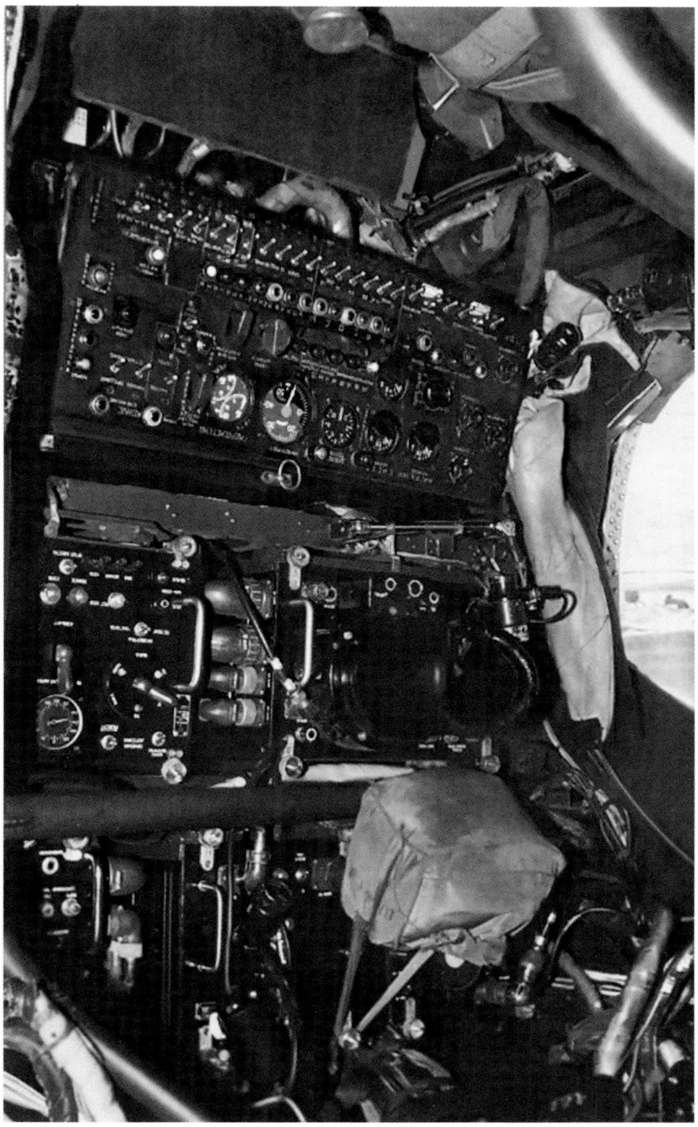

Above: The instrument panel and map table in the navigator's workstation of a 3M tanker

Below: The dorsal gunner's seat and observation blister on a 3M tanker

Avionics and equipment: The 3M is fully equipped for all-weather day/night operation.

a) navigation and piloting equipment: The navigation suite is revised to include SD-1 distance-measuring equipment (*samolyotnyy dahl'nomer*), a DISS-1 Doppler speed/drift sensor, an additional ARK-42 ADF, and an SP-1 star tracker mounted on top of the flight deck, offset to port. The 3MD is fitted with an SPI-3 navigation receiver/display (*samolyotnyy preeyomoindikahtor*).

To prevent stalling, the 3M and 3MD are equipped with an ASKR-3 automatic critical flight mode indicator (*avtomaht signalizahtsiï kriticheskikh rezhimov*), alerting the pilots that the angle of attack is too high 1°–1.5° before the AoA becomes critical.

The pilots' and navigator's instrument panels feature AGB-2 artificial horizons (*aviagorizont bombardirovshchika* – artificial horizon for bombers); KUS-1200 airspeed indicators (*kombineerovannyy ookazahtel' skorosti* – combined ASI); VD-20 two-needle barometric altimeters (*vysotomer dvookhstrelochnyy*); VAR-30-3 vertical speed indicators (*variometr* – VSI); an SSN-3 dynamic pressure indicator (*signalizahtor skorosnovo napora*); EUP-53 turn and bank indicators (*elektricheskiy ookazahtel' povorota*); MS-1 Mach meters; an AM-10 accelerometer; KI-12 and KI-13 induction-type compasses (*kompas indooktsionnyy*); a DAK-DB-5 celestial compass (*distantsionnyy astrokompas dahl'nevo bombardirovshchika* – remote celestial compass optimised for long-range bombers); an AK-53P celestial compass; 13ChP-20ChP, AChKh, and AVR-M chronometers; TNV-15 ambient air thermometers (*termometr naroozhnovo vozdukha*); a VS-46 high-altitude pressure-warning indicator (*vysotnyy signalizahtor*) telling the crew when it's time to don the oxygen masks; and a TUE-48 thermometer (*termometr ooniversahl'nyy elektricheskiy* – multipurpose electric thermometer).

The following engine instruments are installed: ITE-1 remote-sensing rpm gauges (*indooktsionny takhometr elektricheskiy* – induction-type electric tachometer) for the engines (and, on the 3MS, TE-45 rpm gauges for the turbostarters); 2TVG-366 exhaust gas temperature gauges (*termometr vykhodyashchikh gahzov*); a TTsT-13 thermometer; an EMI-3R three-needle electrical indicator (*elektricheskiy motornyy indikahtor*) showing fuel pressure, oil pressure, and oil temperature; an EDMU-3 electrical, remote, standardised pressure gauge (*elektricheskiy distantsionnyy manometr oonifitseerovannyy*); DIM-80T induction-type remote pressure gauge; RTS-16A-35 fuel flow meters; and fuel gauges making up part of the SETS-150 fuel-metering kit.

b) communications equipment: The communications suite is revised by replacing the RSIU-4V UHF command radios with R-802V (RSIU-5V) command radios.

d) IFF system: SRO-1P (*izdeliye* 62-01) IFF transponder.

Armament:

a) offensive armament: The 3M is equipped to carry nuclear or conventional free-fall bombs internally. Possible payload options originally included:

• six 3,000 kg (6,613 lb.) FAB-3000 M-54 HE bombs on two KD4-M bomb cassettes

• two 5,000 kg (11,020 lb.) FAB-5000 M-54 HE bombs on a KD6-M bomb cassette on the starboard side of the bomb bay

Right: The DB-33A dorsal turret of a 3MS-2, with appropriate blast plates. The port 'towel rail' communications aerial and the SP-1 star tracker ahead of it are also visible.

Below: Close-up of the dorsal turret and the barrels of the twin AM-23A cannons

Below right: The tail gunner's station and DB-35AM tail turret of a 3MS. Note the distinctive triangular red blade aerial of the SRO-1P IFF transponder above the turret.

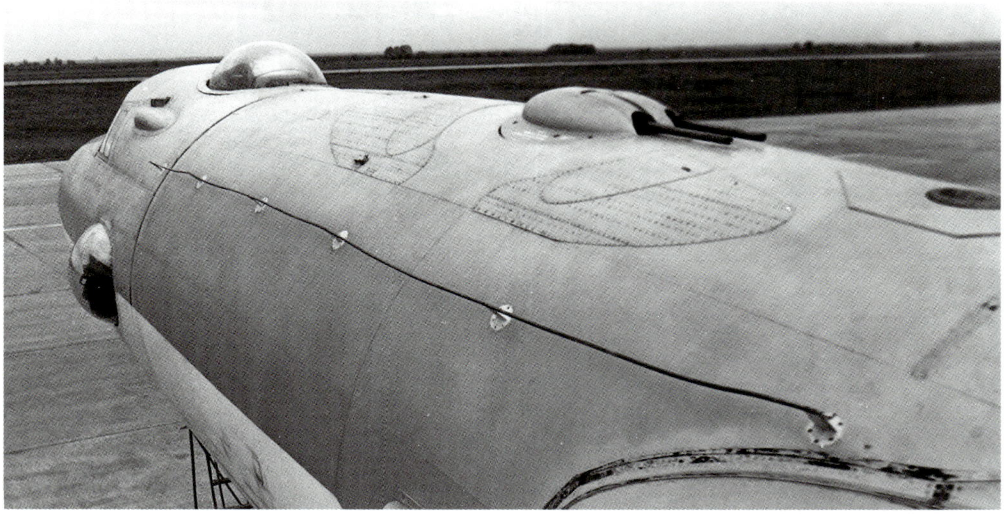

• six BRAB-3000 armour-piercing bombs on two KD4-M bomb cassettes

• four 6,000 kg (13,230 lb.) BRAB-6000 armour-piercing bombs on two KD6-M bomb cassettes

• two 9,000 kg (19,840 lb.) FAB-9000 M-54 HE bombs on two KD6-M bomb cassettes

• one FAB-9000 M-54 bomb in a nose-down position on an MBD6-M gantry-type rack.

With two long-range fuel tanks in the upper part of the bomb bay, the options were:

• two FAB-3000 M-54 bombs side by side on an MBD6-M rack with two frontal braces

• one FAB-5000 M-54 bombs on an MBD6-M rack with one frontal brace

• two BRAB-3000 bombs side by side on an MBD6-M rack

• two BRAB-6000 bombs side by side on an MBD6-M rack

The maximum ordnance load determined by the bomb bay capacity was reduced from 24 to 18 tons (from 52,910 to 39,680 lbs.) when the 6,000 kg bombs were withdrawn from the inventory.

FAB-3000 M-54 bomb carriage on an MBD6-M rack on the 3M bomber

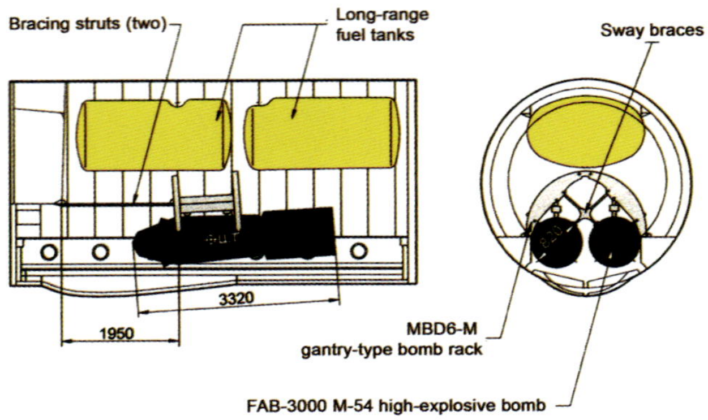

FAB-3000 M-54 bomb carriage on KD4-M racks on the 3M bomber

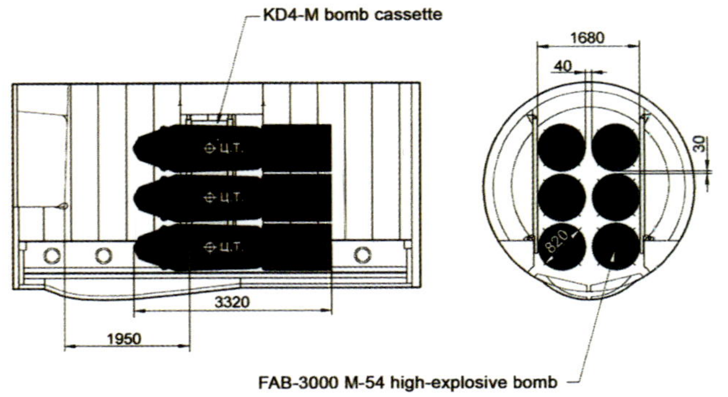

FAB-5000 M-54 bomb carriage on an MBD6-M rack on the 3M bomber

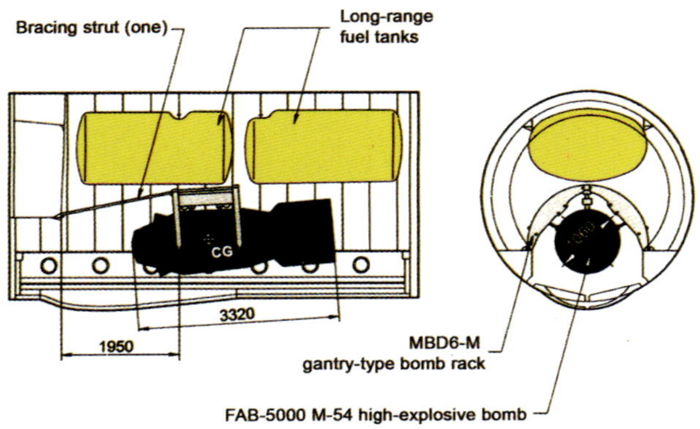

FAB-5000 M-54 bomb carriage on a KD6-M rack on the 3M bomber

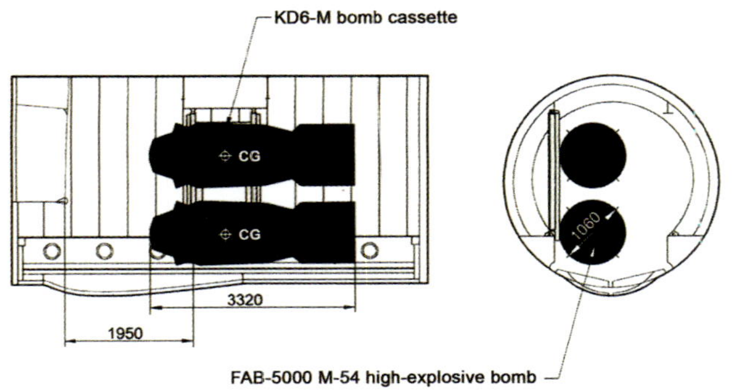

BRAB-3000 bomb carriage on an MBD6-M rack on the 3M bomber

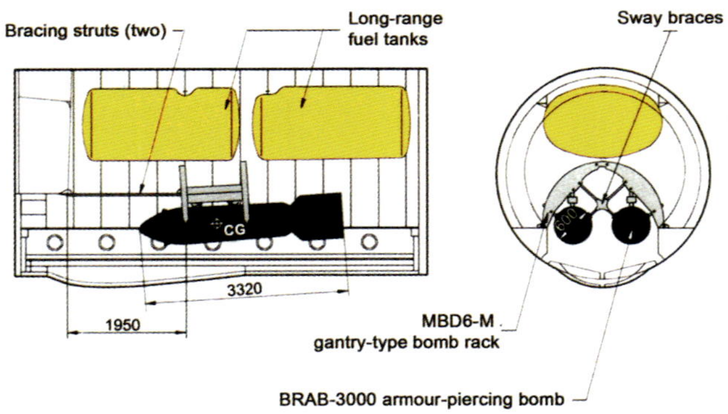

BRAB-3000 bomb carriage on KD4-M racks on the 3M bomber

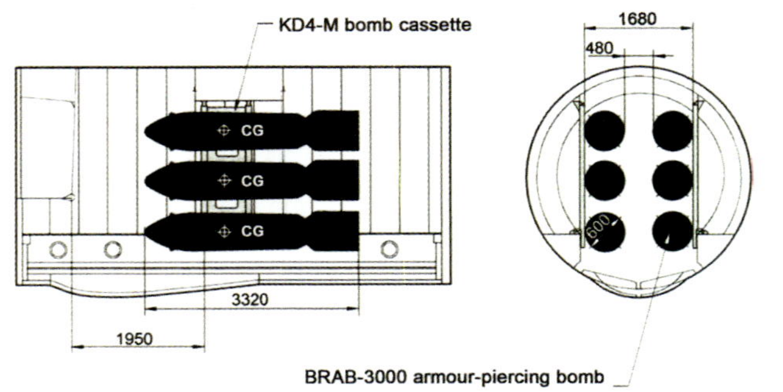

BRAB-6000 bomb carriage on an MBD6-M rack on the 3M bomber

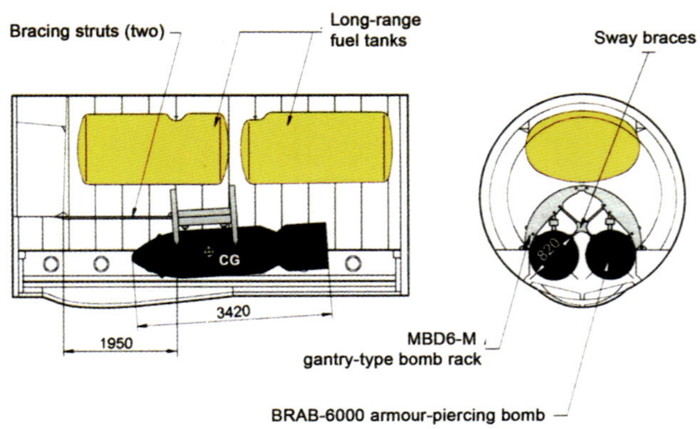

BRAB-6000 bomb carriage on KD6-M racks on the 3M bomber

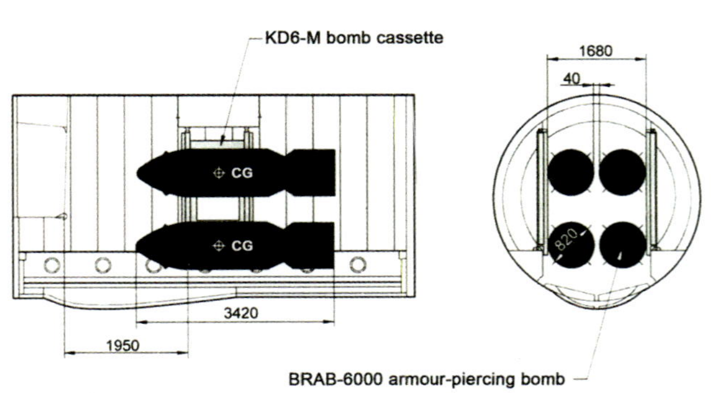

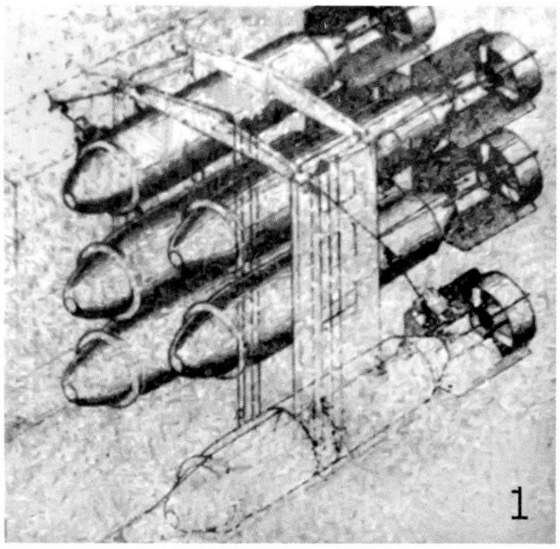

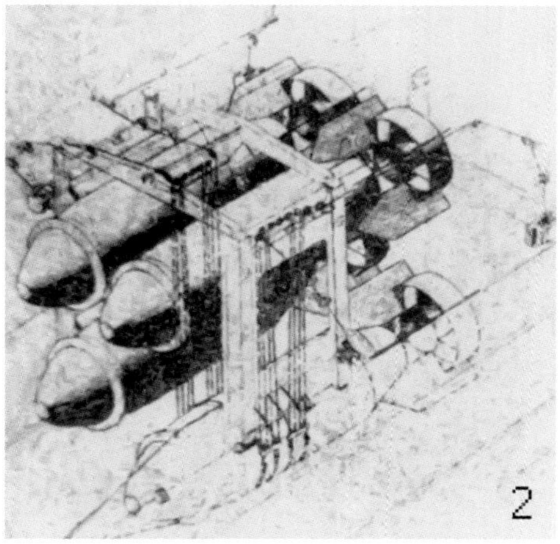

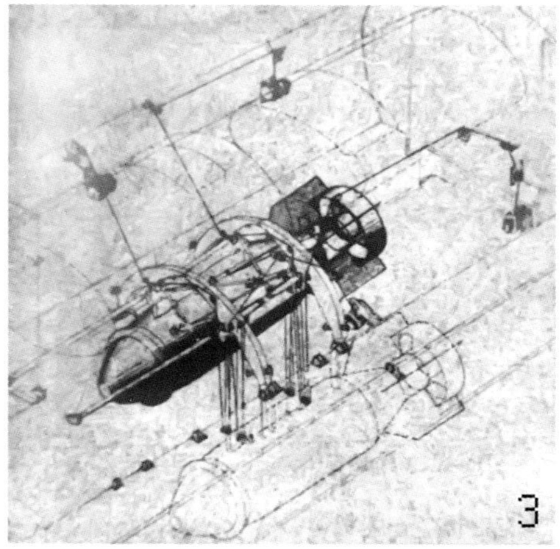

1. The carriage of six BRAB-3000 armour-piercing bombs on two KD4-M bomb cassettes (one bomb is shown being lifted into position by a bomb hoist)
2. The carriage of four BRAB-6000 armour-piercing bombs on two KD6-M bomb cassettes
3. The carriage of two BRAB-6000 bombs on an MBD6-M gantry-type rack; again, one bomb is shown being lifted into position.
4. The carriage of two FAB-9000 M-54 HE bombs on KD6-M bomb cassettes
5. The carriage of a single FAB-9000 M-54 bomb on an MBD6-M rack

A bomb trolley with a FAB-9000 M-54 is positioned for loading below the bomb bay of a 3M. This photo shows how the bomb bay doors could be opened through 180° on the ground, allowing the bomb carts to pass freely.

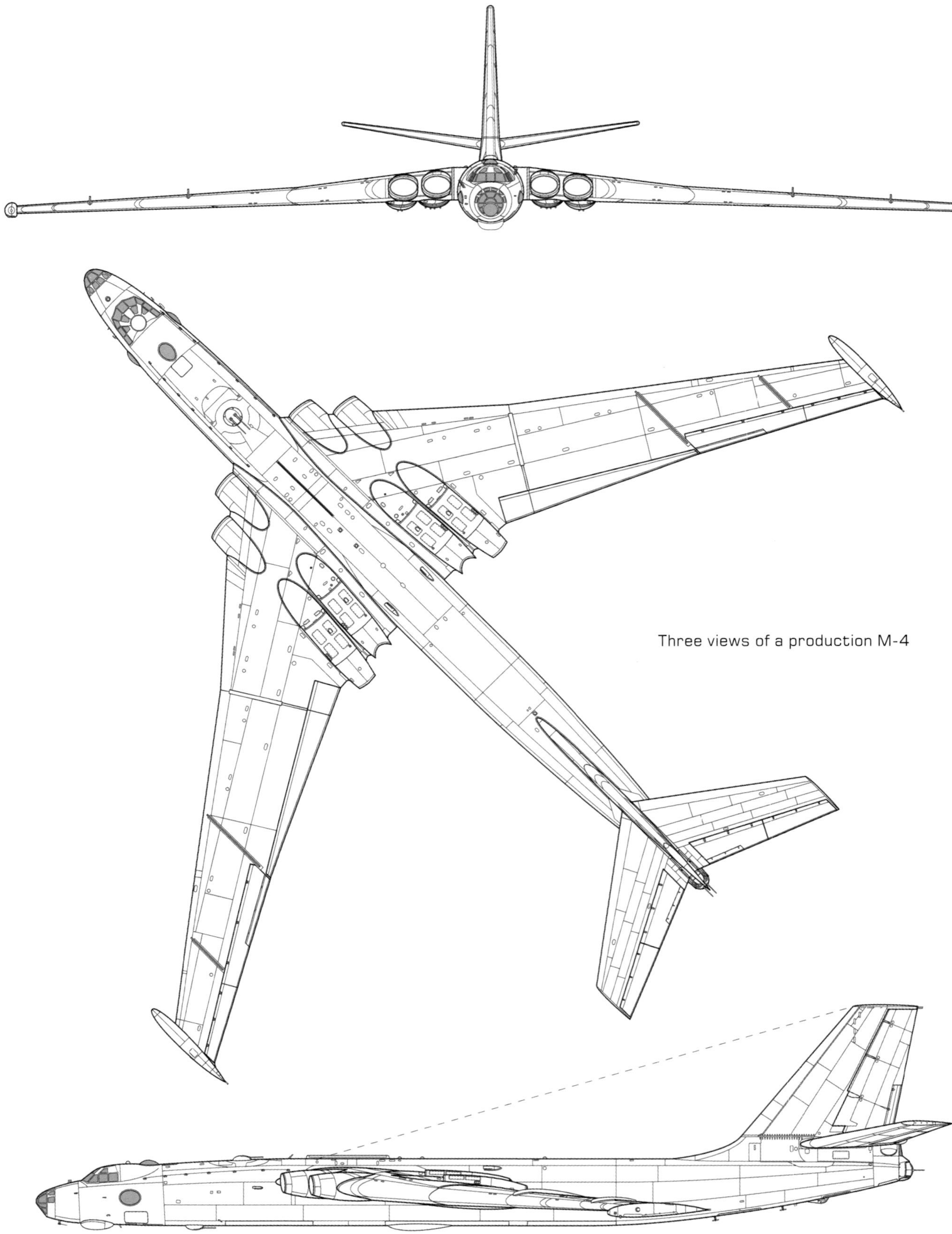

Three views of a production M-4

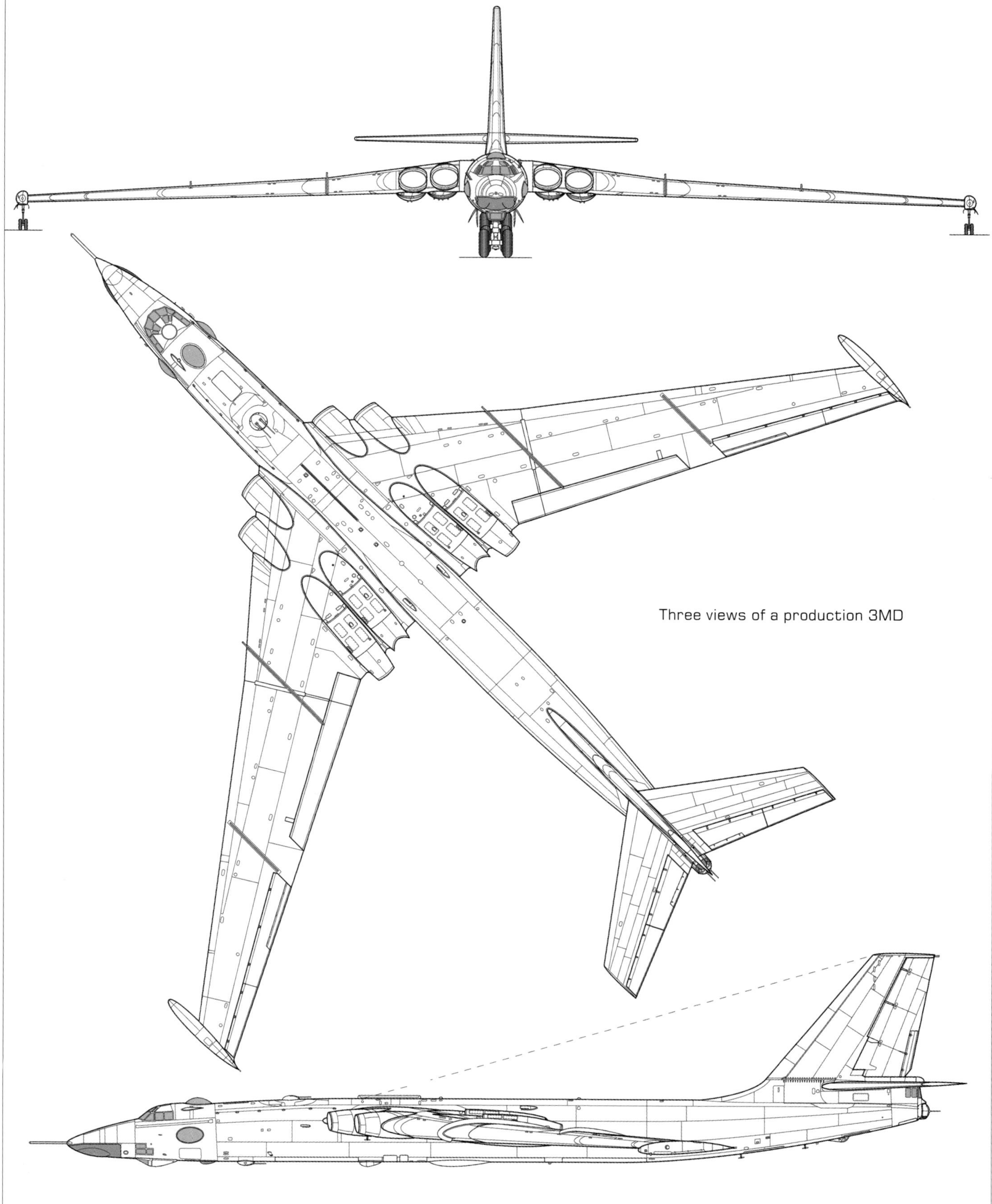

Three views of a production 3MD

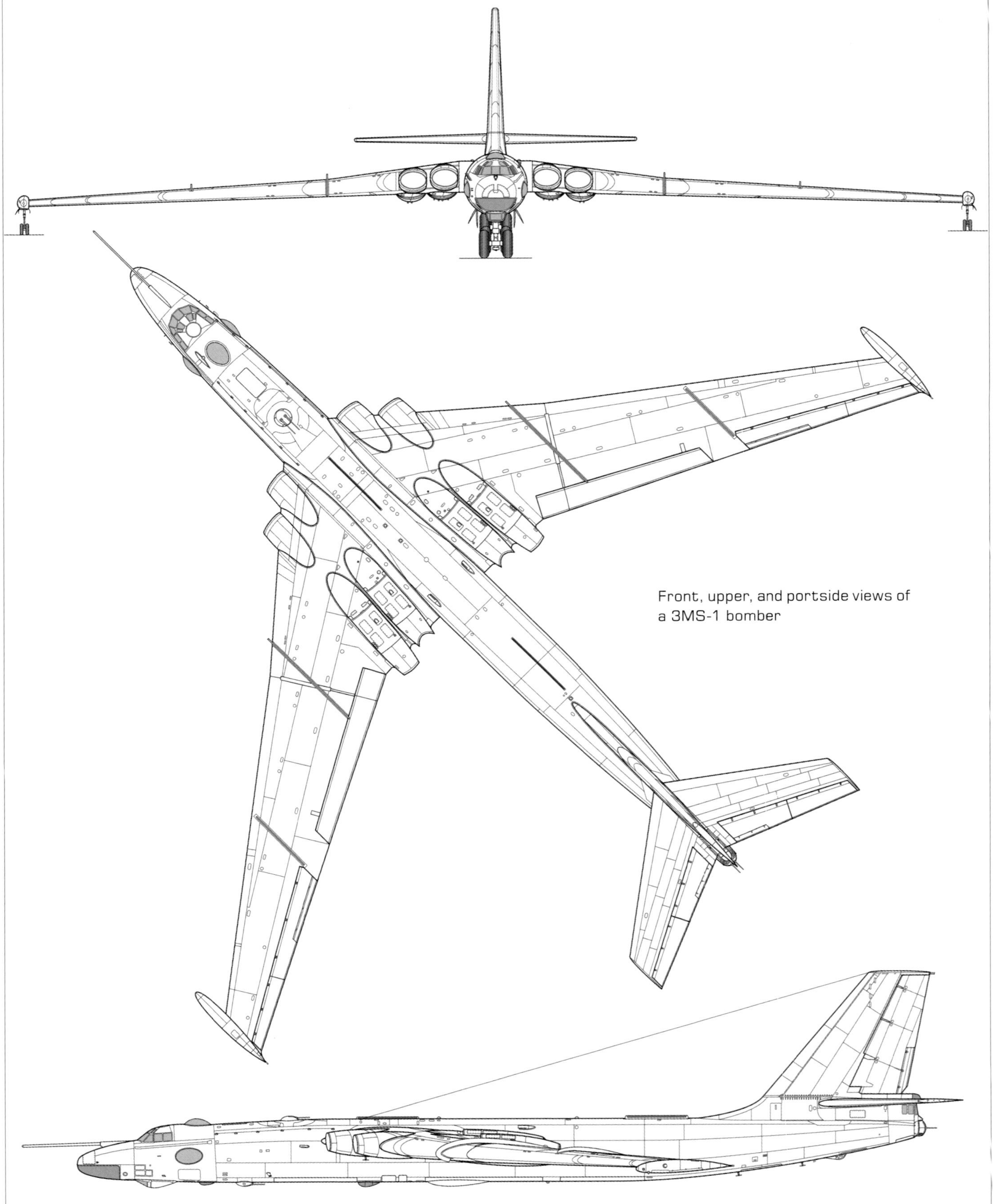

Front, upper, and portside views of a 3MS-1 bomber

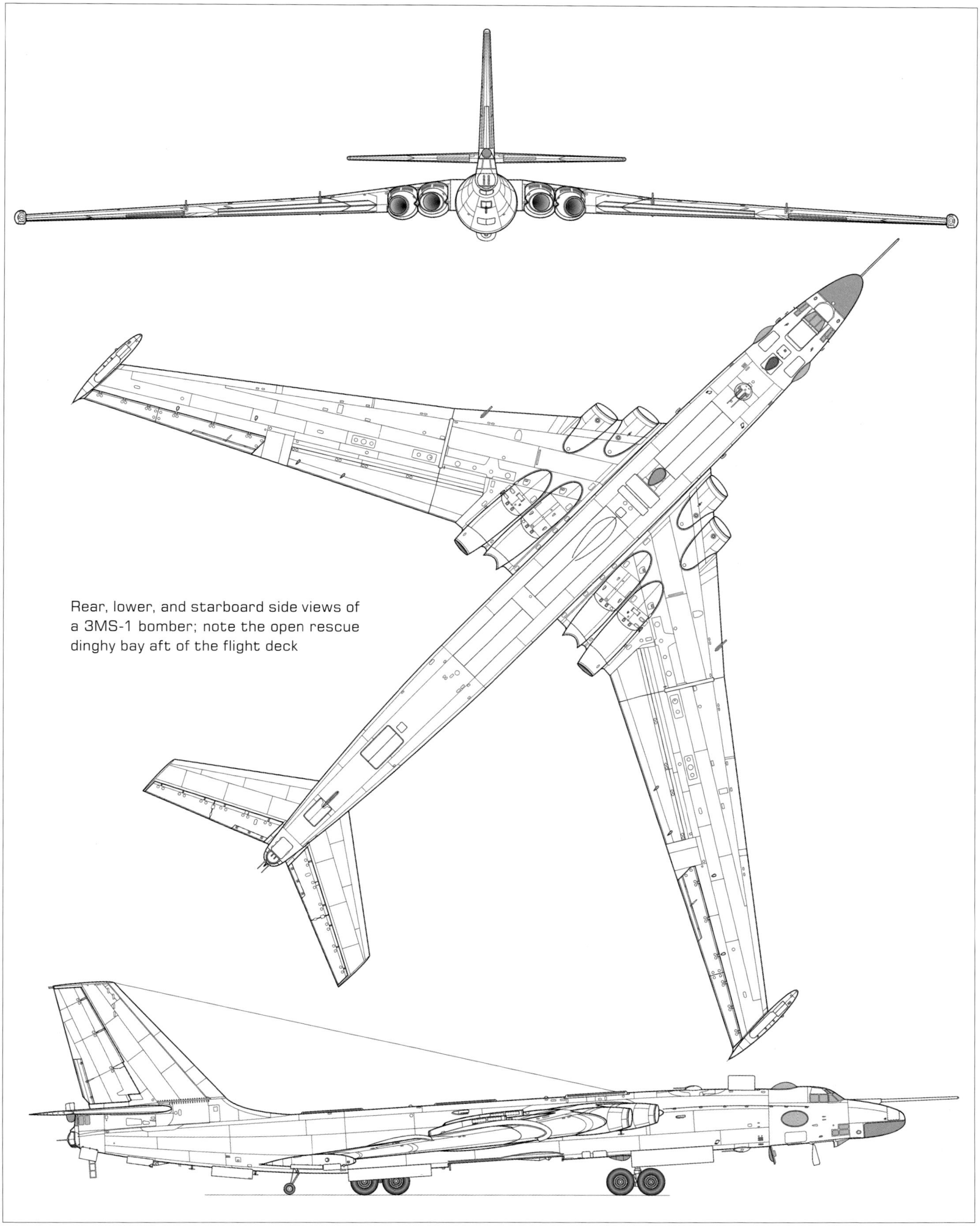

Rear, lower, and starboard side views of a 3MS-1 bomber; note the open rescue dinghy bay aft of the flight deck

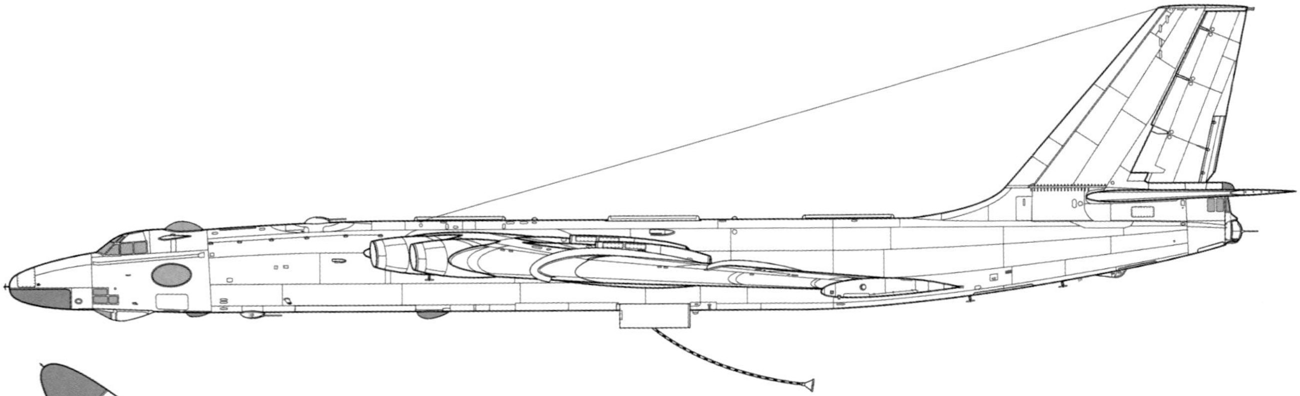

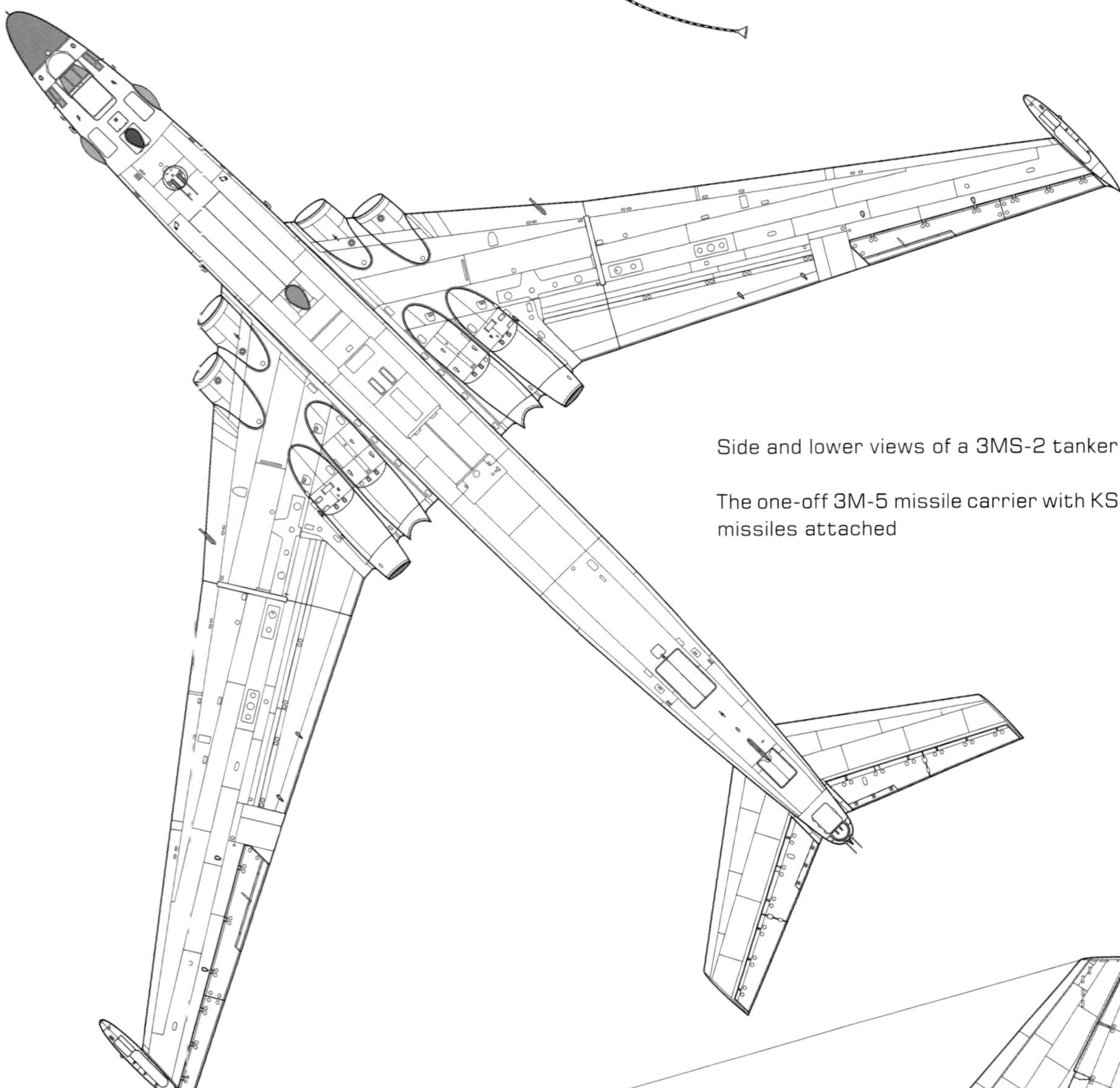

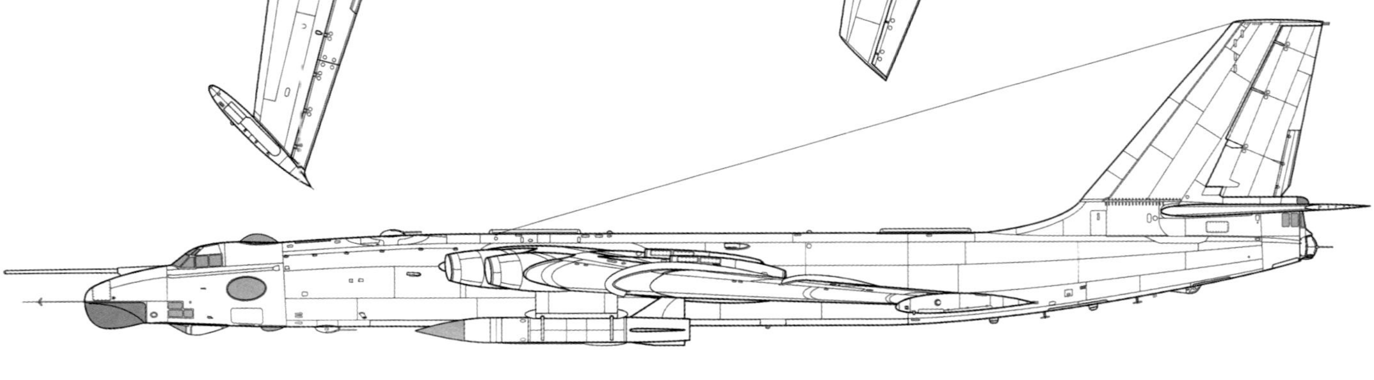

Side and lower views of a 3MS-2 tanker

The one-off 3M-5 missile carrier with KSR-5 missiles attached

Chapter 5

The *Bison* in Service

The Long-Range Aviation started getting to grips with the M-4 in 1954, as soon as the bomber entered production. The type first entered service with the newly established 201st TBAD (*tyazholaya bombardirovochnaya aviadiveeziya* – Heavy Bomber Division, roughly equivalent to a bomb wing [heavy] in USAF terms) headquartered at Engels, which had been formed on 4 September 1954 with Maj.-Gen. Serafim K. Biryukov as CO, reporting to the 50th VA (*vozdooshnaya armiya* – air army, roughly equivalent to a numbered air force in the USAF), headquartered in Smolensk. The division's first unit was the 1096th TBAP (*tyazholyy bombardirovochnyy aviapolk* – heavy bomber regiment, roughly equivalent to a USAF bomb squadron) at Engels-2 AB; the colocated 1230th TBAP was added in 1955. Each regiment was to comprise two squadrons of M-4s and a squadron of Tu-16s; the Badgers were meant primarily for proficiency training so as to save wear and tear on the *Bisons*.

The air force personnel got their first taste of the M-4 directly at plant No. 23 and at the Myasishchev OKB's flight test facility in Zhukovskiy, where conversion training initially took place. The first M-4 delivered to the air force arrived at Engels-2 AB on 28 February 1955, followed by the second aircraft on 2 March.

The new bomber looked impressive as hell, especially in comparison with the Tu-4, to say nothing of the wartime bombers, which many of the division's pilots remembered very well. At last the DA had an aircraft with true strategic capability! However, the pilots were quick to discover that the M-4 had its peculiarities.

M-4 '03 Red' (c/n 5302528) seen immediately after liftoff, with the gear just beginning to retract. Note the strake aerials on top of the centre fuselage and on the rear fuselage underside.

Left: An early M-4 coded '31 Red' passes over Moscow's Red Square with fighter escort during one of the May Day parades. Note the absence of star insignia on the fuselage, indicating that the photo was taken after 1955 (possibly in 1957).

Below left: The same bomber seen from below, with all four escorting MiG-17s *sans suffixe* in the picture

Above right: A formation of three M-4s makes a flypast during the Aviation Day parade at Moscow-Tushino airfield in 1956.

Right: Close-up of the lead ship in the formation

Below right: This view of M-4 '24 Red' (c/n 5301114) shows that the aircraft still wears pre-1955 insignia and huge tactical codes on both the nose and the fin.

Bottom right: Another view of the same trio as it departs Tushino after the flypast

The aircraft was demanding to fly and unforgiving of pilot errors, and the crew workload on takeoff was very high. Being accustomed to aircraft with a tricycle landing gear, the pilots had a tough time adapting to the bicycle-gear M-4, with its totally different takeoff technique. The ingrained habit of hauling back on the control column to initiate rotation, which was necessary on a tricycle-gear aircraft, proved harmful since it could cause the M-4 to pitch up excessively and stall; this necessitated revisions to the control system.

When an M-4 stalled and crashed on takeoff for the first time, an MAP and air force board hurriedly convened to deal with the bomber's dangerous behaviour. The devised remedy was to reduce the elevators' upward travel from 14°–15° to 6°–8°, apply special markings to the flight deck sides, and install special indicators on the control columns, helping the pilots check the elevator deflection angle. After liftoff, the aircraft was required to fly level at 5–10 m (16–32 ft.) for a few seconds until it accelerated to sufficient speed to gain full control authority.

Speaking of which, the landing gear was a constant source of problems in the early days. Occasionally the gear would collapse or refuse to extend. In one case, this resulted in a major incident when an M-4 captained by Maj. Yuriy P. Zasedatelev made a textbook wheels-up landing on the unpaved reserve runway at Chagan AB, near Semipalatinsk, Kazakhstan, at night (!). The aircraft suffered only minor damage and was returned to service.

(As an aside: Believe it or not, a while later, an M-4 captained by the same Yuriy P. Zasedatelev made a belly landing at Ookraïnka AB, in the Amur Region of the Soviet Far East, in identical circumstances. Again, the aircraft suffered only superficial damage, but the cause of the gear malfunction defied belief. The flight engineer (Lt. Gorbachov) had had an argument with the unit CO, claiming the landing gear would not deploy if even a single fuse in the electric circuitry was 'fried.' His immediate superior suspended him from flying, replacing him with another officer, but Gorbachov had managed to remove one fuse from the fuse box – just to prove his point. Well, prove he did, but the security service did not take long to identify the culprit. Lt. Gorbachov was court-martialled and charged with sabotage, which could well mean a death penalty; eventually he was sentenced to six years' imprisonment.)

The wide landing-gear track created problems in the winter – the outrigger struts were liable to damage during ground handling (pushback) if the taxiways were not cleared of snow to a sufficient width. The wheel brakes had an antiskid system, but this novel (and useful) feature was unreliable, with the wheels often locking on landing, with many a burst tyre as a result.

The insufficient AoA margin was equally dangerous on landing because reducing the AoA a bit too much would cause the nose to dip and the nose gear to touch down first, resulting in a bouncy landing. It was important to touch down at the very beginning of the runway and delay brake parachute deployment until the speed bled off to 130 km/h (80 mph); otherwise the parachutes could fail. Runway excursions were therefore not uncommon, since the pilots had trouble slowing the aircraft in time. Small wonder that the service pilots were wary of the M-4 at first.

The environmental-control system drew a lot of criticism – the crew cabin heating was unsatisfactory, the temperature dropping below zero at high altitude. The fur-lined flight attire actually made the problem worse in the summer – the crewmen would be drenched with sweat at low altitude and thus liable to catch cold at high altitude. A chronic bad cold was one of the M-4 crews' professional diseases!

The hydraulic actuators in the flight control system were unreliable. It should be noted that while the Myasishchev OKB had foreseen this and taken precautions by incorporating a manual-control mode, flying the M-4 in this fashion was no piece of cake. Even after the surviving M-4s and 3Ms had undergone a midlife update in 1957–58 to facilitate flying in manual mode, the forces on the aileron control wheels at 450–500 km/h (280–310 mph) exceeded 40 kgf (88 lbf). The artificial-feel mechanism in the control system also left a lot to be desired, the control forces being excessively high and causing pilot fatigue.

As if that weren't enough, the M-4's manufacturing standard was very variable; each aircraft had its individual peculiarities, which complicated conversion training. The aircraft gave the ground personnel a hard time too; for example, adjusting the control system was very labour intensive. Even such a simple thing as a wheel change on the main gear bogie required the aircraft to be trestled, using six huge jacks – and defuelled first.

A lot of surprised reaction, or even downright rejection, on the part of the pilots was caused by the wingtips flexing up and down in flight – sometimes a dozen feet or more. This aeroelasticity sometimes startled the crews. On one occasion, when the bomb armament was being tested during the evaluation period at Engels, the aircraft unexpectedly dumped the entire bombload instead of dropping just one bomb as planned; the force acting on the aircraft was so great that the wings bent, propelling the bomber upward like a springboard! All attempts to get to the core of the problem

on the ground failed, and the crew, being rather shaken, did not want to push their luck.

GK NII VVS chief navigator Col. Aleksandr N. Rekunov rushed off to Engels to take the matter in hand. After analysing the incident, he took the decision to fly a sortie. The heavily loaded M-4 took off, heading for the local weapons range. The first drop went normally, but on the second try the scenario was repeated – the bomber's wings creaked threateningly under the sudden stress as the entire remaining load of heavy-calibre bombs tumbled earthward. It was obvious that continuing the trials was too dangerous. The defect was eventually traced to a seemingly minor error – the wires in the bomb release control circuit had been cross-wired to the wrong connectors. This again proves the old adage: in aviation there are no trifling matters!

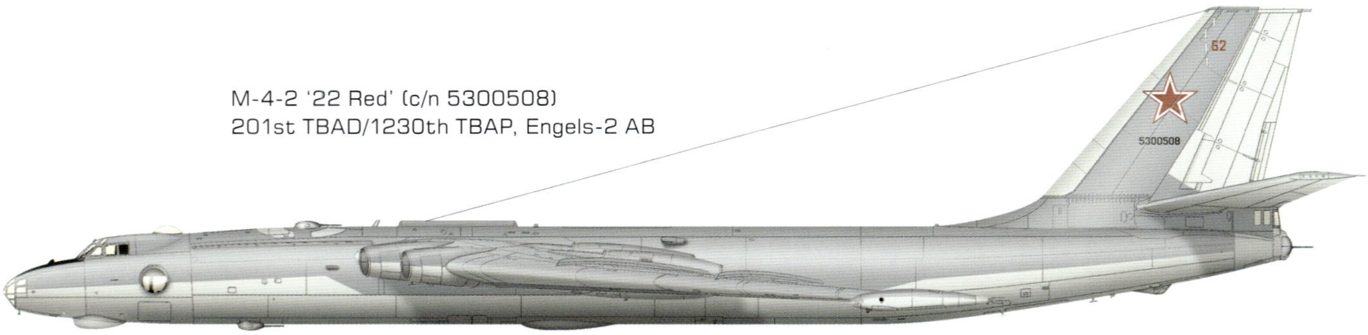

M-4-2 '22 Red' (c/n 5300508)
201st TBAD/1230th TBAP, Engels-2 AB

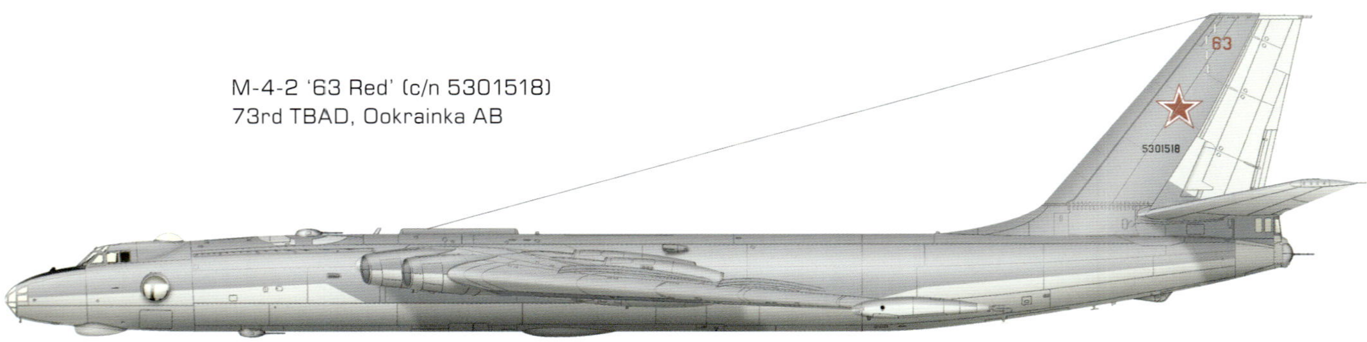

M-4-2 '63 Red' (c/n 5301518)
73rd TBAD, Ookrainka AB

Left: M-4 c/n 5300508 in its early days, still with no antiflash paint job and no tactical code; the c/n is still repeated on the nose.

Right: M-4-2 '62 Red' takes on fuel from a TZ-22 articulated refuelling bowser (note the hoses leading into the nosewheel well). The tanker wears an 'Excellent aircraft' maintenance award badge on the nose gear door.

Below right: Routine maintenance on an M-4; the maintenance platform is noteworthy. The placard in front reads, 'Aircraft under electric power. All work must proceed under the supervision of the technician in charge.'

As M-4 production continued, enough aircraft were available to equip a second division. In November 1956, 10 M-4s were ferried to the aforementioned Ookraïnka AB (a.k.a. Seryshevo) near Belogorsk and delivered to the 79th TBAP of the 73rd *Ternopol'skaya* TBAD (the latter was headquartered at the same base), reporting to the 5th VA headquartered in Blagoveshchensk (Amur Region). Both the division and the regiment had recently been transferred to Long-Range Aviation; the division's honorary appellation was awarded for its part in taking the town of Tarnopol' (later renamed Ternopol') during the Great Patriotic War. The unit's bombers became operational in early 1957.

The 3M achieved initial operational capability in 1957, the first production bombers arriving at Engels-2 AB in May. The introduction of the *Bison-B* in the 201st TBAD proceeded more quickly and smoothly; not only did the division have its previous experience with the M-4 to fall back on, but the 3M was much more refined. Still, the service entry of both the M-4 and the 3M was by no means trouble free and was accompanied by a good deal of accidents and incidents. For instance, on the *Bison-A* there were two cases of the elevator skin ripping away in flight; fortunately, on both occasions the crew managed to land the aircraft in one piece. The 3M proved to be just as prone to pitching up and stalling on takeoff if you weren't careful. Its new VD-7B engines brought problems of their own, and unscheduled engine changes were all too frequent.

Because of the M-4's teething troubles, by the end of 1957 the Myasishchev OKB apron at Zhukovskiy was clogged with up to 30 *Bison-As* requiring modifications for flight safety reasons. The situation was so grave that on 30 December 1957, the Council of Ministers issued grounding orders for the M-4 until all defects had been rectified. For example, the abovementioned elevator integrity problem was cured by altering the design and stiffening the elevator control rods.

Technicians make the necessary adjustments to the port tailplane of a 3MN.

On 31 December 1957, a second unit, the 40th TBAP, was established at Ookraïnka AB, reporting to the same 73rd TBAD; this new unit was equipped with M-4s and 3Ms.

In spite of the initial troubles, the crews rated the 3M quite highly. Many pilots who flew the *Bison-B* had fond recollections of the aircraft and its pleasant handling – the 3M was, to use an old cliché, a 'pilot's aeroplane.'

Pretty soon, the M-4s and 3Ms started venturing over international waters; these missions were a 'show of force,' demonstrating the Soviet Union's military power to the West. The *Bisons* and other Soviet bombers frequently 'harassed' NATO naval task forces and skirted the maritime borders of NATO nations and Japan; in so doing, they were habitually shadowed by NATO and Japanese Air Self-Defence Force fighters. Apart from the 'harassment' bit, these sorties involved an extremely important photoreconnaissance mission, including nighttime reconnaissance. To this end, the M-4 featured a tilting camera mount with AFA-42 day cameras or NAFA-MK night cameras aft of the mainwheel well, while a fairly large bay farther aft housed flare bombs.

M-4 operations placed high demands on the crews' skill. As a rule, each crew flew two 10-hour night sorties weekly, logging about 1,200 flight hours per annum. As often as not, the flights took the aircraft over the North Pole. An M-4 captain was usually in the rank of major and had a Pilot 1st Class skill rating; the copilot was

A 3MS-2 tanker is prepared for jacking up prior to a landing-gear check. Two of the six hydraulic jacks are visible here. Note the open engine cowlings and the engine numbers on the red-painted nozzle covers.

Another 3MS-2 undergoing maintenance, with a parked 3MD in the background. The domed fairing of the DB-33A dorsal barbette is removed, exposing the two AM-23 cannons and their ammunition belt sleeves.

in the rank of captain, with at least a Pilot 2nd Class rating. After flying the M-4 for three to five years, some pilots gave up and arranged for a transfer to the Tu-16 or quit flying altogether.

The *Bison*'s main mission was the destruction of key military and industrial ground targets. The 201st TBAD covered the northern theatre of operations, being meant to strike across the North Pole at the USA; the 73rd TBAD's area of responsibility was the Far Eastern TO. It should be noted, however, that production M-4s were overweight, and in the absence of IFR capability their range with a 5-ton (11,020 lb.) bombload released halfway through the mission did not exceed 8,000 km (4,968 miles). This meant the chances of making it back to own territory were rather slim; as a last resort, the crews were supposed to guide the aircraft to a designated ocean area and eject in the hope of being rescued by friendly ships or submarines.

Standard armament options included two free-fall *special munitions* (a euphemism for nuclear bombs), two 9,000 kg (19,840 lb.) FAB-9000 high-explosive bombs, or four 6,000 kg (13,230 lb.) FAB-6000 HE bombs. Smaller bombs, naval mines, or torpedoes could also be carried on cassette- or gantry-type racks. If one FAB-9000 was carried, the bomb bay also accommodated two long-range fuel tanks – the so-called group 12 tanks.

Speaking of nukes, originally the M-4 was meant to carry the RDS-4 *Tat'**ya**na* first-generation nuclear bomb (8U69 or

An armourer loads 23 mm ammunition into the ammo boxes for the DB-35AM tail turret via the brake parachute bay doors. The wooden crates are marked BZT (*brone**boy**no-zazhi**gah**tel'no-tras**see**ruyushchiy* – armour-piercing/incendiary/traced).

Left: A 3MS makes a banked turn, showing off the zero-dihedral tailplanes and the tandem ADF strake aerials on the fuselage. The aircraft may appear to be camouflaged but it's a trick of the light.

Right: A 3MN-1 bomber moments before touchdown with the flaps fully down. The VD-7B engines were characterised by their smoky efflux.

Below and below left: The captain of a 3M clad in typical flight gear of the period, including a leather helmet.

izdeliye 244N), with a yield of 42 kilotons. RDS supposedly stood for *re'aktivnyy dvigatel' spetsiahl'nyy* ('special reaction engine' or 'special jet engine'), a name used for the first Soviet nuke for security reasons. Later this acronym generated alternative explanations, such as *re'aktivnyy dvigatel' Stalina* ('Stalin jet engine'), *Rosseeya delayet sama* ('Russia makes [the atomic bomb] on her own'), etc. Subsequently the relatively compact 3-megaton RDS-37 bomb (*izdeliye* 37D) was regarded as the main weapon; however, it was handicapped by a whole set of imperfections, including a limited shelf life.

In 1961 the aforementioned 71st Test Range (the nuclear test outfit based at Bagerovo AB) organised three detachments of Tu-16, Tu-95, and 3M bombers, which participated in aerial nuclear tests at the proving ground on the Novaya Zemlya ('New Land') archipelago, located in the Arctic Ocean not far from the Kola Peninsula. Three such tests were held successfully within 13 days (20 September–2 October 1961); two detachments of 3Ms participated in the 2 October test. Five more aerial nuclear tests were held at in Novaya Zemlya in 1962, the aircraft operating from Olen'ya AB, a bomber base near Olenegorsk (Murmansk Region) on the Kola Peninsula.

In so doing, special tests were held to determine the effect of the nuclear blast on the aircraft; to this end, the bombers were outfitted with up to 40 sensors, 30 instruments, and five to seven oscillographs. This instrumentation made it possible to monitor and control the nuke's systems (en route to the target and after release) and monitor the temperature in the environmentally controlled bomb bay. The aircraft's speed and altitude were recorded; the bomb's descent and the ensuing mushroom cloud were photographed and filmed. Measurements were made of the engines' operating parameters and the airframe deformation when the blast wave of the explosion caught up with the aircraft, as were the bomber's stability and control characteristics at this moment; the temperature of the aircraft skin when exposed to the flash of the explosion was measured to check the efficacy of various protective coatings.

Right: An air-to-air of a 3MN-1 (note the rectangular orifices on the engine cowlings identifying the VD-7B engines). Interestingly, the aircraft lacks the usual 'antiflash' colour scheme, only a small section of the belly being painted white.

Below: A 3MN-1 or 3MS-1 completes its landing run with the brake parachutes deployed.

New conventional munitions were also envisaged, including armour-piercing and concrete-piercing bombs. The hitherto current 1946-model and 1950-model bombs with designations suffixed M46 and M50 were supplemented by the latest 1954-model bombs with designations suffixed M54; these included thick-walled HE bombs with increased penetration capability, HE/fragmentation, and incendiary bombs.

Now we come to the big question: Could the *Bison* penetrate the air defence system of the 'potential adversary' (NATO)? The numerous radar pickets, SAM sites, and supersonic interceptors stationed in western Europe and North America left few chances of getting through, even in a first-strike scenario. The six cannons of the M-4 would hardly give it enough protection. The only hope the bomber's crew had lay in the ECM suite and the armour plating protecting the crew against shell fragments.

The notorious shoot-down of Francis Gary Powers's high-flying Lockheed U-2A spy plane (56-6689) by an S-75 SAM (NATO code name SA-2 *Guideline*) near Sverdlovsk on 1 May 1960 had made it obvious that strategic bombers were no longer invulnerable at high altitudes. However, it soon became clear that their perceived extreme vulnerability to enemy air defences was somewhat exaggerated. The USA quickly devised an 'anti-SAM tactic' – from 1962 onward, the B-52s were supposed to penetrate the enemy's SAM barriers at low altitude, where the missile guidance radars could not track them. Exercises showed that low-level/ultra-low-level penetration could give a fair chance of success. The Soviet Union

answered in kind; between January and August 1964, a research programme was held to investigate the possibility of maximum-range flights of Tu-95 and 3M bombers at 50–200 m (164–660 ft.) in the daytime, when obstacles and terrain were visible, and at 200–300 m (660–980 ft.) at night. This low-level tactic significantly increased the chances of air defence penetration by single aircraft (formation flying at low level was ruled out – for safety reasons, if nothing else). Of course, the turbulence at low altitudes imposed additional strain on the airframe, and this was where the Myasishchev bomber had an advantage over the Tu-95 thanks to its flexible wings, which could cope better with the turbulence. Another important advantage was the *Bison*'s lower radar cross-section in the frontal aspect because it was jet powered; the Tu-95's propellers markedly increased its radar signature.

However, these low-level flights never got further than the experimental stage; operational use of this tactic required the aircraft (including the 3M) to be modified in order to withstand the turbulence at low altitude. MAP declined to undertake this work, which eventually had to be performed by the Ministry of General Machinery (MOM). The Council of Ministers tasked Viktor N. Bugaiskiy, who headed Branch 1 of MOM's OKB-52, with getting the 3M low-level combat operations programme included into the task list of KB-90, headed by V. Gusarov; the work was to proceed in January–February 1966. The bomber's range on a 'hi-lo-hi'

Above: 3MS-2 '85 Red' (c/n 7300504) shows the tanker's characteristic fuselage illumination lights, as well as the lack of the bomb bay door bulge identifying the 3MS-2s modified as nonconvertible tankers.

Above left: A 3MN-1 comes in to land, leaving a characteristic smoke trail. This view shows the position of the nose and main gear bogies in no-load condition.

Left: An air-to-air of '52 Red,' a 3MN-1 bomber powered by VD-7B engines; again, this particular aircraft lacks full 'antiflash' colours on the fuselage.

Right: A 3M drops a FAB-9000 M54 bomb. The huge size of the stars is noteworthy.

Left: This 3MS-2 tanker coded '21 Red' retains the ability to be reconverted into a bomber, as indicated by the unmodified bomb bay doors. The underside of the inner wing torsion box is unpainted; note also the smaller star insignia.

Below: 3MS-2 '90 Red' (c/n 7300505) with the landing gear in midretraction

Right: Seen through the port beam blister of a 3MS-2, a 3MN-1 bomber approaches the tanker.

Below right: Here the 3MN-1 is seen from the tail gunner's station as it takes on fuel.

mission profile decreased dramatically, but this was an acceptable price for the reduced vulnerability during air defence penetration.

Another point to be considered is that even though the M-4 and 3M had their reliability issues, they were far more reliable than contemporaneous intercontinental ballistic missiles. Also, unlike ICBMs, strategic bombers were highly mobile, having greater mobility than the ballistic missile submarines that were just coming on the scene; not only could they be rapidly redeployed, but they could be assigned a new target in midmission if the need arose.

Well, actually the *Bisons* were initially authorised to use only a handful of bases with long runways – Engels-2 AB, Ookraïnka AB, Dyagilevo AB (near Ryazan', in central Russia), Siauliai (pronounced '***Sha****ooliay*'; in Lithuania), and Chagan AB (in Kazakhstan). Incidentally, Dyagilevo AB hosted the DA's 43rd TsBP i PLS (*Tsentr boye****voy*** *podgo****to****vki i pere****oo****chivaniya* ***lyot****novo sos****tah****va* – Combat Training & Aircrew Conversion Centre), which was soon assigned responsibility for M-4 crew training. However, a technique was devised of taking off with a reduced fuel load that allowed the bombers to be dispersed to virtually any Soviet air base if the possibility of a preemptive strike loomed. The service pilots quickly discovered that with a reduced TOW, the M-4 really leapt into the air.

Here we have to go back a little. In 1960, Long-Range Aviation was reorganised. The 50th VA was transformed into the 6th OTBAK (*otdel'nyy tyazholyy bombardirovochnyy aviakorpus* – Independent Heavy Bomber Corps, HQ Smolensk), while the 5th VA became the 8th TBAK (HQ Blagoveshchensk). These retained control over the 201st TBAD and the 73rd TBAD, respectively.

In the autumn of 1960, OKB-23 was disbanded; yet, its progeny continued flying and scaring the Soviet Union's 'potential adversaries.' 3MSs and 3MNs flew long-range reconnaissance missions over international waters on a regular basis, shadowing NATO warships, and participated in military exercises right up to the 'ministerial' level (that is, those commanded by the Soviet minister of defence). To fool the West into thinking the Soviet air force had more *Bisons* than it actually had, the tactical codes carried on the fin and the nose gear doors were changed from time to time (codes in the 1* and 2* ranges were replaced by codes in the 6* and 7* ranges). In reality, however, the VVS suffered from a shortage of these heavy aircraft.

According to the Air Force, *Bison* accident attrition was caused mostly by serious design shortcomings, which came to light in the course of operational service. In all, 31 of the 125 M-4s and 3Ms *et seq.* built were lost in accidents (see appendix 3). True, the Myasishchev OKB, which was reborn in 1967, did its best to improve the operational safety and reliability of the M-4 and 3M by making changes to the flight control system, the gust locks, the fire suppression system, and the ejection seats and by improving the crews' working conditions.

Tanker Operations

As the 1950s drew to a close and a new decade began, certain factions in the VVS and the aircraft industry grew increasingly vocal, questioning the advisability of having two heavy bomber types in the DA's inventory, especially now that the Tu-95KM *Bear-C*, an IFR-capable missile carrier, was available. This sparked the decision to convert a substantial proportion of the *Bison* fleet for the IFR tanker role; these tankers would support the operations of the Tu-95KM and the remaining 3MS-1 and 3MN-1 bombers. The conversion programme began in 1961 as a joint effort by aircraft factory No. 23, the air force's ARZ No. 360 (*aviaremontnyy zavod* – aircraft repair plant) at Dyagilevo AB, which refurbished the type, and the tech staff of the 201st TBAD at Engels. Although the conversion jobs were a bit crude, the tankers showed an adequate flight safety record and fairly high performance – they could transfer up to 46 tons (101,400 lbs.) of fuel while having a combat radius of 5,500 km (3,420 miles).

In 1961 the M-4s of the 73rd TBAD were withdrawn from the 40th TBAP and 79th TBAP and, after being converted to M-4-2 tankers, were pooled in the newly established 251st OAESZ (*otdel'naya aviaeskadril'ya samolyotov-zaprahvshchikov* – Independent Aerial-Refuelling Squadron), which was likewise stationed at Ookraïnka AB. It had the unofficial name *Letoochiye myshi* (Bats). It may as well be said now that the squadron proved to be short lived – it was disbanded in 1965 and the tankers were distributed between the bomber regiments. Another M-4-2 squadron, the 179th OAESZ of the 6th OTBAK, was deployed at Siauliai and nicknamed *Normahndiya* (Normandy).

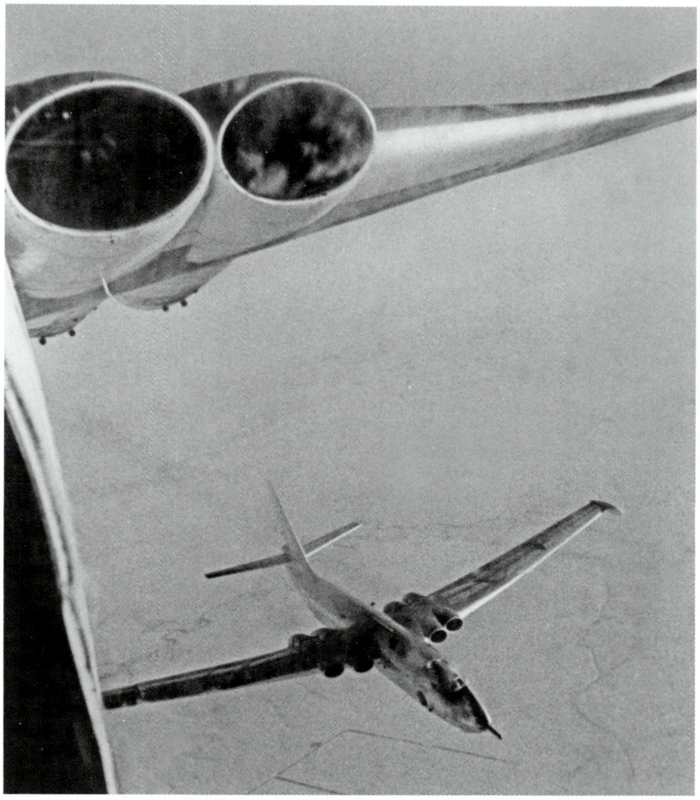

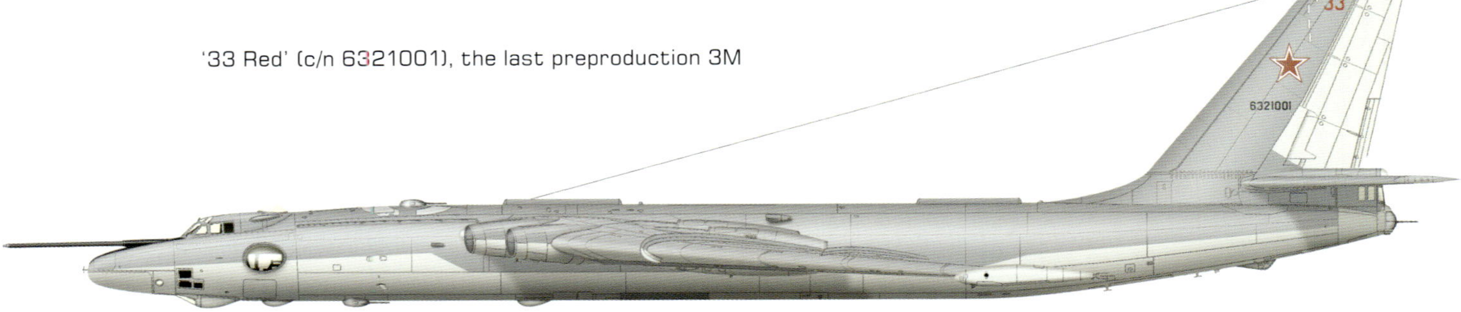

'33 Red' (c/n 6321001), the last preproduction 3M

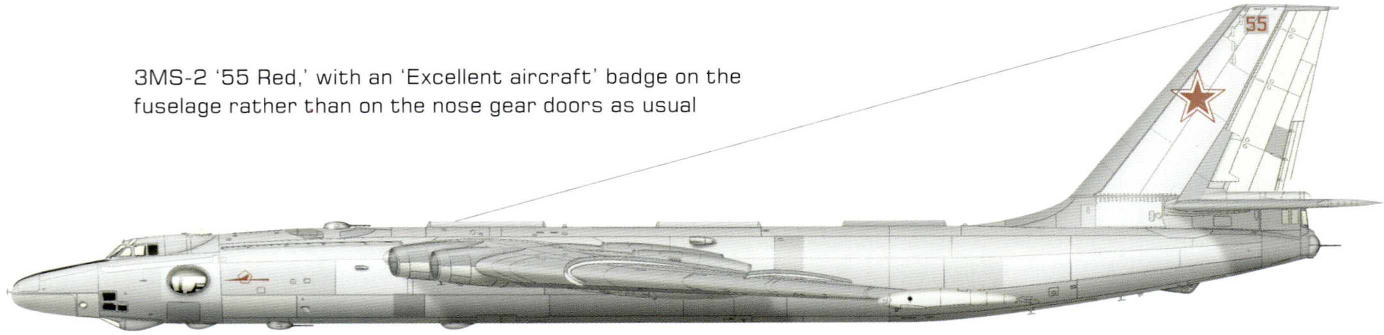

3MS-2 '55 Red,' with an 'Excellent aircraft' badge on the fuselage rather than on the nose gear doors as usual

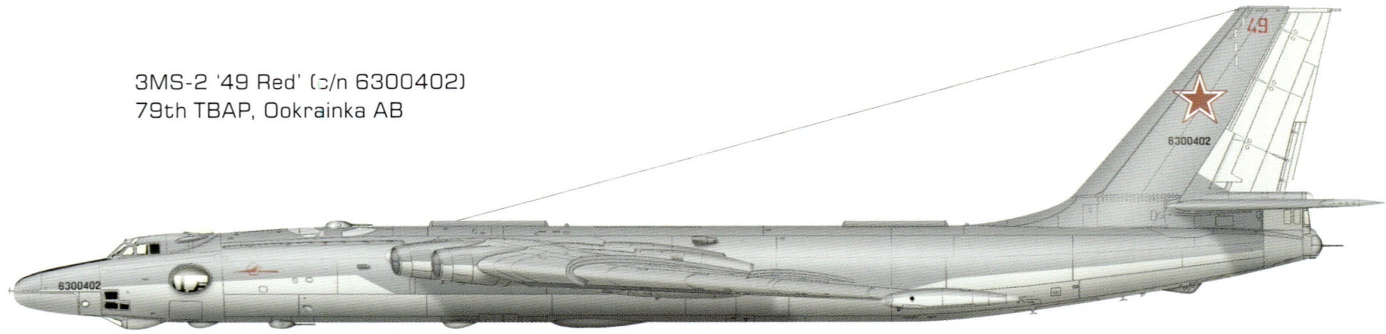

3MS-2 '49 Red' (c/n 6300402)
79th TBAP, Ookrainka AB

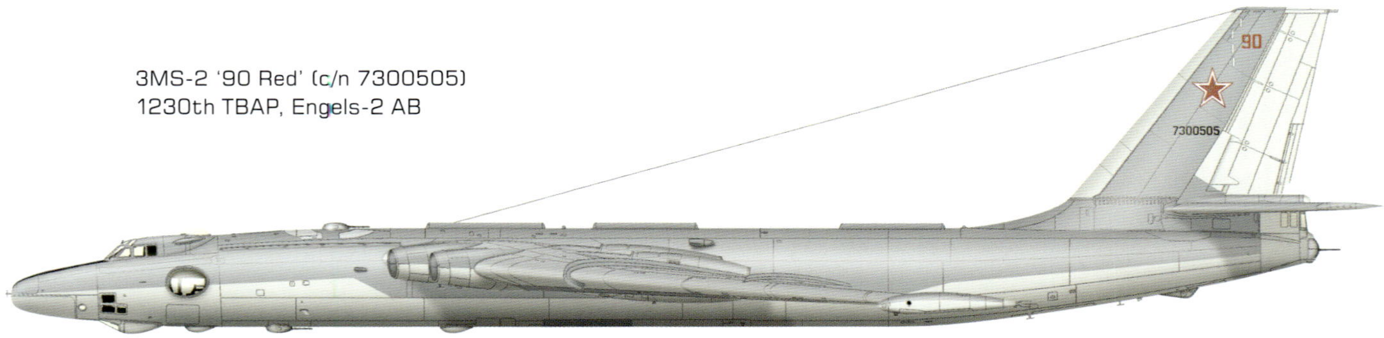

3MS-2 '90 Red' (c/n 7300505)
1230th TBAP, Engels-2 AB

Siauliai also hosted a fighter unit, and the fighters kept getting in the way of the M-4s, whose pilots kept complaining about having 'these cigarette butts underfoot.' Of course the fighter pilots took offence, promising they would 'shoot down' the M-4s during the next exercise. On one occasion, detecting an intercepting fighter in timely fashion during such an exercise, the gunner of an M-4 alerted the captain; when the fighter had closed in for a simulated cannon attack, the M-4 made an abrupt manoeuvre and the fighter hit the bomber's wake turbulence, entering a spin!

In contrast to the M-4, all *Bison-B*s and 3MDs had IFR capability as standard. The first in-flight refuelling of an in-service 3M was performed as early as 1 March 1959; the bomber was captained by Maj. Nazimkin. By the end of the year, the crews of the 201st TBAD had successfully performed 303 fuel top-offs; in June 1960 a 3M captained by Maj. Bulavin flew a maximum-range mission with a single fuel top-off, remaining airborne for 17 hours, 25 minutes.

To be honest, DA crews had a bit of trouble mastering the probe-and-drogue IFR technique, which turned out to be not that

Top: Front view of a 3MS-2 operated by the 1230th APSZ at Engels-2AB

Above: 3MS-2 '26 Red' rests on the hardstand at Engels-2AB in the evening light.

Right: The tankers of the 1230th APSZ await the next mission on a bleak winter's day. Note the obligatory fire extinguisher placed in front of the nearest aircraft; not that it would be much use in the event of a major fire, but this is standard operational procedure.

Above: A mix of 3MS-1s and 3MS-2 on the flight line at Engels; '56 Red,' in the foreground, represents the bomber version.
Below: Preproduction 3M c/n 6320801 was delivered to the VVS, eventually becoming 3MN-2 '24 Red,' seen here landing at Engels.

Above: Another tanker returns to base after a sortie.
Below: 3MS-2 '15 Red,' landing at Engels, shows the new cruciform brake parachutes that replaced the original circular ones.

Left: '83 Red,' another 3MS-2 tanker based at Engels

Below left: A somewhat unusual perspective of 3MS-2 '14 Red,' giving details of the outrigger strut fairing and the outer wing underside

Right and far right: This is how the pilots of the Tu-95 saw the Myasishchev tanker. Note the six lights around the circumference of the drogue used at night.

Far right, below: View from the tanker's port blister, showing a 3MD being refuelled

simple after all. On occasion, bomber crews returned to base without making contact, reporting at the debriefing that they had followed the prescribed 'approach, point, and shoot' procedure, but the probe simply would not lock in to the drogue. Careful analysis of the missions, however, showed that the pilots had been overcautious, 'firing' the probe at excessively long range. Small wonder it did not lock into place!

Indeed, tanker operations were so intensive in those days that the rubber industry could not supply enough replacement fuel transfer hoses for the HDUs as they ran out of service life. The VVS command had no choice but to extend the hoses' service life; at times this led to dangerous incidents involving Tu-95s being refuelled. On at least three occasions the hose broke and became entangled around the receiver aircraft, nearly causing it to lose control. One such hair-raising episode occurred on 24 June 1986, when a 1006th TBAP Tu-95MS *Bear-H* missile carrier from Uzin AB coded '42 Red' and captained by Maj. Aleksandr G. Firstov was being refuelled by a 3MS-2 tanker. The rotten hose broke at the drum of the HDU, whipping around the bomber; 'We've had it,' the captain thought. Flailing wildly in the slipstream, the hose

repeatedly struck the elevators, making the aircraft pitch up and down; Firstov could barely keep it under control. All at once the seesawing subsided; the tail gunner reported that the end of the hose had become wedged between the cannon barrels. With all due care, the pilots brought the bomber home for a safe landing. Similar incidents occurred in 1973 and 1983.

There was a case when the drogue was ingested into the propeller of a Tu-95KM, ruining it completely; the bomber barely made it back to base. On another occasion, when the Tu-95KM pilots lost concentration, the receiver aircraft overshot and collided with the tanker, losing part of the vertical tail and seriously damaging the tanker; luckily, both aircraft landed safely.

Bison-A/B tankers were used a lot to support long-range operations of the DA's Tu-95s in the 1970s and 1980s as well. Col. Vladimir V. Yegorov (Ret.), a former Long-Range Aviation pilot who flew various versions of the Tu-95, recalled that in 1980 he served with the 1023rd TBAP at Semipalatinsk as a Tu-95 copilot. *'That was when my first exercise in division strength began. In all, 38 out of 40 Tu-95s operated by the division's two regiments were involved (the other unit – the 1226th TBAP – operated Tu-95K and Tu-95KM missile carriers). In the course of the exercise the crews of both regiments performed in-flight refuelling, practice launches of Kh-20M missiles, and bombing attacks with all compatible conventional bombs at two target ranges (in Kazakhstan and in the Far East). The missions spanned over ranges in excess of 10,000 km [6,213 miles], lasting up to 14 hours. It has to be said that this was the normal scenario for a division-strength exercise in the DA at the time; each regiment would put up 18–20 aircraft – that is, all available machines. This characterises the Soviet Long-Range Aviation's operational readiness and state in those days.*

In addition to proficiency training and live weapons training, the crews of our division routinely flew signals intelligence (SIGINT) missions along the borders of the potential adversary – the NATO nations – and tracked the movements of US Navy carrier task forces in the Pacific, using our SIGINT, photo, and radiation reconnaissance equipment. As a rule, such missions

were flown by a Tu-95MR-2 [reconnaissance aircraft] accompanied by a Tu-95KM from our "brother unit," the 1226th TBAP. They included refuelling from M-4-2 or 3MS tankers [3MS-2 – *Auth.*] *on the outbound and inbound legs; the tankers were operated by the 73rd TBAD / 401st TBAP, based at Blagoveshchensk* [*sic*; the unit was actually the 40th TBAP! – *Auth.*). *Such reconnaissance missions could take as long as 24–27 hours; each crew performed three or four such ultralong missions annually.'*

The IFR-capable Tu-95MR-2 *Bear-E*s routinely flew SIGINT/PHOTINT sorties along the borders of NATO nations and over the Pacific and Arctic Oceans, shadowing US Navy carrier task forces led by the aircraft carriers USS *Midway* (CV-41) and USS *Constellation* (CV-64). Vladimir V. Yegorov recalls: *'The climax of the mission was the moment when the aircraft broke through the overcast after cruising for hours on end, and we would suddenly find ourselves over a convoy comprising an aircraft carrier escorted by 10 or 12 other warships. After this we would photograph the ships at low altitude, circling over the convoy for 20–30 minutes; then we climbed to 10,000 m [32,810 ft.] or so and headed for the spot where we were to rendezvous with the tanker. After topping off the tanks with 35 tons [77,160 lbs.] of fuel we headed back to base, covering the remaining 5,000 km [3,100 miles] to Semipalatinsk almost in a straight line in autopilot mode.'*

M-4-2 tankers (or, in later years, 3MS-2/3MN-2 tankers) were also used to support the operations of the *Bear*'s naval versions – the Tu-95RTs *Bear-D* maritime reconnaissance/over-the-horizon targeting aircraft (*razvedchik-tsele'ookazahtel'* – reconnaissance/target designator [aircraft]) and the Tu-142 *Bear-F* ASW aircraft. Having no hose-and-drogue tankers of its own, the Soviet Naval Aviation (AVMF) cooperated with the Air Force, which supplied the tankers. On 16 April 1967 two *Bear-Ds* from the North Fleet Air Arm's 392nd ODRAP (*otdel'nyy dahl'niy razvedyvatel'nyy aviapolk* – independent; that is, direct reporting long-range reconnaissance air regiment) flew the first reconnaissance sortie involving in-flight refuelling; they were captained by Lt.-Col. Ivan F. Gladkov (the unit's CO) and Lt.-Col. V. I. Khayarov. Probe-and-drogue IFR proved to be as demanding for the naval airmen as it was for their air force colleagues – it was no easier than the wing-to-wing technique used by the Tu-16. By June 1968, ten 392nd ODRAP crews had received their IFR procedure ratings. In 1969

Left: A Soviet bomber pilot in winter attire, complete with heavy fur-lined boots, poses for a publicity photo with a 3MD. The latter is apparently coded '25 Red' (ladders and other support equipment were often marked with the aircraft's tactical code to discourage theft). Note the drop tank hardpoints between the engines; the drop tanks were never carried by in-service 3Ms.

Below left: A 3MD negotiates a turn on the taxiway.

Right: A *Bison-C* makes a smoky take-off. The 3MD is powered by the same VD-7B engines as the 3MN, hence the sooty efflux.

Right: A 3MD 'hits the tanker'. . .

Below right: . . . and is doused with residual fuel as it disengages from the drogue due to a faulty valve.

Far right: Another *Bison-C* is refuelled before a sortie. This 'bird's-eye' perspective shows well the 3MD's characteristic 'cheekbones.'

Far right, below: The crew of a 3MD commanded by A. N. Markov

Left: A typical Soviet-era publicity shot, showing the crew of a 3MD conferring before a nighttime winter sortie. The pilots are studying a map that the navigator has produced from his 'brain bag.'

Below and below left: Ground crewmen do a quick check on the same aircraft. Note that on the 3MD, the outer landing lights are housed in the outrigger strut fairings, not in the wing roots as on all earlier variants.

Above right: An atmospheric night shot of a 3MD, showing the sculpted radome and the buried drogue illumination lights flanking the IFR probe

Right: Another 3MD in a similar setting. Again, the aircraft's tactical code ('47 Red') is painted on the crew access ladder.

the unit began flying routine reconnaissance sorties over the Atlantic Ocean involving in-flight refuelling.

In addition to its primary mission, the Tu-95RTs was used to support the Soviet space programme. The 392nd ODRAP *Bear-Ds* were called on to locate Soviet military space vehicles that splashed down in the Indian Ocean after reentry, transmitting their coordinates to Soviet navy ships; it was imperative that the latter reached and retrieved the space capsules before the US Navy did. Such missions were code-named ***Ellips*** (ellipse); their extreme range and duration necessitated support by *Bison-A/B* tankers. For demonstration purposes, on 1 August 1968 two *Bear-Ds* captained by Ivan F. Gladkov (by then promoted to colonel) and squadron commander Maj. A. I. Startsev flew a 20-hour reconnaissance mission over the Indian Ocean, refuelling over Iranian territory.

The first real-life Ellips mission took place on 21 September 1968 and was led by Col. Ivan F. Gladkov; the vehicle to be detected was a Soviet lunar probe. Eight *Bears* took off from Mozdok (in what was then the Chechen-Ingush Autonomous SSR) in pairs at one-hour intervals; a 3MS-2 tanker would take off first, followed two minutes later by a Tu-95RTs. Knowing the space probe's estimated time of arrival (that is, the scheduled recovery parachute deployment time), the first pair's departure was scheduled so that the Tu-95RTs would be on location at the southernmost point of the search area, which was about 10 hours' flight time from Mozdok. After takeoff, each pair headed for Tbilisi, Georgia; thence the aircraft proceeded to Djul'fa, a town on the Iranian border, where the international airway exiting Soviet airspace began. Next, the pair made a port turn and headed for Tehran and then via Mashhad, Birjand, and Zahedan to the coast of the Indian Ocean. The refuelling procedure began over the coast; by then the *Bear* had used up about 20 tons (44,090 lbs.) of fuel – the maximum amount that the tanker could give and still make it back to base. The procedure took 10-12 minutes, then the tanker headed home and the *Bear* continued its flight all alone over the ocean – first in darkness, then in daylight, and then again in darkness. All in all, 392nd ODRAP crews flew more than 10 such missions from Mozdok.

Another 'customer' for the tankers was the Air Force's 67th Independent Airborne Early-Warning Squadron, based at Siauliai and operating a handful of Tu-126 *Moss* AEW aircraft.

The tanker crews' job was not easy either. On one occasion, a 3M bomber had trouble making contact with an M-4-2 tanker, and when the pilots finally succeeded and topped up their tanks, the tanker was much farther away from home – and with much less fuel remaining – than planned. To reach the base, the M-4-2 needed to fly as high as possible because the fuel consumption was lower at high altitude. Yet, the HDU jammed when the hose had been rewound only halfway; the emergency guillotine cutting the hose also malfunctioned. This meant the bomb bay doors could not be closed, increasing drag and hence fuel consumption.

Barely making it back to base, the tanker ran into a storm front on the landing approach; the turbulence subsided only just before touchdown. As the aircraft rolled along the tarmac, the crew was startled by a sudden loud bang inside the fuselage. It turned out that the HDU had been blocked by ice building up on the drum; now this ice had melted, and the HDU, which was still set to rewind, had pulled in the drogue with a thump!

Left: On 9 July 1967 the 3MD had its public debut when an example coded '49 Red' was in the static display of the air show held at Moscow-Domodedovo airport. Note the data placard in front of the nose gear unit.

Right: A small crowd of spectators flock around the 3MD at the 1967 Domodedovo air show.

Below right: Another view of '49 Red' at Domodedovo before the really big crowds came in. The 3MD almost lacked the 'antiflash' white colouring of the fuselage underside characteristic of the *Bison-A/B*.

In another incident, on 29 February 1972 the port observation blister of a 3MS-2 captained by Maj. A. G. Ignat'yev cracked and failed at 11,000 m (36,090 ft.), causing instant decompression of the flight deck. The pilots, who had neglected to don their oxygen masks, passed out due to hypoxia, and the tanker spiralled earthward with left bank, out of control. After trying repeatedly but unsuccessfully to call the pilots over the intercom, the tail gunner ejected at 4,800 m (15,750 ft.), believing that a crash was imminent. The flight engineer saved the day; putting the oxygen masks on the pilots, he helped them regain consciousness and take control. Only after landing did they discover that one crewman was missing!

On the Brink of War

In 1962 the VVS inventory included seventy-eight 3M-whatever bombers and 25 M-4-2 tankers – 103 of the 125 *Bisons* built. Most of them had by then been in service for more than six years, exceeding their provisional service life of 1,000 hours. On 15 April 1962, Pyotr V. Dement'yev, chairman of the State Committee for Aviation Hardware (GKAT – *Gosu**darst**vennyy komi**tet** po aviatsi**on**noy **tekh**nike*) and former minister of aircraft industry, reported thus to the Communist Party Central Committee: *'Due to the air force's demands to ensure safe operation of these aircraft up to 1970, it is advisable to perform certain design work and modifications aimed*

3MD '25 Black'

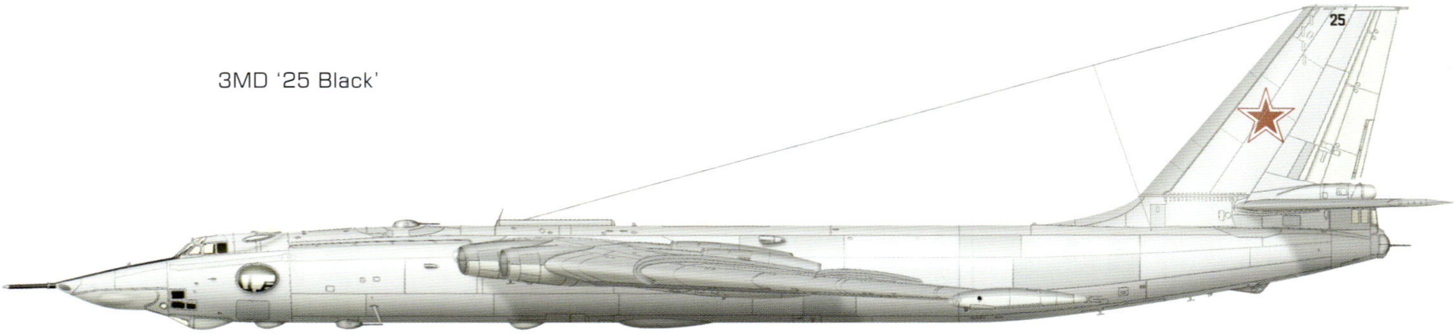

at keeping the 3M and M-4 aircraft up to current operational and tactical standards. To this end, the KB-201 design bureau responsible for the 3M and M-4 aircraft has now been formed at the branch office of OKB-52. The plant named after [Mikhail V.] Khroonichev [that is, the former aircraft factory No. 23 – *Auth.*] *has been assigned responsibility for all manufacturing work associated with 3M and M-4 operations and conversion to refuelling tankers.'*

(Note: In December 1957, MAP was 'demoted' to a state committee along with several other ministries due to changing government policies during the Khrushchov era. In 1965, however, their names and status were restored after Nikita S. Khruschchov had been unseated and replaced by Leonid I. Brezhnev as the Soviet leader in 1964.)

Above left: A 3M coded '53 Red' cruises at high altitude.

Left: A technician replaces the air intake covers of a 3MD (again, duly marked with the engine numbers); note the taxi light on the starboard nosewheel well door.

Above right: A 3MD coded '56 Red' is prepared for a sortie on a sunny winter day. There seems to be an accumulation of snow on the fuselage after a recent snowfall.

Right: Before the introduction of the 'Excellent aircraft' badge (a stylised aircraft silhouette superimposed on the 'State Quality Mark' pentagon), the maintenance award for faultless operation was presented verbatim in the form of the legend *Otlichnyy* (Excellent), as seen on 3MD '55 Red' here.

In the meantime, the brief period of détente between the East and the West was succeeded by a new round of confrontation. Despite the failure of the Bay of Pigs invasion (Operation *Zapata*) between 14 and 19 April 1960, the USA did not give up the idea of toppling the Cuban Communist government of Fidel Alejandro Castro Ruz. No longer relying on mercenaries and Cuban 'freedom fighters' backed by the CIA, the US administration was planning a full-scale military operation. Knowing that the small Cuban Revolutionary Armed Forces were no match for the US war machine, Castro requested military assistance from the Soviet Union. An agreement was soon reached whereby a 40,000-strong Soviet military contingent was deployed on Cuba, including three regiments of R-12 (*izdeliye* 8K63, NATO SS-4 *Sandal*) intermediate-range ballistic missiles and two regiments of R-14 (*izdeliye* 8K65, NATO SS-5 *Skean*) IRBMs. It was these weapons systems that triggered the infamous Cuban Missile Crisis of September–November 1962, which put the world on the brink of the Third World War.

Faced with the increasing probability of a hot war with the USA, the Soviet MoD regrouped its forces hastily. Among other things, the 1096th and 1230th TBAPs, making up the 201st TBAD, were redeployed to a forward-operating location at Siauliai in order to reduce their target approach time. This was the only occasion when 3M bombers on quick-reaction alert (QRA) duty actually sat armed with nuclear bombs, ready to take off at short notice. (In contrast, USAF B-52s prowling along the Soviet borders always had nukes on board in those days.)

Two squadrons of each regiment were to strike at the continental US; the third squadron was equipped with hastily outfitted 3MS-2 tankers to support the operation. The bomber crews were issued sealed flight bags with route maps indicating their targets, and sealed envelopes with the nuclear bombs' activation codes.

Top: Ground crewmen use an SPO-15 boom lift on a ZiL-164 4×2 general-purpose lorry chassis to inspect the vertical tail of a 3MD.

Bottom: A 3MD is waiting to be refuelled by two TZ-22 fuel bowsers.

Opposite: A pair of FAB-3000 M46 bombs on their bomb trolleys stand ready for loading into a 3MD.

They never got to see the nukes proper, accepting the already bombed-up aircraft with security seals on the bomb bay doors. The Soviet air force's tactical-aviation units and the DA's Tu-16P *Badger-J* ECM aircraft (P = *posta**nov**shchik* ***pomekh*** – ECM aircraft) were to help the bombers get unscathed past the NATO bases in western Europe, but later the *Bisons* would be on their own. The airmen were subjected to immense psychological strain, knowing that in the event of a nuclear war their own families would surely perish. Still, morale was high; 'Do what you have to do, and come what may!'

However, realising that an all-out war with the USA would be a war of destruction, the Soviet leaders had the common sense to back down and seek a political solution. On 26 October, Nikita S. Khrushchov acknowledged the deployment of Soviet missiles to Cuba (which he had hitherto denied) but stated they would never be used against the USA. He offered to US president John F. Kennedy to withdraw all Soviet offensive weapons from Cuba in exchange for a pledge of nonaggression against Cuba; he also demanded that the US ballistic missiles deployed in Turkey, which were causing great concern of the Soviet government, be withdrawn. On 27 October, JFK replied that the USA guaranteed nonaggression and that the blockade of Cuba would be lifted on condition that the withdrawal of the Soviet offensive weapons be monitored by the United Nations. A day later, the Soviet gov-

ernment accepted this offer, making an appropriate statement. Between 5 and 9 November, all 42 IRBMs were shipped back to the USSR, and the launchpads were demolished. Hence the state of heightened alert was cancelled, and the bombers of the 201st TBAD returned to their home base in Engels – much to the relief of everyone concerned.

Starting in the summer of 1962, the surviving M-4s had their outer wings reinforced as a precaution against fatigue cracks (several cases of fatigue cracking had been discovered on operational aircraft). The design work was performed by a team of 10 KB-201 employees under deputy chief designer Yakov B. Nodelman (consider that the new 'Bison Bureau' had a total of only 30 people on its payroll). Obtaining an M-4 airframe, TsAGI held a new series of static tests. Together with the VVS, the engineers set up a system of monitoring the current condition of operational *Bisons*, detecting fatigue cracks as they appeared. The tests at TsAGI and SibNIA were backed by theoretical research, which made it possible to pinpoint the places in the highly stressed airframe structure that were most prone to fatigue damage during lengthy operation; this allowed the aircraft's service life to be appreciably extended in the long run. Later, Vladimir M. Myasishchev and his team were awarded honorary medals by the VDNKh fairground

Above: 3MS-2 '80 Red' (c/n 6300401) is poised for takeoff at Engels-2 AB.

Below: A 3MN-2 tanker (note the slits in the engine cowlings, characteristic of the VD-7B engines) flies over the Volga River.

Above: 3MN-2 '55 Red' sporting the 'Excellent aircraft' badge heads a long row of sister ships on the flight line at Engels.

Below: 3MS-2 '83 Red' (c/n 7300501) on a rain-soaked hardstand at Engels-2 AB

(*Vystavka dostizheniy narodnovo khoziaystva* – National Economy Achievements Exhibition) in Moscow for this research effort.

When a new period of détente began in the early 1970s, the Soviet leader, Leonid I. Brezhnev, promised the new US president, Gerald Ford, that the Soviet strategic bomber fleet would be reduced by permanently converting twenty 3MSs into tankers by deleting all bomb armament and the IFR probes and sealing the bomb bay, leaving only small clamshell doors for the HDU. The Long-Range Aviation command supported the idea, believing that tankers were more valuable than obsolescent bombers. The conversions were performed by ARZ No. 360; the programme was coordinated by an MAP commission under the then minister of aircraft industry Ivan S. Silayev, whose short temper had earned him the nickname Ivan Groznyy (Czar Ivan the Terrible). Eventually, however, not all of the envisaged 20 aircraft were thus modified.

Above: M-4-2 '72 Red' (c/n 5300710) is placed into storage – never to fly again (note that it is parked on a grass field, with only a few concrete slabs under the nose and main gear to prevent it from sinking into the mud later).

Left: The end of the road. This dismembered 3MN-1 shared the sad fate of nearly all *Bisons*, being scrapped in compliance with arms reduction treaties. Note the VD-7B engines lying beside the aircraft.

Above right: The gutted hulks of three 3MDs, with radomes cut open and hatches gone. Note the ILS aerial normally hidden from view inside the radome.

The end of the decade, however, saw the outbreak of a new war – the Afghan War (1979–89); for the first time since 1945, the Soviet armed forces were officially involved in an armed conflict on foreign soil. Long-Range Aviation was also involved, albeit on a limited scale, and occasionally the 'oldies' turned out to be more appropriate for the current task than newer types. On 15 April 1979, several months before the war, rebellion broke out in the town of Herat, the capital of Afghanistan's Herat Province; it included both a popular uprising caused by the socialist reforms implemented by the Democratic Republic of Afghanistan government (notably the agrarian reform) and a mutiny of Afghan army troops. The DRA government requested help from the Soviet Union, and for a while the Soviet government considered the possibility of air strikes against the insurgents. To this end, the 3Ms of the 1096th and 1230th TBAPs, based at Engels-2, were prepared for the mission along with other types; they were fitted with cassette racks for carrying 52 FAB-250 M54 HE bombs, weighing 12,640 kg (27,870 lbs.), or 28 FAB-500 M54s, weighing 13,280 kg (29,280 lbs.). Eventually, however, the Herat uprising was quashed by loyal government troops on 20 April and the air raid was called off. Much later, the DA did participate in the Afghan War, but only Tu-16s and Tu-22M3 *Backfire-C* bombers were involved. Thus, the *Bison* was never used in anger after all.

Eclipse

In 1980 the DA's organisation was changed again. The 6th OTBAK was replaced by the 46th VA VGK (SN); that is, *voz**doosh**naya* ***ar**miya Ver**khov**novo **glav**noko**mahn**dovaniya (strate**gich**eskovo naz**na**cheniya)* – 46th (Strategic) Air Army of the Supreme High Command (HQ Smolensk); the 8th TBAK was succeeded by the 30th VA VGK (SN), headquartered in Irkutsk. These took over control of the 201st TBAD and 73rd TBAD, respectively.

Gradually the *Bison* was displaced by newer types in the DA's inventory. By the late 1960s, new versions of the Tu-95 had taken over the 3M's role as a strategic strike aircraft. By the early 1980s, nearly all surviving *Bison-B*s had been converted into tankers, forming the backbone of the Soviet air force's tanker element until the advent of the Ilyushin Il-78 *Midas*.

The units at Ookraïnka AB were the first to give up the type. In 1984 or 1985, the 40th TBAP and 79th GvTBAP phased out the M-4 and 3M, exchanging them for Tu-95s. In particular, the former unit received the Tu-95M bombers and Tu-95MR reconnaissance aircraft of the aforementioned 1023rd TBAP; the latter reequipped with Tu-95MS missile carriers, assisting the 40th TBAP in converting from the Myasishchev tankers to the *Bear-A/E*. The 79th TBAP reequipped with Tu-95KM *Bear-C* and Tu-95K-22 *Bear-G* missile carriers; all Tu-95K-22s were now concentrated in this unit.

True enough, in the mid-1970s an effort was launched to extend the 3M's service life. As part of this effort, a high-time 3MS (c/n 7300703) was ferried to Novosibirsk and tested by the Siberian Aviation Research Institute, named after Sergey A. Chaplygin (SibNIA – ***Si**beer**skiy** naoochno-is**sled**ovatel'skiy insti**toot** avi**ah**tsiï*), at loads approaching the actual operational load limits. Recommendations issued by SibNIA would allow the *Bison-B/Cs* to remain operational for another 10 years or more. A structural upgrade programme was developed in keeping with these recommendations and implemented by ARZ No. 360; yet, again only part of the fleet was eventually upgraded.

Above: The so-called 'Sixth Field' in Engels – an unpaved storage area – with stored 3MS-2s awaiting disposal

Left: Curiously, the nearest three *Bisons* here are parked in the sequence of their tactical codes – '54 Red', '55 Red,' and '56 Red.' The cannons have been removed from the turrets.

Right and above right: Carnage at the 'Sixth Field,' with the mingled remains of scrapped 3Ms and Tu-95s

The Dobrynin-powered 3MNs and 3MDs were phased out in the 1980s and prepared for scrapping, though a few aircraft were placed in flyable storage. The Mikulin-powered 3MSs served on thanks to their more reliable engines.

After the breakup of the USSR in late 1991, the newly independent Ukraine laid claim to all ex-Soviet military assets on its territory. Among other things, it appropriated the 184th GvTBAP (*Gvar**deys**kiy tya**zh**olyy bombardi**rov**ochnyy **a**via**polk*** – Guards Heavy Bomber Regiment) in Priluki (Chernigov Region), which had all of the Soviet air force's operational Tu-160 *Blackjack* swing-wing supersonic missile strike aircraft, and the 409th APSZ (***a**via**polk** samo**lyo**tov-za**prahv**shchikov* – Aerial-Refuelling Regiment) at Uzin AB, near Kiev, with Il-78 tankers, which supported the Tu-160s' operations. Left without up-to-date strategic aircraft, the Russian MoD had to take remedial action. As *Blackjack* production continued, the 1096th TBAP started converting from the 3M to the Tupolev bomber, taking delivery of its first brand-new Tu-160 on 14 February 1992 (some sources say 16 February); by May this number had increased to three. The remaining 3MS-2 tankers powered by low-time RD-3M-500As were concentrated in a single unit, the 1230th TBAP, which was reorganised as the 1230th APSZ. They were a useful asset; the breakup of the Soviet Union left the Russian air force with only 13 modern Il-78/Il-78M *Midas-A/B* tankers, which meant the 3MS-2 could not be retired just yet.

Still, the *Bison*'s days were numbered, since the Soviet Union had pledged to scrap a substantial number of these bombers under the Strategic Arms Reduction Treaty (START 1), signed on 31 July 1991 – even though the VVS had many fewer heavy bombers than the USAF. The last 3Ms soldiered on until 1994, by which time the unit at Engels had mastered the Tu-160 and increased its *Blackjack* fleet to five. On 10 February 1994, all remaining 3Ms were struck off charge by an MoD order. Thus, the 1230th APSZ was left with Il-78/Il-78M tankers only, retiring the last 3MS-2s.

By 1996 the remaining *Bisons* were concentrated at the so-called Sixth Field (Engels-6), a vast unpaved storage area near Engels-2 AB, awaiting disposal; by August 1997 most of them had been scrapped. Looking back on it now, this was a premature decision.

By the end of its career, the M-4/3M family had been in service for nearly 40 years. One of the reasons for its early eclipse was undoubtedly the negative attitude of the nation's political leaders because the *Bison* had never been 'missilised.'

Three M-4-2 tankers survive in Russia, but none are readily accessible to the aviation enthusiast. One aircraft coded '60 Red' (c/n 6302831) is immaculately preserved in the DA Museum at Dyagilevo AB, Ryazan'. Another tanker, '63 Red' (c/n 5301518), is preserved at its home base, Ookraïnka AB, which has a small museum of its own; unfortunately, it is in less-than-ideal condition – the lateral sighting blisters had been smashed, and the apertures simply had been closed with sheet metal. The third *Bison-A*, '66 Red' (c/n 5302528), was flown to Syzran' (Samara Region) on 21 January 1991 and preserved at the premises of the Syzran' Military Pilot College (SVVAUL – ***S**yzranskoye* ***vys****sheye* *voy**en**noye* *aviatsi**on**noye* ***o**o**chil**ischche* ***lyot**chikov*). In 2004 it was disassembled and transported to Moscow for preservation at its birthplace, the present-day Khroonichev Scientific & Production Centre. Alas, something went wrong and the aircraft was not reassembled; as of October 2019 the major airframe components were still in open storage at the plant, suffering the attentions of vandals.

A single 3MS-2, '14 Red' (c/n 7300805), is on display at the *other* DA Museum at Engels-2 AB. This aircraft had the distinction of making the last flight of a 3M, on 21 March 1994 when it was ferried from Dyagilevo AB to Engels by a crew under Maj. V. Goora.

The next-to-last 3MD, '30 Red' (c/n 0301804), resides in the Central Russian Air Force Museum in Monino, south of Moscow. It was flown there from Engels on 8 July 1986 by a crew under Maj. V. Sirotkin, with copilot Lt.-Col. V. Pavlyukov, navigator Maj. S. Chikunov, radar operator Capt. G. Sysoyev, flight engineer Capt. G. Filippov, gunner / radio operator Ens. V. Katkov, and tail gunner Ens. M. Hisyametdinov. The landing on the short dirt runway at Monino was almost foiled by a recently erected high-rise construction crane but went well. On landing, the bomber deployed the brake parachutes, jettisoning them at the end of the landing run; by the time the crew shut down the engines and climbed out, the parachutes had been stolen!

A sad sight – more than a dozen dismembered 3MN-1 bombers at the storage and disposal area in Engels

Chapter 6

The Spacecraft Carriers

The mid-1970s were characterised by notable achievements in space exploration, which looked set to continue at an ever-increasing pace. Further use of conventional disposable space launch vehicles (SLVs) appeared to be wasteful; this sparked a need for reusable launch vehicles. It was presumed that such vehicles would be more efficient economically if used with sufficient frequency.

Hence, as early as 1968, the USA began accelerated development of a reusable space transport system; the programme was defined in 1970 as the project that became the famous space shuttle. Not to be outdone, the Soviet Union soon followed suit: a decision was taken to develop an indigenous reusable space transport system with similar characteristics; it was officially named Baikal after the world's deepest lake, located in East Siberia. On 17 February 1976 the Council of Ministers and the Communist Party Central Committee issued directive No. 132-51, which was backed by MAP orders No. 81, of 24 February 1976, and No. 106, of 15 March 1976. The directive assigned the newly established NPO ***Mol****niya* (Lightning), headed by Gleb Ye. Lozino-Lozinskiy within the MAP framework as the main designer of the Soviet shuttle orbiter, with Myasishchev's Experimental Machinery Plant being included into NPO Molniya as codesigner. (This was not a matter of chance – the Myasishchev OKB had worked on the M-19 aerospace plane project in 1974–75. Even though the M-19 was never built, this experience would now be put to good use.)

The same directive tasked the Ministry of General Machinery's NPO ***Ener****giya* (Energy) with developing a monstrous SLV,

CCCP-01502, the second VM-T, takes off with a mockup version of 'payload 1GT' (the oxidiser tank of the Energiya SLV).

the likes of which had not been seen before. Initially known as *Rassvet* (Dawn) but subsequently renamed Energiya, this was intended for placing large satellites, manned space station modules, and space shuttles into orbit.

The Soviet space shuttle received the name *Buran* (Blizzard). Two very different versions were considered at the technical proposal stage: a lifting-body design with small dihedral – no, *very* dihedral wings, designated *izdeliye* 305-1, and a more conventional vehicle with larger cranked-delta wings resembling the space shuttle, which was designated *izdeliye* 305-2. The latter version won (more on this follows a little later). The *izdeliye* 305-1 resembled a scaled-up version of the Mikoyan OKB's *Spirahl'* (Spiral) aerospace plane project; this was no coincidence – Gleb Ye. Lozino-Lozinskiy had supervised the design effort on that project in the days when he worked at the Mikoyan OKB.

Now a big problem arose. The programme envisaged construction of space vehicles of hitherto unseen size and weight whose various components were to be manufactured by hundreds of enterprises located in the central regions of the USSR. The initial idea of assembling the SLV and the space shuttle at the famous Baikonur Space Centre in Kazakhstan was quickly dropped, since this would require large-scale manufacturing facilities to be constructed there. It was deemed more expedient to have the principal subassemblies of the SLV and the payload it was to launch (not necessarily the space shuttle) manufactured by the main contractors and delivered to Baikonur for final assembly.

Thus, it was necessary to transport outsized items, some of which had a length of 60 m (197 ft.) and a diameter of 8 m (26 ft.), over a distance of 1,500–2,000 km (932–1,243 miles). Besides, although the Buran's envisaged primary landing site was Leninsk (a.k.a. Yubileynyy) airfield located next door to the Baikonur Space Centre, the shuttle could land elsewhere if an early deorbiting became necessary. Alternate airfields for the Buran were located far apart – at Simferopol' in the Crimea, Ukrainian SSR, and at Khorol' AB in the Far East. From there it would have to be taken back to Baikonur, where it would be prepared for the next launch.

The big question was how to transport the Energiya SLV and the shuttle orbiter to Baikonur. Transporting these items by road, rail, or river/sea was impracticable – because of the sheer bulk and weight of the cargoes, it would be necessary to build new roads and upgrade existing ones, rebuild bridges, widen tunnels, raise high-voltage power transmission lines, etc. Imagine the costs involved! Also, the cargo would be subjected to prolonged vibrations and rocking during the journey, which could damage the space vehicle. Air transport was the only realistic means of delivery.

The next question was, What type of aircraft should be used? The Mil' Mi-26 *Halo* heavy-lift helicopter, which was then under development, came into consideration, since it was to carry slung loads of up to 18.5 tons (40,780 lbs.). The idea cropped up of loading sections of the SLV on a purpose-built truss-type platform that would be lifted by four (!) Mi-26s. To check the validity of the method, LII manufactured a subscale model – a chunk of fuselage from a retired Il-18 airliner – which was lifted in this fashion. However, during one of the flights, the external load started rocking dangerously after encountering turbulence, and the helicopter pilots were forced to jettison it. The method was considered unsafe – not to mention the fact that the helicopters' short range would require many refuelling stops on the route and hence require construction of many helipads, which was uneconomical.

An even more exotic proposal envisaged the use of airships, but these would have to be developed from scratch. Thus, fixed-wing aircraft were the only option. MOM placed an order with MAP for the development of an airlift system that would comprise a special outsized-cargo transporter aircraft, appropriate loading equipment, and ground support equipment.

In 1975, MAP addressed the Kiev-based OKB-473, headed by Oleg K. Antonov, with a request to develop a dedicated aircraft for the carriage of space system subassemblies. At the time, the Antonov OKB was working on the project of the An-124 *Ruslan* heavy military transport (NATO reporting name *Condor*), with a load-carrying capacity of 120 tons (264,600 lbs.); MAP decided this was a good basis for the specialised transport system.

Here we have to say a few words about the Soviet shuttle orbiter to give an idea of what the aircraft was to carry. In 1975, when it was at an early development stage, the future Buran looked almost like a straight copy of the American space shuttle, featuring three large main rocket motors akin to the SSME (space shuttle main engine) and riding a fuel/oxidiser tank of the same shape as the space shuttle's external tank (ET). The difference was that the tank was flanked by four reusable liquid-propellant rocket boosters instead of two solid-propellant boosters and that the rear fuselage featured two jettisonable launch escape system (LES) rocket packs, which would carry the shuttle out of harm's way if the launch went awry. The launch weight excluding the LES was nearly 120 tons (264,550 lbs.); hence the shuttle was designated OS-120 (*orbitahl'nyy samolyot* – 'orbital aircraft' or space plane). Dimensionally it was almost identical to the space shuttle, having a length of 37.5 m (123 ft., 3/8 in.), a height of 17.4 m (57 ft., 1 3/64 in.), and a wingspan of 22.0 m (72 ft., 2 9/64 in.). The dry weight was 68 tons (149,910 lbs.) – almost the same as for the space shuttle. Therefore the An-124 was envisaged as the shuttle carrier aircraft for the Buran.

However, the An-124's original project configuration, known as *izdeliye* 200, was too conservative and did not make it possible to achieve the required level of design perfection. In 1976, Oleg K. Antonov took the decision to rework the project completely, the An-124 in its ultimate form becoming the *izdeliye* 400. This aircraft, which looked set to be the OKB's 'bread and butter' for many years to come, had to be finalised and tested first, so the spacecraft transporter project was shelved for the time being.

Concurrently the designers at NPO Molniya concluded that the OS-120 was too heavy, and by January 1976 the shuttle had been redesigned to become the OK-92 (*orbitahl'nyy korahbl'* –orbiter). The figures reflected the greatly reduced launch weight of 92 tons (202,820 lbs.) less launch escape system. The concept was fundamentally changed in so doing. The OK-92 lacked the main rocket motors – these were relocated to the core module of the RLA-130 SLV, which the shuttle was now to ride instead of a simple fuel/oxidiser tank; their location was occupied by the LES. On the other hand, the OK-92 had two turbofan engines in pylon-mounted pods on the rear fuselage sides, which also housed the rear thrusters for orbital manoeuvres. Upon reentry, the fairings closing the turbofans' air intakes and exhausts were jettisoned, allowing the shuttle to fly under its own power and facilitating landing. In May 1976 the launch weight was increased to 98 tons (216,050 lbs.); hence the originally envisaged 12,000 kgp (26,455 lbst) Solov'yov D-30KP nonafterburning turbofans borrowed from the Il-76 *Candid* transport were replaced by 12,500 kgp (27,560 lbst) Lyul'ka AL-31F afterburning turbofans from the Sukhoi Su-27 *Flanker-B* fighter.

The OK-92 was shorter and lower than the previous version, with a length of 34.5 m (113 ft., 2$\frac{17}{64}$ in.) and a height of 15.8 m (51 ft., 10$\frac{3}{64}$ in.); the wingspan and the size of the payload bay were the same. Dry weight without the LES was 51 tons (112,430 lbs.), which meant the shuttle could be airlifted by the most capable extant Soviet transport aircraft – the An-22 ***Antey*** (Antaeus; NATO reporting name *Cock*); a version of the An-22 carrying the OK-92 piggyback was contemplated but not built.

The OK-92 was not built either, evolving into the Buran as we know it (*izdeliye* 11F35) by March 1978. The definitive Buran had no jet engines (with the exception of the BTS-002 flight test article) and was heavier, with a launch weight of 105 tons (231,480 lbs.) and a dry weight of 62 tons (136,680 lbs.); it had a length of 36.37 m (119 ft., 3$\frac{57}{64}$ in.), a height of 16.35 m (53 ft., 7$\frac{45}{64}$ in.), and a wingspan of 23.92 m (78 ft., 5$\frac{47}{64}$ in.). Concurrently, the Energiya SLV was also reworked – the core module now had four rocket motors instead of three (albeit smaller ones) and was made longer to hold more fuel and oxidiser; the thrust of the booster packs was also increased.

Later, in 1981, when development of the baseline An-124 (*izdeliye* 400) was nearing completion, the Antonov OKB undertook a design study of a specialised version provisionally designated An-124KT (*kosmicheskiy transportirovshchik* – space transporter, or spacecraft carrier). Since the Ruslan's freight hold was not large enough to accommodate the Buran space shuttle or components of the Energiya SLV, there was no option but to carry them piggyback. This prompted a redesign of the tail unit; the single vertical tail was first enlarged and then replaced by a twin-fin tail unit, which would be unaffected by the turbulence generated by the external cargo. However, the customer wanted the An-124KT to be suitable for suborbital launch of spacecraft; the designers of NPO Molniya came to the disheartening conclusion that the weight of the load to be separated from the aircraft acting as the launch platform could be as much as 270 tons (595,350 lbs.), not the 170 tons (374,850 lbs.) envisaged originally. As a result, it became obvious that the An-124 was unsuitable for use as the first stage in a suborbital launch system, as well as for carrying the Buran and components of the Energiya. A new versatile transport aircraft with even-greater lifting capability had to be developed. Eventually, responding to insistent demands from MOM, in 1983 the Antonov OKB began developing a tailor-made 'rocket transporter' aircraft – the mighty six-engined, twin-tailed An-225 ***Mriya*** ('dream' in Ukrainian; NATO reporting name *Cossack*) derived from the An-124.

However, we are jumping ahead of the story. Much time was wasted in cajoling Antonov; the hypothetical space transporter version of the An-124 seemed years away. Yet, the need for such an aircraft grew increasingly acute because the prototypes of the Energiya SLV and the Buran orbiter were going to be manufactured soon and would need to be transported from Moscow to the Baikonur Space Centre.

In this situation, the Myasishchev OKB offered to adapt the 3M bomber for the space transporter role. This bold proposal came at just the right moment. The idea was not as wild as it may seem; the 3M could cope with the envisaged loads, whose maximum weight was 32 tons (70,550 lbs.). (True, the standard bomber had a maximum payload of 24 tons [52,910 lbs.], but this was limited by how many bombs you could cram into the bomb bay; theoretically the aircraft could lift more, but you can't put a quart into a pint pot!)

Actually the OKB had considered the 3M as a 'shuttle carrier' even as the technical proposal for the *izdeliye* 305-2 orbiter was drawn up. Several wind tunnel models representing an otherwise standard 3M with auxiliary endplate fins on the stabilisers and the space shuttle riding piggyback were manufactured and tested. When the shuttle was fitted with a tail fairing covering the rocket motor nozzles for a ferry flight, the bomber's normal vertical tail would be removed to avoid buffeting, leaving the aircraft with a twin-tail empennage. Experiments and calculations performed by the OKB confirmed the feasibility of such a conversion.

Interestingly, general designer Vladimir M. Myasishchev was opposed to the idea at first. One by one he summoned and sternly questioned aerodynamicist Aleksandr A. Brook, project engineer V. V. Lyubakov (who was responsible for the general arrangement), structural strength engineer A. Kh. Karimov, the control system engineers, electrical engineers, operations department specialists, and economists. All of them unanimously spoke in support of the project, persuading them that the idea was feasible and should be proceeded with. Myasishchev thanked each interviewee in a restrained manner and let him go.

Nevertheless there were considerable risks involved. The 3M had been in service for 20 years by then and was considered obsolete. Moreover, the 3Ms had a lot of hours on the clock, and many people were not sure if the conversion was worthwhile, considering the short remaining service life – even as a stopgap until the An-225 appeared. Since the idea involved adapting an existing aircraft rather than developing a purpose-built one, there was little scope for modification. It was not known how the outsized cargo would affect the aircraft's stability and handling and the airflow around the tail surfaces. Likewise, it was not certain that the drag caused by the cargo would not be excessive – even though the 3M's excellent lift/drag ratio offered some reserves in this respect.

The idea of using the 3M as a shuttle carrier sparked bitter opposition both from the bosses of NPO Molniya (of which the Myasishchev Experimental Machinery Plant was part at the time) and the scientific leaders of LII and TsAGI. TsAGI's director, Gheorgiy P. Svischchev, and first vice director, Gheorgiy S. Büschgens, both called on Myasishchev, urging him to drop this 'crazy idea.' Nevertheless, Vladimir M. Myasishchev stuck to his decision to go ahead with the project.

3M-T Outsized-Cargo Transporter (Second Use of Designation; *izdeliye* 3-35, First Project Form)

On 21 November 1977 the Council of Ministers and the Communist Party Central Committee issued joint directive No. 1006-323, which was followed up by MAP order No. 512, dated 30 December 1977.

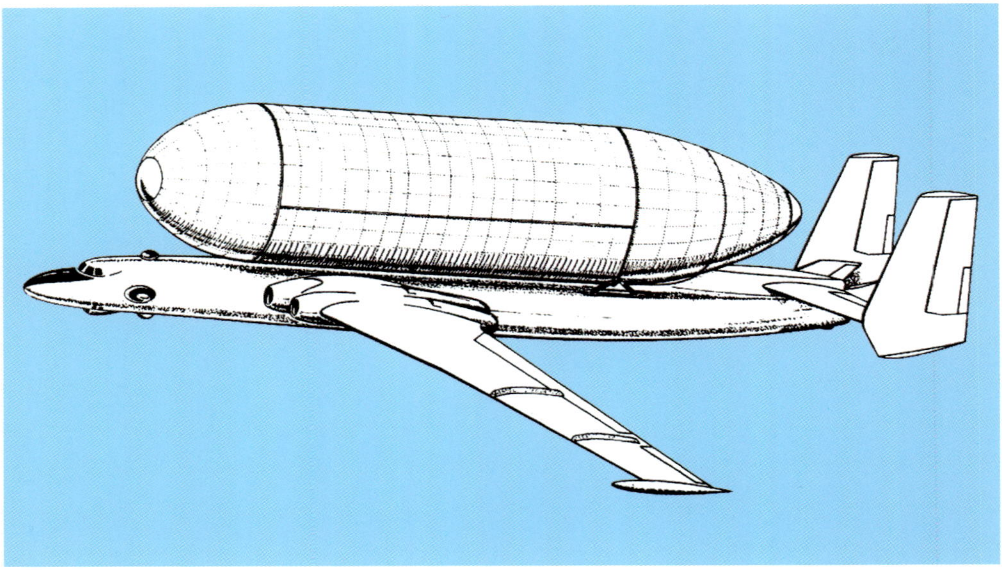

An artist's impression of the 3M-T space transporter's early project version, with a detachable dorsal payload pod, which was developed into the definitive VM-T. Note the swept horizontal tail with no dihedral and the pentagonal vertical tails.

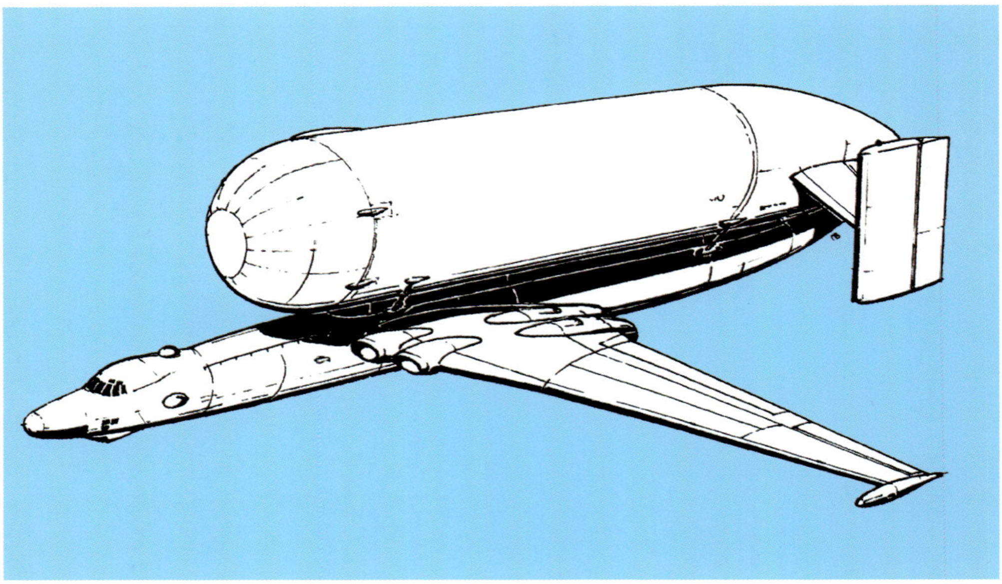

An artist's impression of the 3M-T's so-called 'Guppy' project version, with a permanently installed cylindrical payload pod carrying the tail unit with rectangular vertical tails. The pod's front end opens for loading/unloading.

'The Dragon and His Tail': this view illustrates the rear fuselage curving up around the underside of the pod, as well as the strongly anhedral horizontal tail and the vertically placed fins.

These documents officially tasked the Myasishchev OKB with developing a specialised cargo aircraft for hauling outsized loads, specifying that it was to carry components of the SLV's core module from Kuibyshev to Tyuratam (an alternative name for the Baikonur Space Centre) and the shuttle orbiter from Zhukovskiy to Tyuratam via Engels. (The Energiya SLV was to be manufactured by MOM's Progress plant – the former MAP aircraft factory No. 1 – in Kuibyshev; the orbiter would be built by MAP's Tushino Machinery Plant (TMZ – ***Tushinskiy mashinostroitel'nyy zavod***), formerly plant No. 82, in Moscow, and delivered to Zhukovskiy for onward transportation.) True, this aircraft was seen as an interim solution until a more capable aircraft was developed by the Antonov OKB. Still, the high level of the authorising documents speaks for itself: the shuttle carrier was included into the national space programme, ergo ultra-important.

At the Myasishchev OKB the programme was code-named ***tema*** 3-35; this designation figured in the CoM directive and the MAP order. The aircraft itself was designated 3M-T, inheriting the designation of a stillborn IFR tanker project developed in 1960; in this case the T stood for ***trahnsportnyy [samolyot]*** – transport aircraft.

Boris I. Zorin was appointed the 3M-T's chief project engineer; after his death in harness, V. A. Korchagin became project chief, assisted by project engineer V. M. Kirillin. The OKB's involvement in the multipurpose reusable space transport system programme – even as a 'subcontractor' – stretched its resources to the limit, requiring a substantial reinforcement of the design staff. Several experienced engineers were hired, along with numerous graduates of MAI and the Kuibyshev Aviation Institute. Additionally, the OKB's experimental shop and flight test facility in Zhukovskiy had to be thoroughly reorganised and reequipped with the latest hardware; new ground test rigs were constructed and cooperation ties were established with numerous enterprises within the MAP and MOM frameworks.

The issue of choosing the right layout for the 3M-T now loomed large. Of course, the Myasishchev OKB took due notice of what Western companies were doing in this field. Starting in the early 1960s, the US company Aero Spacelines had developed six outsized-cargo carrier aircraft based on heavily modified Boeing C-97 transports for carrying spacecraft components. These bore the common name Guppy and, so to say, got steadily guppier (Pregnant Guppy, Very Pregnant Guppy, Super Guppy, etc.). The last of the line was the Colossal Guppy, which had an upper fuselage lobe diameter of 12.2 m (40 ft.) and a 90-ton (198,400 lb.) payload.

Initially, the Myasishchev OKB likewise contemplated internal cargo carriage. The first project version of the 3M-T was conceived as a huge cargo pod to which the crew section and wings (complete with engines) of the 3M were mated, together with a new twin-fin tail unit. For all intents and purposes, this was a new aircraft that would require too much time (and money) to develop, and this layout was soon dropped.

3M-T Outsized-Cargo Transporter (*izdeliye* 3-35, Second Project Version / 'Guppy')

The second project version was also designed with internal cargo carriage in mind but had far-greater structural commonality with the 3M. It was no less bizarre, though. The forward/centre fuselage, wings, and landing gear were taken wholesale from the standard bomber. Everything else was new. A huge cylindrical pod of 10 m (32 ft., 9⁴³⁄₆₄ in.) diameter was mounted above the fuselage on a short, thick pylon that ran all the way from the wing leading edge to the fuselage tail cone. The pod was braced by two pairs of sloping struts; its rear end was ogival, with the tapered rear fuselage curving upward to follow its lower contour and eerily resembling a dinosaur's tail. This ogival rear fairing carried the tail unit, which consisted of strongly anhedral (!), gently swept stabilisers tipped with rectangular fins disposed vertically. The parabolic front end of the pod was hinged, swinging open for loading and unloading. This version of the 3M-T earned the in-house nickname 'Guppy,' obviously by analogy with the Aero Spacelines Guppy.

The 'Guppy' had a number of major shortcomings. First, once again such an extensive conversion would take too long to manufacture. Second, the Buran could be carried only with the wings detached; this involved removing part of the quartz heat insulation tiles at the wing/fuselage joint, which was undesirable because the tiles could get damaged. A damaged tile could lead to catastrophic structural failure on reentry caused by local overheating (as would later be demonstrated by the crash of space shuttle *Columbia* on its 28th flight on 1 February 2003). As for the Energiya SLV, its assembled sections could fit inside the pod, having a diameter of 8 m (26 ft., 2³¹⁄₃₂ in.). Third, loading and unloading would be very inconvenient. Hence this version was not proceeded with.

3M-T Outsized-Cargo Transporter (Definitive Version; *izdeliye* 3-35, VM-T Atlant)

The next configuration suggested by Vladimir M. Myasishchev involved a much less extensive conversion. Once again, the bomber's forward/centre fuselage, wings, and landing gear were unchanged; they were mated with a new, longer rear fuselage carrying a new twin-fin tail unit. The latter consisted of shoulder-mounted swept stabilisers with no dihedral and pentagonal fins at the tips, with one-piece inset rudders terminating well short of the bottom. The payload would be carried in a detachable cargo pod mounted above the fuselage, which was similar in shape and size to the one in the 'Guppy' project. To enable carriage in this pod, the core module of the Energiya SLV was broken down into three subassemblies at this stage – the so-called module Ts1, the fuel tank that was part of 'module Ts2,' and the liquid oxygen (LOX) tank, intermediate casing between the tanks and the engine bay fairing of 'module Ts2.' Hence the new version of the 3M-T space transporter was unofficially dubbed '3-blok Ts'; the letter "Ц" was a reference to *tsen**trahl'**nyy blok* – centre or core module of the SLV.

(It should be noted that like their Soviet colleagues, American aeronautical engineers considered external carriage of bulky items by heavy transport aircraft. For instance, the Lockheed Company had similarly tried to adapt the C-5A Galaxy military transport for carrying SLV stages in a special container atop the fuselage under a NASA contract – an idea that eventually came to nothing. The Boeing Company converted a former American Airlines Boeing 747-123 into the Boeing 747SCA (shuttle carrier aircraft) for NASA in 1976; this aircraft was used for free-flight gliding tests of the space shuttle prototype, *Enterprise*, in 1977. Moreover, in spite of the huge costs involved, NASA commis-

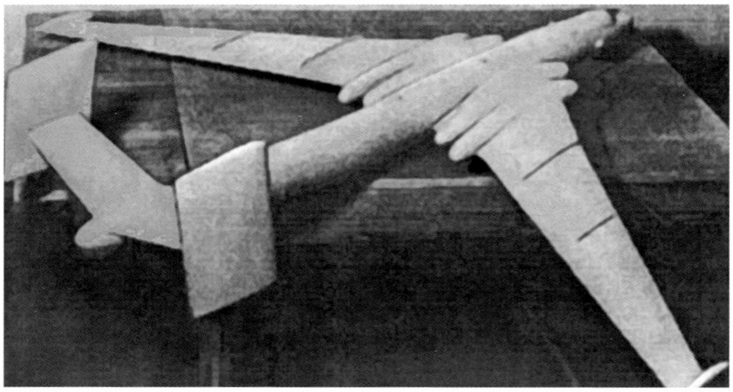

Top and above: Pictures from a TsAGI bulletin, showing a model of the 3M-T in its final version (without payload)

Right: A separate model of the 3M-T's nose gear unit for wind tunnel test, with a matchbox for scale

A wind tunnel model of the 3M-T carrying 'payload 2GT' (the SLV's fuel tank and service bay)

sioned development of several highly unorthodox aircraft tailored for carrying rocket system components externally (the Virtus, Twin Body, etc.); none of them materialised.)

It was envisaged that Myasishchev's space transporter would retain the baseline bomber's IFR capability, which would obviate the need for refuelling stops en route. The designers considered the possibility of free-flight tests in similar manner to the Boeing 747SCA / space shuttle combination – the *izdeliye* 305-2 orbiter would be taken aloft by the modified 3M, gliding to earth after separation.

Changes were made as the design work progressed – first and foremost to the tail surfaces. In the final version of the project, the fins were of simple rectangular shape, just like on the 'Guppy' version. However, they were at right angles to the stabilisers, which were now unswept and featured marked dihedral, so that the fins were canted inward. Sweep was not necessary, since cruising speed with an external load would not exceed 450 km/h (280 mph), while the strong dihedral reduced the torsional loads generated by the tail unit. Also, the larger area of the stabilisers compared with the 3M increased the lift and hence the pitch-down force generated by them, compensating for the pitch-up force created by the dorsally mounted load. The tail unit area amounted to 26% of the wing area; making it any smaller was impossible because the powerful vortices generated by the bulky external load would cause tailplane buffet – a highly dangerous phenomenon that might result in structural failure.

Since the new twin-fin tail unit generated much-higher structural loads, the airframe had to be reinforced and the rear fuselage redesigned completely. The new rear fuselage was 7 m (22 ft., $11^{19}/_{32}$ in.) longer and upswept to provide an adequate ground clearance – a necessity due to the aircraft's characteristic nose-up ground angle. The outer wings had a new three-spar design and new lower skin panels; the forward fuselage skin panels aft of the pressurised flight deck, the centre fuselage skin panels aft of the wings, and some internal structural elements were all new. The bomb bay was faired over; the skin panel replacing the doors absorbed part of the structural loads and incorporated a small maintenance hatch for access to the equipment housed in the bay. The dorsal and ventral cannon barbettes were also deleted, of course. Suitably stressed attachment points for the external cargo were provided on the rear fuselage sides and on top of the forward fuselage. A system pressurising the external load was added, using engine bleed air; this was necessary to ensure that the external load maintained its proper shape and was not damaged by the pressure of the slipstream.

The nose and main landing-gear units were stock 3M items, but the outrigger struts had to be reinforced because of the greater loads absorbed in the event of a banked landing in a crosswind. The flight control system was new, featuring an ASUU-335 stability augmentation system (*avtomaticheskaya sistema oostoychivosti i oopravleniya* – automatic stability and control system for *izdeliye* 3-35); irreversible hydraulic actuators were used in all three control circuits instead of reversible ones. This ensured flight safety and adequate stability and handling throughout the flight envelope both in no-load and loaded configuration.

The fuel system was modified, allowing fuel to be transferred in order to change the CG position according to the weight of the cargo carried. For a while, the OKB considered installing VD-7M afterburning turbojets taken from the Tu-22 *Blinder* bomber, with an afterburner rating of 16,000 kgp (35,270 lbst). However, these were unduly bulky, protruding nearly 3 m (9 ft., $10^{7}/_{64}$ in.) beyond the wing training edge. Eventually the choice fell on uprated Dobrynin VD-7MD nonafterburning turbojets delivering 10,750 kgp (23,700 lbst) for takeoff. In a nutshell, it would not be doing the Myasishchev OKB justice to say that the 3M-T was merely an adaptation of the 3M for a new role; the changes were so extensive that the result was, in effect, a new aircraft.

Another important change at this stage was that the OKB dropped the idea of using the external cargo pod, which added empty weight and reduced the payload; now the cargo would be carried above the fuselage and fitted with fairings as appropriate to reduce drag. Four payload configurations were envisaged. The heaviest of these was the so-called payload 0GT – the Buran space

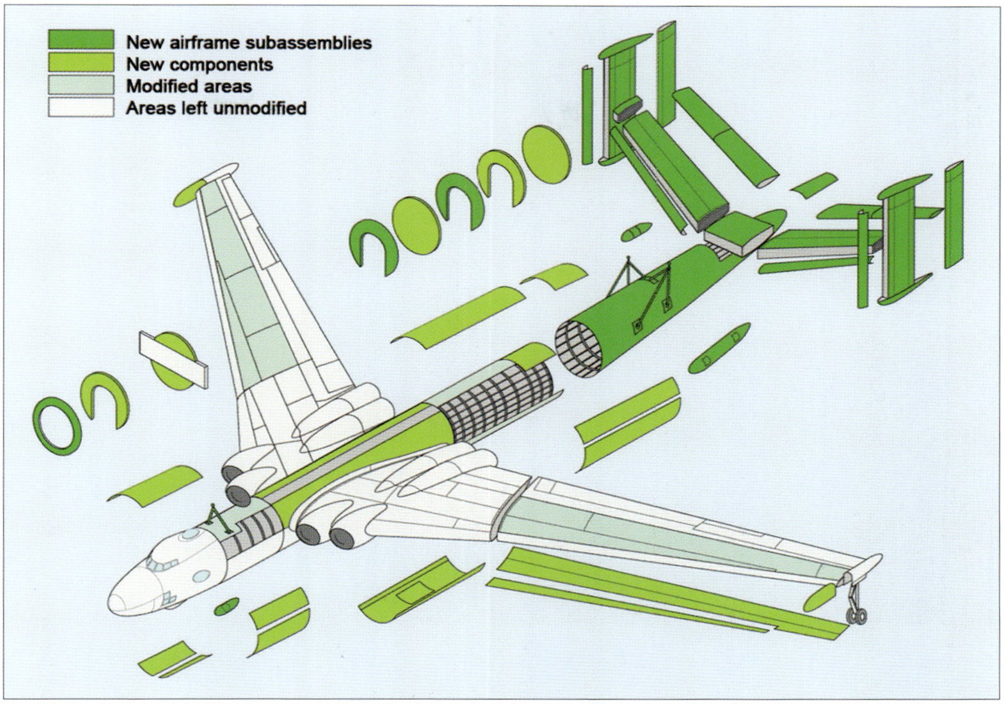

Right: A drawing showing the changes in the 3M-T's airframe structure in comparison with the 3MN

Below: The first flying example of the 3M-T – or rather VM-T, as it was known by then (c/n 9301402) – nears completion in the Myasishchev OKB's experimental shop in 1980. The fuselage and inner wings of the static-test article (c/n 9301504) can be seen in the background

shuttle (*izdeliye* 11F35), the letters denoting *grooz tekhnolo***gich***eskiy* (technological cargo); it would be carried in a slightly nose-up attitude.

'Payload 1GT' was the cylindrical liquid hydrogen (LH_2) tank of the SLV's core module (stage 2); it was 40 m (131 ft., 2⁵¹⁄₆₄ in.) long and nearly 8 m (26 ft., 3 in.) in diameter. During transportation, the tank was pressurised with nitrogen to maintain its proper shape, and the ends were provided with fairings pressurised by the slipstream by means of air scoops (the front fairing was hemispherical and the rear one was ogival).

'Payload 2GT' represented a combination of the SLV's proper nose fairing (featuring the same ogival shape), the LOX tank, the equipment module, and the engine module; this assembly was mounted back to front and provided with the same hemispherical forward fairing as in the previous case. Finally, 'payload 3GT' was the draggiest of all – a huge 'egg' made up of the fuel tank's front and rear fairings joined together so that they could be returned to the factory for the next mission (thus it was occasionally called 'return payload'). There was also a planned payload version with a so-called UAK-1 cargo pod (*ooniver**sahl'**nyy aviatsi**on**nyy kon**tey**ner* – versatile aviation container), which was similar in cross-section to 'payload 2GT' but was shorter, accommodating various cargo items, but this never materialised.

Having very different dimensions and aerodynamics, these external loads would obviously affect the rear fuselage and tail unit of the 3M-T differently. This necessitated an additional round of wind tunnel tests. As already mentioned, several models had been tested in TsAGI's wind tunnels to prove the possibility of using the

Left: A schematic model of the 3M-T with 'payload 1GT' in the Myasishchev OKB's MAGDA integrator

Below: This partial model of the 3M-T tested in TsAGI's T-104 wind tunnel is fitted with a model of 'payload 2GT.' Note the payload attachment points and the wing 'endplates.'

Opposite page: A model of the 3M-T (VM-T) in the OKB's T-1 wind tunnel. The flattened pod was proposed for transporting the French Hermes shuttle orbiter; note the wool tufts for flow visualisation.

3M as a 'shuttle carrier.' The Myasishchev OKB's aerodynamics department further insisted that a quarter-scale model of the aircraft with various payload configurations be tested in TsAGI's oldest and largest wind tunnel – the T-101. Many questioned the advisability of manufacturing such an expensive model; it was only the support of Vitaliy G. Mikeladze, the then director of TsAGI's Section 2, that tipped the scales in favour of these tests. It was worth the effort: the test results eliminated the irregularities experienced in the smaller wind tunnels, allowing the aircraft's control system to be kept simple. The quarter-scale model was designated PM-335 (*pro**doo**vochnyy ma**ket*** – wind tunnel mockup of *izdeliye* 3-35).

Another major success achieved by the Myasishchev OKB was the resolution of the aerodynamic interference problems associated with carrying huge external loads. Of course, this work could not have been performed without close cooperation with TsAGI. When the placement of all cargoes envisaged by the project had been tested in wind tunnels and by mathematical extrapolation (in a remarkably short period, it should be noted), the OKB was able to select the optimum parameters for each payload configuration. Additionally, a schematic model of the 3M-T with 'payload 1GT' was tested in the Myasishchev OKB's own integrator, known as MAGDA (*mag**nit**naya **a**ero**ghid**rodina**mich**eskaya ana**log**iya* – magnetic aerohydrodynamic analogy), at Zhukovskiy.

The promising results yielded by the wind tunnel tests and mathematical analysis under the *izdeliye* 3-35 programme triggered the development of numerous transport systems designed for airlifting outsized loads. This was prompted by the fact that TsAGI and the State Institute of Transport Problems (a division of Gosplan – the Ministry of State Planning) had by then accumulated a wealth of statistical data on where and how such loads were transported.

Above: The first *Bison* to visit Baikonur was the 3M-5 missile carrier prototype (c/n 7300503), which landed at Yubileynyy airfield in March 1982 to check the base's suitability for the 3M-T. Hence the original caption to this photo called it 3M-T. The missile pylons and rear ECM fairing were removed by then, but the nonstandard bulbous radome unique to this version is very evident.

Left: The officers of the Buran's automated control system group (including group commander Lt.-Col. Anatoliy A Oosik on the left) pose for a photo beside the 3M-5 at Yubileynyy.

Opposite page: Three photos of the first VM-T, CCCP-01402, on final approach to Zhukovskiy in no-load configuration, showing clearly the aircraft's dachshund-like fuselage. Note the ventral ADF strake aerials with Doppler speed/drift sensor array between them. The IFR probe added in late 1984 is clearly visible.

In the case of the 3M-T, the loaded aircraft was a preposterous picture, since the fuel tank / main module of the Energiya SLV was 2.3 times larger in diameter than the bomber's fuselage and nearly as long as the aircraft itself! (To be precise, the length of 'payload 1GT' made up more than 80% of the fuselage length – including the stretch made in the process of conversion for the transport role!) Nevertheless, after receiving the go-ahead from the scientists, the designers set to work.

The cargo was secured by a system of struts forming an inverted-V brace at the front and a pair of pyramidal trusses at the rear. Each strut was adjustable lengthwise and featured joints at both ends to alleviate bending loads. To allow for possible errors in placing the anchoring points on the cargo, two of the struts forming the starboard rear truss incorporated spring-loaded length compensators, making it possible to attach the cargo without levelling the aircraft properly. The crew was reduced to six, the dorsal and tail gunners being unnecessary on a transport aircraft.

If developing the 3M-T was a complex task, then developing the associated cargo-handling equipment was no easy thing either. A huge gantry crane designated PKU-50.01 (*po**d**yo**m**no-kozlo**vo**ye* *oo**stro**ystvo* – gantry-type lifting device) was designed in-house, the figures reflecting the lifting capacity of 50 tons (110,230 lbs.).

Above: The third non-space-f ying test article of the Buran (OK-KS, c/n 0.03) minus vertical tail is loaded onto a VM-T by a PKU-50.01 gantry crane in February 1383. Note the relative size of the two aircraft. A BelAZ-7420 tug is preparing to tow the aircraft.

Below: VM-T CCCP-01402 carrying another nonflying Buran, the OK-ML1 (c/n 0.04), again minus vertical tail for lateral stability reasons

Above: Placing the cargo (in this case a dummy 'payload 2GT') on the aircraft's back required strength and precision – and the PKU-50.01 gantry crane had both. The VM-T is dwarfed by the huge crane. A ChTZ S-80 bulldozer acts as the tug in this instance.

Below: A dummy 'payload 2GT' is loaded onto the VM-T at Zhukovskiy, with fire engines (including a fire chief van) in attendance. This photo shows how the two halves of the PKU-50.01 could be slid apart to suit different payload lengths.

Above and below: The 'payload 1GT' was more than twice the fuselage diameter and nearly as long as the aircraft itself! Small wonder some people were opposed to the idea of the VM-T flying with cargoes! Fortunately the sceptics were wrong.

It was to load and offload the components of the Energiya/Buran space transport system, the heaviest of which weighed 32 tons (70,550 lbs.) – and handle them gently; the accuracy margin was to be no more than a few millimetres, and the crane was to operate at wind speeds up to 15 m/sec. (30 kts.). This huge machine was more than 50 m (164 ft.) long, 60 m (196 ft., 10 in.) wide, and 20 m (65 ft., 7 in.) high. Two such cranes were built – one in Zhukovskiy and one at Yubileynyy airfield near Baikonur. (Later, in 1987, a much more capable and truly colossal gantry crane designated PUA-100 (*po**dyom**no-oosta**no**vochnyy agre**gaht*** – lifting/installation device) with a capacity of 100 tons (220,450 lbs.) was built at Yubileynyy airfield for loading/unloading the VM-T and the An-225.)

The advanced development project of the 3M-T was submitted in mid-1978 and approved by the industry's R&D establishments and the customer; the OKB started issuing manufacturing drawings and preparing to convert . . .

Above and below: VM-T CCCP-01402 in its early days (without IFR probe) with 'payload 2GT'; the fluted texture of the Energiya SLV's cylindrical lower section is noteworthy. The aircraft ostensibly wore the then-current Aeroflot livery, with additional Aviaspetstrans titles.

To convert *what*? Good question. Obtaining the required three *Bison-Bs* for conversion into two flyable space transporters and a static-test airframe proved to be a major problem. The help of merited military pilot Maj. Gen. Nikolay F. Kuznetsov, Dr. Mil. Sc., who was then director of the Cosmonaut Training Centre, proved pivotal in this matter. By the greatest good luck, he and the DA commander, Col.-Gen. Vasiliy V. Reshetnikov, were buddies, having served in the same air force unit during the Great Patriotic War. After arranging a meeting with Reshetnikov, whom he had not seen for a long time, Kuznetsov arrived not alone but with Myasishchev OKB representatives in tow. The two veterans gave each other a great bear hug and stated recalling days gone by. A while later, when everyone was in a good mood, the OKB's spokesman stated his case, summing up with the request to transfer three bombers to the Myasishchev Experimental Machinery Plant. Reshetnikov, whose assent was all-important, was unenthusiastic, to say the

Above: CCCP-01502, the seccnd VM-T, makes a typically smoky takeoff with 'the Great Plywood Rocket' (a dummy 'payload 1GT').

Below: Even with a huge payload like Energiya's fuel tank, the VM-T displayed surprising agility, as demonstrated by this tight turn during an Aviation Day flypast at Moscow-Tushino, where the aircraft made its public debut.

Above right: CCCP-01502 flies with 'return payload 3GT,' the draggiest of all the VM-T's payloads.

Right: As-yet probeless CCCP-01402 with 'payload 2GT.' Note the two pairs of faired attachment points for the payload retention struts.

Below right: Another air-to-air of VM-T CCCP-01402 in postupgrade days (after 1984)

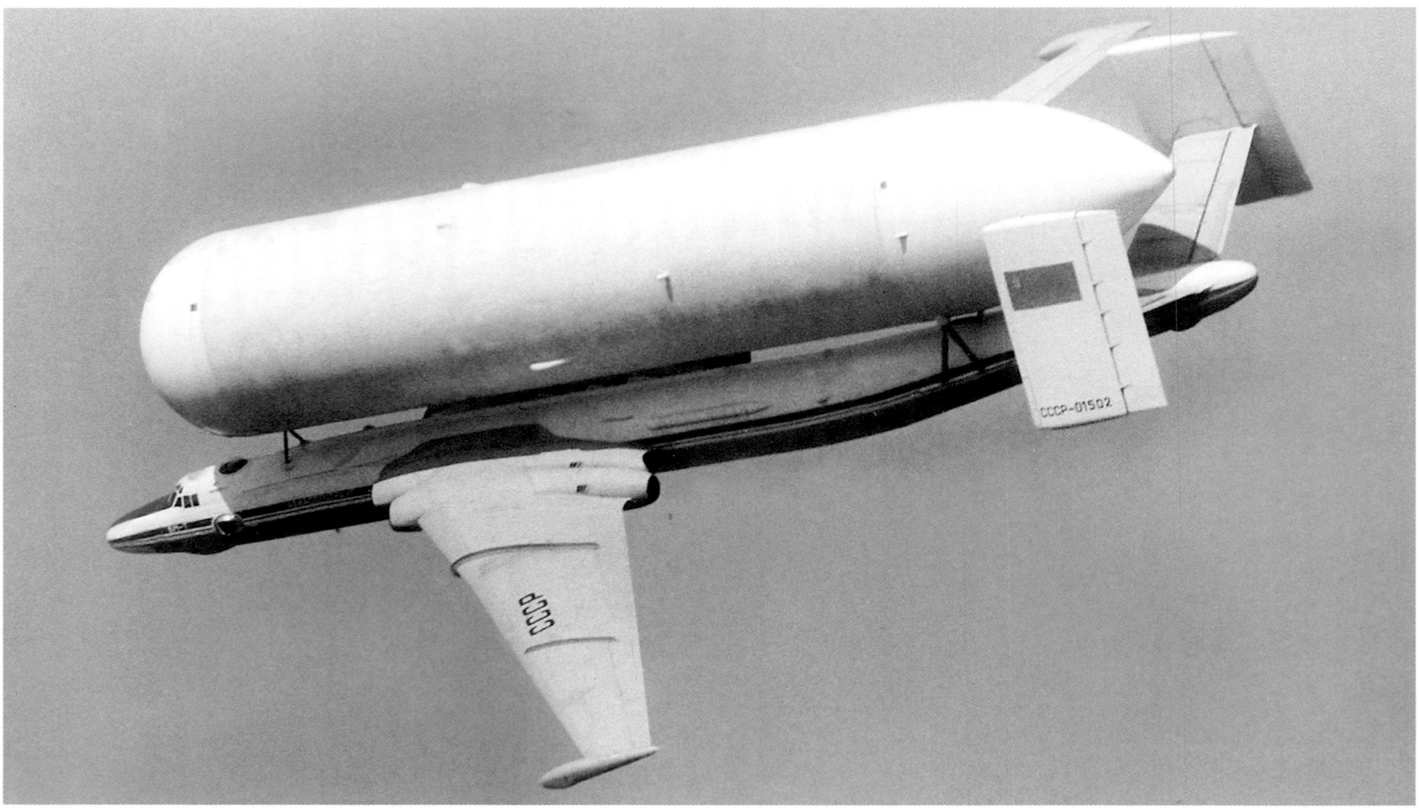

CCCP-01502

CCCP-01402

01402
CCCP-01402

least. *'Are you kidding? We're short of aircraft as it is!* he said. – *There's only two regiments left of them, one in the Far East* [at Ookraïnka AB – *Auth.*] *and one in Engels, and here you are demanding three aircraft!'* Still, he promised that the request would be duly considered.

Eventually, permission was granted and three low-time 3MN-2 tankers in good condition (tactical codes unknown, c/ns 9301402, 9301502, and 9301504) were turned over to the OKB in 1979. The airframes were thoroughly checked for fatigue cracks and deformations, and any defective parts were replaced; this work was performed by TsAGI and the Siberian Aviation Research Institute (SibNIA – *Sibeerskiy naoochno-issledovatel'skiy institoot aviahtsii*) in Novosibirsk under the supervision of the Myasishchev OKB.

Conversion work started in 1979. Airframe c/n 9301504 was earmarked as the static-test article, while the other two were to be completed to flying condition. The conversion job was a large-scale cooperative effort undertaken by MAP and the Ministry of Defence Industry (MOP – *Ministerstvo oboronnoy promyshlennosti*). The other participants were the Moscow Machinery Plant (named after Mikhail V. Khrunichev), OAPO *'Polyot'* ('Flight' Omsk Aircraft Production Association, formerly MAP plant No. 166), the *'Strela'* (Arrow) Orenburg Machinery Plant, the Kuibyshev aircraft factory (MAP plant No. 18), and the Tashkent Aircraft Production Association, named after Valeriy P. Chkalov (MAP plant No. 84). These enterprises supplied ship sets of internal structural members and skin panels for the VM-T's rear fuselage and tail unit, some long components of the centre fuselage structure, lower wing skin reinforcement panels, rear fuselage and tail unit assembly jigs, etc. The main part of the work (reinforcement of upper wing skins and wingtip outrigger strut fairings,

Above: CCCP-01502 taxies over the 'piano keys' of a wintry runway at Zhukovskiy, with Buran c/n 0.03 (the OK-KS systems test article), a.k.a. 'payload OGT,' on top. The OK-KS had a natural-metal finish, being devoid of heat insulation (the white-and-black ceramic tiles and the reinforced carbon-carbon composite nose cap and wing leading-edge fairings).

Left: The crew of VM-T CCCP-01402 poses after the aircraft's maiden flight on 29 April 1981. *Left to right*: I. Semookhin, V. Padukov, N. Gheneralov, A. Kucherenko, S. Sokolov, B. Ayzatoollin.

systems integration, design and manufacture of the jigs for the centre fuselage, outrigger strut fairings and the static-test airframe's outer wings, manufacture of some crucial airframe components, and final assembly of all three airframes) was the responsibility of the Experimental Machinery Plant.

Sadly, Vladimir M. Myasishchev did not live to see the 3M-T materialise, passing away on 14 October 1978 at the age of 76. In memoriam, the 3M-T was redesignated VM-T (VM for Vladimir Myasishchev), while the Experimental Machinery Plant was named after its founder. (There is also an alternative explanation that the aircraft was given the new designation because the original one plainly indicated its bomber origins, and the counterintelligence department put its foot down.) Later the VM-T received a popular name, ***Atlant*** (the Atlas of Greek mythology), as a reference to its impressive lifting capacity. The name was a fitting one: Atlas was doomed to hold up the sky till the end of time, while the VM-T was 'doomed' to uplift vehicles designed to go skyward.

After Myasishchev's death, the enterprise continued working under the direction of acting chief designer Valentin A. Fedotov, who was officially appointed chief designer by an MAP order on 23 May 1979. Fedotov was a disciple of Myasishchev, and within the first three months he showed clearly that his choice as the new chief designer was a good one. He managed to keep up the traditional Myasishchev approach to aircraft design, characterised by creative endeavour and engineering daring; this allowed the extremely complex transport aircraft project to be completed within an incredibly short time.

Final assembly proceeded in Zhukovskiy. Interestingly, modification work on the centre fuselage structure and wing assembly was performed without jigs, allowing the first flying example to

Above: CCCP-01502 takes off from Zhukovskiy, taking a Buran to Baikonur – possibly the one that eventually flew the space mission (c/n 1.01). Unlike the OK-KS, this one has most of the heat insulation tiles in place (though the tip of the nose and the wheel well doors remain bare). This photo gives a better view of the detachable tail fairing fitted to the Buran when carried by the VM-T.

Right: A Myasishchev OKB test crew discusses a flight in front of VM-T CCCP-01402 sometime after 1984.

The VM-T with no payload

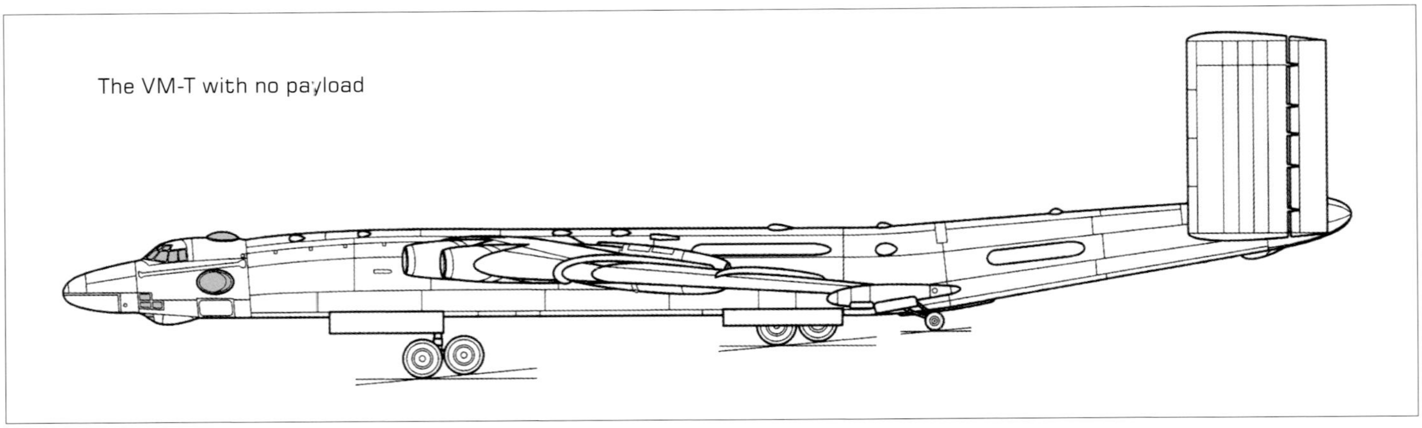

enter test 18 months after the commencement of the work and cutting conversion costs considerably (this immediately brings to mind how the M-4 bomber was created in just 22 months). The intensive and dedicated work of all the OKB's departments allowed the first prototype and the static-test article to be completed in late 1980, whereupon the fuselage and tail unit of c/n 9301504 were delivered for destructive testing to TsAGI, and the wings to SibNIA. The static tests took place in 1981.

Aptly registered CCCP-01402 and wearing the full 1973-standard blue/white livery of Aeroflot Soviet Airlines with additional *Aviaspets**trans*** (Air Special Transport) titles, the first flying VM-T (c/n 9301402) was completed in December 1980 and underwent a complete cycle of ground system tests with no payload (in the so-called autonomous configuration) between December 1980 and April 1981. The aircraft moved under its own power for the first time on 30 March; this was followed by a series of high-speed taxi runs. Finally, on 29 April 1981, the VM-T successfully made its first flight; the test crew comprised captain Anatoliy P. Kucherenko, copilot Nikolay N. Gheneralov, navigator S. Sokolov, conavigator I. Semookhin, flight engineer B. Ayzatoollin, and radio operator V. Padukov. Interestingly, Viktor V. Ootkin, the then director of LII–cum–chairman of MAP's Methodical Council (a body whose functions included authorising the flight tests of new aircraft types), stated bluntly when the VM-T's first flight clearance was being considered that a flight of this aircraft with an external load would take place only over his dead body!

During its maiden flight the VM-T experienced auto-oscillations of the rear fuselage caused by the rudder control runs not being properly balanced. The remedy was to add a 30 kg (66 lb.) load-carrying control rod in the rear fuselage. By the end of 1981, CCCP-01402 had made 19 test flights – all with no payload.

Tests with external loads were the next item on the agenda. In October 1981 the aircraft underwent ground tests with a mockup of the Energiya SLV's fuel tank / main module attached. Taxiing

The VM-T with payload 3GT

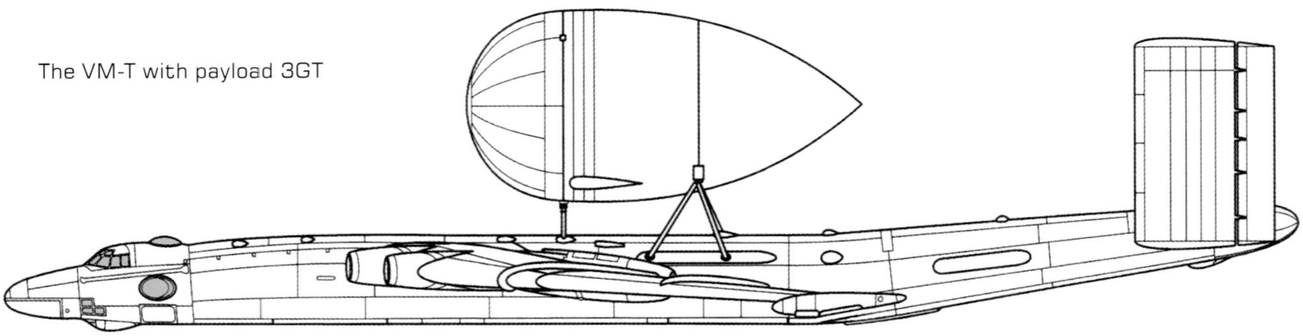

The VM-T with payload 2GT

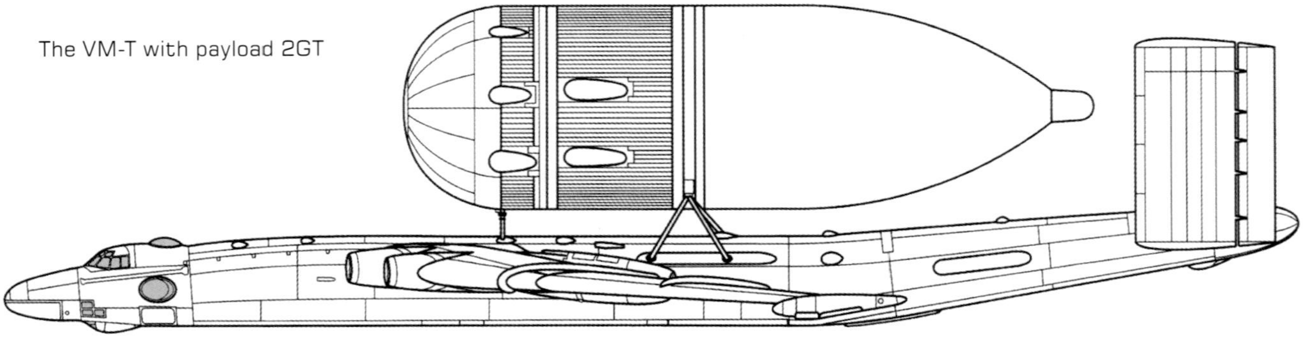

By 1991, VM-T CCCP-01502, which was used for demonstration flights with a dummy 'payload 1GT,' became a 'flying billboard,' advertising for the ASKO joint-stock insurance company.

tests commenced on 5 October; on 11 October and 19 December, CCCP-01402 made a series of runs in which speed was gradually increased. On 18 December, after carefully analysing the materials presented by the Myasishchev OKB, MAP's Methodical Council gave the go-ahead for flight tests in this configuration.

The first flight in loaded configuration was due to take place before the end of the year but had to be postponed several times due to foul weather. The breakthrough finally came on 6 January 1982, when the VM-T successfully flew with a mockup of the Energiya's LH_2 tank – the so-called payload 1GT. The crew was the same one that had performed the type's maiden flight.

Before any flights of the VM-T to Yubileynyy (Leninsk) could be ventured, the 3M-5 missile carrier prototype (c/n 7300503) made several flights to this airfield with a load of ballast in the bomb bay. By then, the aircraft had lost the underwing missile pylons and the ogival rear ECM fairing, but the nonstandard radome bulged at the front was still there; the cannon barbettes were removed. Interestingly, captions to photos of this aircraft at Yubileynyy refer to it as the 3M-T!

Having proved that the VM-T was indeed capable of fulfilling its intended mission, the test crew gradually 'taught it to fly' with all four intended payload configurations. For the early test flights in 'payload 0GT' configuration, the Buran space shuttle's vertical tail was detached to avoid a negative effect on the VM-T's directional stability.

'Payload 2GT' was first carried aloft in mockup form by CCCP-01402 on 15 March 1982; later, Myasishchev OKB test

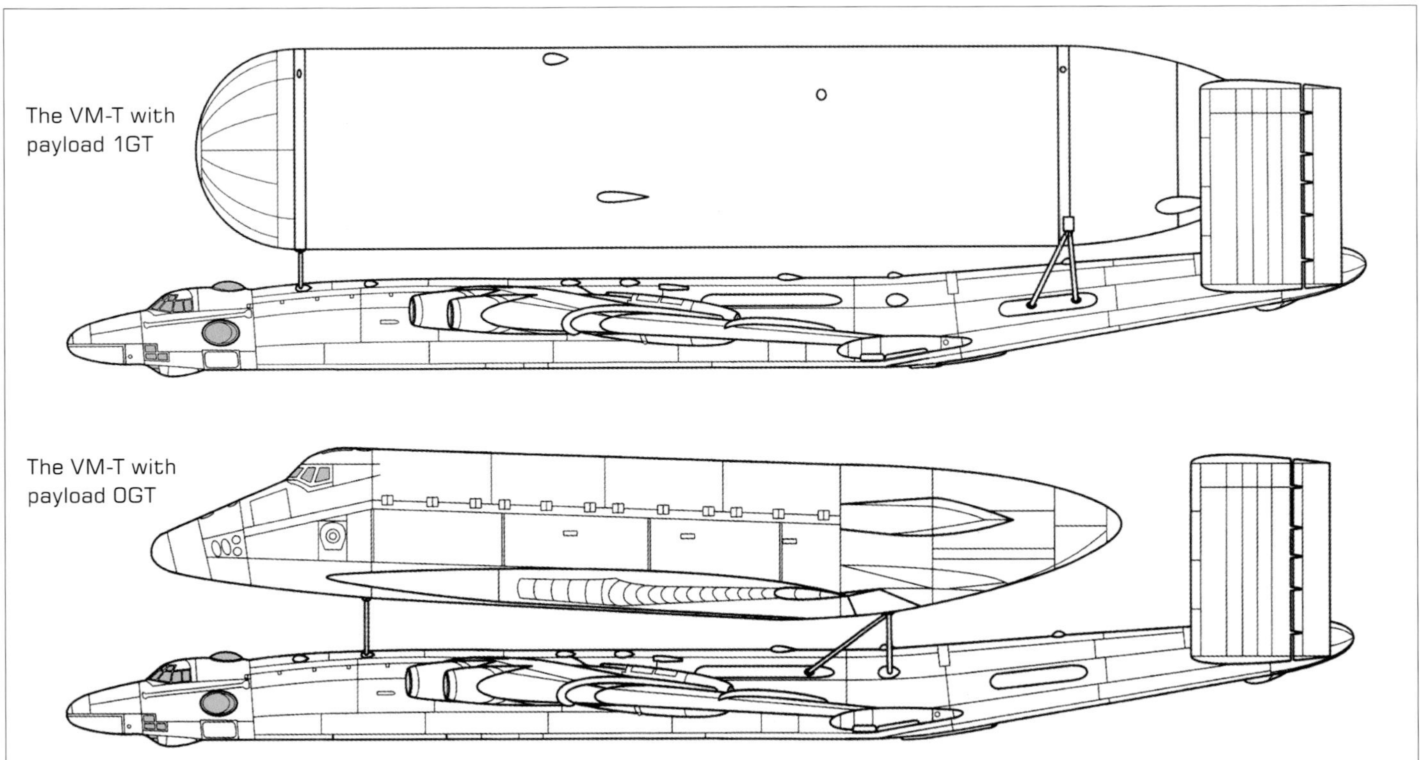

3M-T specifications according to ADP documents

	no load	payload 0GT	payload 1GT	payload 2GT	payload 3GT
Takeoff weight with 50 tons (110,230 lbs.) of fuel, kg (lbs.)	139,200 (306,880)	187,000 (412,260)*	170,500 (375,880)	169,000 (372,575)	152,200 (335,540)
Landing weight, kg (lbs.)	116,200 (256,170)	138,500 (305,335)	129,500 (285,490)	128,000 (282,190)	111,200 (245,150)
Operating empty weight (without fuel or ballast), kg (lbs.)	81,200 (179,010)	126,500 (278,880)	112,700 (248,460)	111,200 (245,150)	96,200 (21,210)
Payload, kg (lbs.)	none	45,300 (99,870)	31,500 (69,440)	30,000 (66,140)	15,000 (33,070)
Ballast fuel, kg (lbs.)	5,000 (11,020)	none	4,800 (10,580)	4,800 (10,580)	3,000 (6,610)
Required runway length, m (ft.)	n.a.	n.a.	3,500 (≈ 11,500)	n.a.	n.a.
Range, km (miles)	3,600 (2,240)	n.a.	n.a.	1,500+ (930+)	n.a.
Speed, km/h (mph)	500–600 (310–372)	n.a.	n.a.	n.a.	n.a.

* with a fuel load of 60 tons (132,275 lbs.)

Weights and dimensions of the 3M-T's payload options

Payload	Weight, kg (lbs.)	Length	Diameter
0GT	45,300 (99,870)*	38.45 m (126 ft., 1 49/64 in.)	23.8 m (78 ft., 1 in.) †
1GT	31,500 (69,440)	44.46 m (145 ft., 10 25/64 in.)	7.78 m (25 ft., 6 19/64 in.)
2GT	30,000 (66,140)	26.41 m (86 ft., 7 49/64 in.)	7.75 m (25 ft., 5 1/8 in.)
3GT	15,000 (33,070)	15.67 m (51 ft., 5 in.)	7.75 m (25 ft., 5 1/8 in.)

Notes:

* Later increased to 50 tons (110,230 lbs.)

† Wingspan

pilots Nikolay N. Gheneralov and Vladimir V. Arkhipenko flew the VM-T loaded with the real thing.

The second flying VM-T – again very appositely registered CCCP-01502 (c/n 9301502) – formally entered test on 25 January 1982. The aircraft was test flown in no-load configuration in March, proceeding to make its first flight with a dummy 'payload 1GT' on 19 April that year. Between August and November, both aircraft were in layup while modifications were being made; upon completion, the upgraded CCCP-01502 flew to Yubileynyy airfield (Leninsk) to undergo ground and flight tests with 'payload 3GT.' Less than a fortnight later, Myasishchev OKB test pilots Anatoliy P. Kucherenko and Eduard N. Chel'tsov took the machine into the air with the 'Easter egg' on top. The tests went without a hitch – only three flights out of the planned eight were required to obtain the desired data. Thus, on 28 December the aircraft could carry a 'payload 3GT' from Leninsk to Kuibyshev (Bezymyanka airfield).

The VM-T's specifications obtained during tests

Crew	6
Power plant	4 × Dobrynin VD-7MD
Engine thrust	4 × 10,750 kgp (4 × 23,700 lbst)
Length overall	51.2 m (167 ft., 11¾ in.)
Height on ground	11.5 m (37 ft., 8¾ in.)
Wingspan	53.14 m (174 ft., 4⅛ in.)
Wing area	351.78 m² (3,782.5 sq. ft.)
Wing loading	387 kg/m² (79.34 lbs. / sq. ft.)
Thrust/weight ratio at takeoff power	0.32
Empty weight	78,700 kg (173,500 lbs.)
All-up weight	136,200 kg (300,265 lbs.)
Fuel load and engine oil weight	53,000 kg (116,840 lbs.)
Payload	50,500 kg (111,330 lbs.)
Top speed	580 km/h (360 mph)
Speed at 4,000 m (13,120 ft.)	550 km/h (340 mph)
Landing speed	210–220 km/h (130–136 mph)
Climb time to 9,500 m (31,170 ft.)	14 minutes
Service ceiling	10,200 m (33,460 ft.)
Range	3,565 km (2,214 miles)
Endurance	Approx. 7 hours
Takeoff run	1,200 m (3,940 ft.)
Landing run	1,800–1,950 m (5,900–6,400 ft.)

In February 1983 the VM-T began ground tests with the third airframe of the Buran space shuttle (c/n 0.03 – that is, batch 0, third airframe in the batch); it was also known as the OK-KS (*orbi**tahl'**nyy ko**rahbl'-kom**pleksnyy stend* – orbiter/integrated test rig) because it was intended for various ground tests. On 1 March, Kucherenko and Chel'tsov performed the VM-T's first flight with the 45.3-ton (99,870 lb.) OK-KS on top. However, the eighth flight in this configuration, on 25 March, ended in an incident when the aircraft ran off the paved taxiway in a crosswind and sank into the soft mud because the nose gear had not been locked neutral; when the ground crew tried to jerk the aircraft free with a tug, the nose gear collapsed, the forward fuselage underside suffering considerable damage. Not until the Buran was lifted clear could the VM-T be towed to a hardstand; the repairs weren't finished until November 1983.

Generally, the tests proceeded at an unprecedentedly brisk pace; yet, even though the VM-T was a derivative of the proven 3M bomber, it did give a few nasty surprises. On one occasion after a generous bootful of rudder, the rear fuselage started oscillating dangerously with an amplitude of up to 250 mm (nearly 10 in.). It should be noted that the length of the control runs had increased as compared to the 3M because of the longer fuselage (as the reader remembers, the VM-T was 7 m longer). This probably caused lags between control inputs and elevator/rudder deflection, leading to oscillations. Had the aircraft retained the bomber's reversible control actuators, the obvious thing to do would have been to switch them off and go to manual mode in order to stop the oscillation. Under the circumstances, however, the only correct thing to do was to kill the speed and be extremely gentle with the controls.

The VM-T as such had a high lift/drag ratio, but the bulky and draggy external load ruined everything, reducing range dramatically. Besides, the large weight of the cargo meant the aircraft could not take off fully fuelled, so as not to exceed MTOW restrictions.

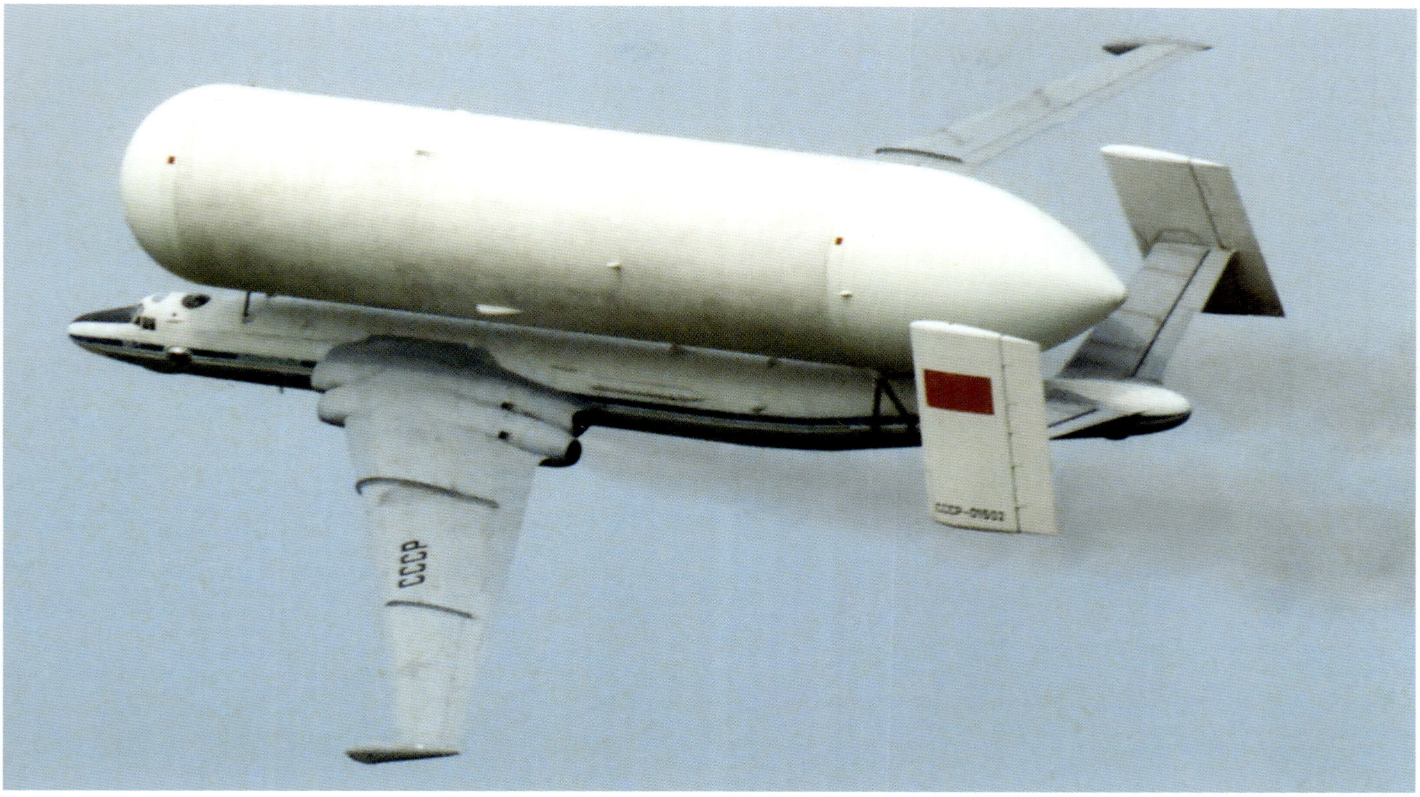

Above and below: VM-T CCCP-01502, carrying a dummy 'payload 1GT,' rolls from right to left during a flypast at an air show.

Above: CCCP-01402, with no payload attached, rests on the Myasishchev apron at Zhukovskiy.

Left and below: At another spot on the same apron, CCCP-01502 is serviced in front of a jet blast deflector, with two SPO-15M 'cherry pickers' on a Ural-375D 6×6 lorry chassis, a trailer-mounted power cart, and a compressed-air cart in attendance.

Top: CCCP-01402 and the OK-KS in flight; makes you wonder who's carrying whom! Seriously, the VM-T's ability to shoulder such loads is truly impressive.

Top right: Now wearing the nonstandard Russian registration RF-01502, the other VM-T sits under the PKU-50.01 crane at Zhukovskiy with a dummy 'payload 3GT' 'egg' attached.

Above and above right: Towed by a K-701 tractor, CCCP-01402 is shown at Yubileynyy airfield, ready to unload an actual 'payload 2GT.' The MAZ-543 prime mover in the background tows a low-loader trailer on which the payload will be lowered after being lifted off the aircraft.

Right: This aspect of the VM-T underscores the contrast between the swept wings and the unswept tail unit.

Hence, the Atlant's range with a 31.5-ton (69,440 lb.) payload was no more than 1,700 km (1,055 miles), necessitating a refuelling stop in Kuibyshev en route from Zhukovskiy to Baikonur. The only way to remedy this was to install an IFR system (the 3MN-2s set aside for conversion had none), and in late 1984 VM-T CCCP-01402 was retrofitted with an IFR probe from a 3M.

The VM-T also performed captive-carry tests with another Buran ground test airframe (OK-M, c/n 0.01). The OK-KS was eventually carried to Baikonur, as was the OK-ML1 airframe (c/n 0.04),

Above: By August 1993, concurrently with the new registration prefix and the Russian flag, VM-T CCCP-01502 gained 'Atlant' titles and the Myasishchev logo.

Below: The same aircraft in the static park of the MAKS-2015 air show

Above right: VM-T RF-01502 at the MAKS-2005

Below right: An aerial view of the same aircraft at Zhukovskiy during the MAKS-99 air show

which arrived in December 1983 for ground equipment integration work, and the OK-MT (c/n 0.15?) used as a prelaunch procedures trainer. The first fully capable Buran (c/n 1.01) arrived at Baikonur in December 1985 atop the VM-T to make its one and only space flight. It was followed on 23 March 1988 by the second fully capable shuttle (c/n 1.02), which, according to some sources, was named ***Boorya*** (Storm). About 300 km (186 miles) from the destination, the VM-T carrying this shuttle developed a fuel leak, forcing the No. 2 engine to be shut down to avoid the risk of a fire. However, it never rains but it pours: when the aircraft was on long finals to Yubileynyy, the generator circuits failed one by one, leaving the DC batteries as the sole source of electric power. Luckily, the landing gear was down by then, but the flaps could not be extended now; to top it all, the No. 1 engine also flamed out as the VM-T sideslipped. The situation was truly catastrophic; the huge freighter with an equally huge 'thunderbird' on top came in to land with the speed of a fighter. Yet, Anatoliy P. Kucherenko and Eduard N. Chel'tsov managed a smooth two-point landing with the utmost precision.

Top row: A comparison of the noses of VM-Ts RF-01502 (*left*) and RA-01402

Second row, left and centre: The nose gear unit and port outrigger strut of VM-T RA-01402

Second row, right: The tail unit of VM-T RA-01402

Above left: The port wing of VM-T RF-01502

Above: The 'payload 3GT' above the fuselage of VM-T RF-01502

Left: The engine air intakes of VM-T RF-01502

Top and top right: The flight deck of VM-T RF-01502

Above and above left: The navigator's station of VM-T RF-01502

Below: The same aircraft completes its landing run at Zhukovskiy.

It should be noted here that six Myasishchev OKB engineers managed the greater part of the work under the *izdeliye* 3-35 programme. They were V. A. Korchagin, V. M. Kirillin, A. I. Nikonov, I. M. Feighel'man, R. V. Zaïtsev, and R. V. Glushkova.

Meanwhile, in 1983 the Antonov OKB finally began designing the unique An-225 freighter tailor-made for supporting the Energiya/Buran multipurpose space transport system. The prototype (CCCP-480182, later reregistered CCCP-82060) was built in 1988 and successfully tested. The Antonov OKB could not pretend the VM-T did not exist, of course, but they often did try to 'play down' the importance of the effort the Myasishchev OKB undertook with this aircraft. For instance, An-225 project chief Anatoliy G. Vovnyanko wrote: *'Another aircraft used for this kind of transport operations was an upgraded 201M (this is how the 3M bomber was called by the Soviet popular press in the days when military censorship was really strict). It delivered large components of spacecraft to Baikonur; yet, this was the best amateurs could do in this field.'* (Amateurs, indeed! It's really disgusting that good aviation engineers never have a word of praise to say for the competition.) Yet, it was the VM-T that made possible both the first launch of the Energiya (loaded with a 100-ton (220,450 lb.) satellite on that occasion) and the first and only flight of the Buran. Between them, the two Atlants made more than 150 trips to Baikonur, carrying components of the Energiya/Buran system.

Incredibly, the entire job of converting three tankers into VM-T transports cost only 60 million rubles in 1980 prices! By comparison, building a factory at Baikonur with the purpose of manufacturing the Energiya SLV on site would have cost billions.

It would seem the new space transport system's unmatched capabilities ensured it a bright future. Yet, even as the programme reached fruition, the clouds began to gather. The first launch of the Energiya on 15 May 1987 ended in the loss of the Skif-DM (Scythian) satellite due to a fault in the latter's electrical circuitry. The second launch and the unmanned flight of the Buran on 15 November 1988 were successful, but then came the political and economic crisis of the late 1980s and early 1990s; the demise of the Soviet Union and the ensuing series of botched economic reforms drove most of the nation's major scientific and defence programmes into the ground. Only those programmes that managed to attract Western interest survived.

Speaking of Western interest, the VM-T was also involved in a western European endeavour to create a semireusable space transport system. In 1976 the French National Centre for Space Studies (CNES – Centre national d'études spatiales) began development of the Hermes aerospace plane in order to avoid excessive reliance on external partners for space exploration. This was a much-smaller vehicle than the space shuttle or the Buran. It was superficially similar in layout to the Boeing X-20 Dyna-Soar 'space glider,' featuring sharply swept delta wings with end-plate fins and rudders. Implementation of the project started in January 1985 – first by CNES and then, from October 1985 onward, by the European Space Agency (ESA). In its final form, the Hermes had a launch weight of 21 tons (46,300 lbs.), a payload of 3 tons (6,610 lbs.), and a crew of three; it was to be placed into orbit by an Ariane 5 SLV. The vehicle had an overall length of 14.584 m 47 ft., 10$\frac{11}{64}$ in.), a fuselage length of 12.875 m (42 ft., 2$\frac{57}{64}$ in.), a fuselage height of 3.094 m (10 ft., 1$\frac{13}{16}$ in.), and a wingspan of 9.402 m (30 ft., 10$\frac{5}{32}$ in.).

For a while, the idea was considered of having the Hermes airlifted to the launch site by the VM-T; the vehicle would be encapsulated in a transport container having a flattened elliptical shape, which made the VM-T eerily resemble an early AEW aircraft (such as the Grumman E-1 Tracer). A wind tunnel model of the aircraft with this payload was built and tested at TsAGI. The Hermes' so-called resource bay (an expendable equipment bay–cum–docking chamber / air lock whereby it was to be attached to the SLV) was to be transported in a separate pear-shaped container. The possibility of using the VM-T for free-flight tests of the Hermes was also considered. Later, however, the ESA decided to use a specially modified Airbus A300 medium-haul airliner as the transporter. Eventually the Hermes programme was cancelled in 1992 – partly due to unachievable cost and performance goals, and partly due to the formation of a partnership with the Russian Space Agency (RKA) within the International Space Station (ISS) programme, which reduced the demand for an independently crewed aerospace plane.

In 1992 the Myasishchev OKB began development of the Demonstrator multistage space launch system, based on the VM-T, as a joint effort with TsAGI and the engine manufacturer NPO Saturn. The aircraft was to carry a hypersonic suborbital launch vehicle powered by two D-57 cryogenic fuel rocket motors; these were, in fact, developed by Arkhip M. Lyul'ka back in the 1960s for the ill-starred N1 lunar rocket. The vehicle was unmanned, having a low-wing, tailless-delta layout with short double-delta wings positioned well aft; strongly canted twin tails; and a flattened fuselage, which had a double-bubble (horizontal figure-eight) cross-section, the two rocket motors being positioned side by side. It carried a Western aerospace plane – the Sänger HORUS (Hypersonic Orbital Upper Stage), ESA Hermes, British Aerospace HOTOL (Horizontal Takeoff and Landing) or the like – or a disposable single-stage launch rocket with a satellite. It could also be used as a launch vehicle for a hypersonic drone or as an engine test bed. By using existing rocket motors, the Demonstrator could be used to investigate such complex problems as in-flight starting of rocket motors running on cryogenic fuel (LOX/LH_2), stage separation dynamics, remote control of an unmanned space system at the orbit entry, deorbiting and automatic landing stages, devising maintenance technologies, etc. Using the VM-T as the carrier aircraft offered substantial cost savings during space transport system and propulsion technology development. At the same time, revenue could be generated by launching small satellites; the Demonstrator could thus compete successfully with the US Orbital Sciences Corporation's Pegasus system (based on the B-52 and the Lockheed L-1011 TriStar airliner) and the proposed Russian Burlak system, using the Tu-160SK – a projected version of the Blackjack optimised for suborbital launch, which never materialised.

Between 11 and 16 August 1992, the second VM-T carrying a dummy 1GT payload was displayed at the MosAeroShow '92 – Russia's first major international air show – at the LII airfield in Zhukovskiy, wearing the altered registration РФ-01502 with the nonstandard prefix RF in Cyrillic characters and Cyrillic 'Atlant' titles instead of Aeroflot/Aviaspetstrans titles. (This unusual prefix denoting *Rosseeyskaya Federahtsiya* [Russian Federation] arose at

Sitting parked on the edge of the apron at Dyagilevo AB, VM-T RA-01402 looks rather eerie on a foggy night like this. Note the fire extinguisher cart in front of the aircraft; now that's some serious fire extinguishers!

A model of the VM-T carrying a 'rocket sandwich' in the form of a winged reusable suborbital launch vehicle with a single-stage rocket (carrying the payload) on top

Here the same model carries a different winged reusable vehicle – apparently a test bed with what looks like a dorsal air intake for an experimental hypersonic ramjet.

A third variation on the same theme – the same suborbital launch vehicle carries a further winged, rocket-powered vehicle. All three versions were developed under the Demonstrator programme.

a time of uncertainty as to what nationality prefix should be used; both Cyrillic and Roman RF versions – or no prefix at all – were tried until the International Civil Aviation Organisation [ICAO] assigned the RA civil prefix. This is not to be confused with the current *quasi-civil* RF prefix introduced for government [that is, military and paramilitary] aviation in 2009.) It was probably at the MosAeroShow '92 that the first attempts were made to secure funds for full-scale development of the Demonstrator system. Initially the latter seemed to attract considerable interest, but no real money was forthcoming; the Russian government either had no money or was not about to earn money that way. Thus ended, sadly, the story of an aircraft that doubtless belongs in Russian aviation's hall of fame.

By 1994, NPO Molniya had completed the preliminary design work on the MAKS reusable aerospace system (*mnogotselevaya aviatsionno-kosmicheskaya sistema* – multipurpose aerospace system). The system comprised the An-325 aircraft (a modified An-225) as a launch platform and a disposable or reusable aerospace vehicle. The latter came in four versions, three of which had two stages. In particular, the MAKS-OS (*orbitahl'nyy samolyot* – 'orbital aircraft' or space plane) was intended for placing a load of 8.5 to 10 tons (18,740 to 22,050 lbs.) into an orbit of 200 km (124 miles) above the Earth. It comprised a manned spacecraft looking like a 'baby Buran,' which had two varieties – OS-P (*piloteeruyemyy* – manned) and OS-B (*bespilotnyy* – unmanned). The sharp-nosed, delta-wing orbiter sat on top of a fuel/oxidiser tank whose shape resembled a rugby ball. For a while, NPO Molniya considered using the VM-T for free-flight tests of the MAKS-OS; however, this spacecraft also remained unbuilt.

Similarly, plans were in hand to use the VM-T for flight-testing the liquid-propellant rocket booster of the Baikal/Angara semireusable space launch system codeveloped by the Khroonichev Space Centre and NPO Molniya. Properly designated U1M (*ooskoritel'* – booster) but popularly known as Baikal, the booster designed by Molniya was meant to fly home after separation from the core module of the *Angara* SLV (named after a Siberian river). To this end, it was equipped with folding wings and tail surfaces, a 5,000 kgp (11,030 lbst) turbojet in the nose, and a retractable tricycle landing gear, as well as an automatic control system. The Baikal was 28.5 m (93 ft., 6 in.) long and 8.5 m (27 ft., $10\frac{41}{64}$ in.) high on the ground, with a wingspan of 17.1 m (56 ft., $1\frac{15}{64}$ in.) in flight mode; dry weight was 17.8 tons (39,240 lbs.) and landing weight 18 tons (39,680 lbs.). However, the Baikal got only as far as the full-size mockup stage.

At the MAKS-93 air show in Zhukovskiy (in this case, the acronym stands for *Moskovskiy aviatsionno-kosmicheskiy salon* – Moscow Aerospace Salon), VM-T êî-01502 was displayed with a dummy payload 3GT 'egg'; since then, this aircraft has been a regular exhibit at the biennial MAKS air shows in this configuration. The first VM-T, now reregistered RA-01402 and still wearing Aeroflot titles (but not the Myasishchev logo), has been in open storage at Dyagilevo AB since 1993.

3M-T Outsized-Cargo Transporter (*izdeliye* 3-35F, Project)

For a while, the Myasishchev OKB considered reengining the 3M-T with Lyul'ka AL-31F afterburning turbofans; this version was known in-house as *izdeliye* 3-35F, the *F* denoting *forsazh* (afterburning). However, this would have required major changes to the wing structure, and the idea was soon dropped.

3M-T Outsized-Cargo Transporter (*izdeliye* 3-35M, Project)

In 1981 the Myasishchev OKB contemplated the possibility of carrying the Buran orbiter on the VM-T in almost completely outfitted condition; this would radically cut the time needed to install various systems and equipment on the orbiter upon arrival to Baikonur. Of course, the fully equipped Buran would be heavier, and the VM-T would need to take off with a reduced fuel supply for TOW limitation reasons; however, this would reduce range unacceptably.

To remedy this, the Myasishchev OKB considered providing it with an IFR system – and a highly unusual one at that. Refuelling from a 3MS-2/3MN-2 tanker in the usual way would be difficult because the external load would be subjected to buffeting after hitting the tanker's wake turbulence, especially because the Buran's vertical tail would be in place. This was not only fraught with damage to the cargo but would endanger flight safety. Therefore, the designers considered an arrangement where the tanker would come up behind the receiver aircraft and pump the fuel up; this would necessitate a different IFR system, but no details are known.

The VM-T equipped in this manner was designated *izdeliye* 3-35M (*modifitseerovannoye* – modified). Soon, however, work in this direction was halted.

VM-T/'53' Air Launch System (Project)

A desktop model exists that shows the VM-T carrying the Myasishchev M-17 *Mystic-A* high-altitude, subsonic aircraft piggyback. The M-17 was conceived for intercepting high- and slow-flying reconnaissance balloons, which were launched *en masse* from western Europe for a while and were hard to kill for conventional interceptors. However, the aircraft carried by the VM-T lacks the dorsal cannon barbette ahead of the wing centre section for destroying the balloons; moreover, the model shows both the VM-T and the M-17 featuring triple winglets! Moreover, the VM-T appears to sport unusual inverted-T aerodynamic surfaces under the wing roots, resembling the hydrofoils of certain ships!

The base of the display stand is enigmatically marked '53,' suggesting the designation M-53. This designation had been allocated in 1959 to one of several supersonic airliner projects by OKB-23, which never materialised; however, the designation may have been reallocated, of course.

3M2 Twin-Fuselage Superheavy Transport Aircraft (3M2-1 through 3M2-4, Projects)

When the Myasishchev OKB started development of the M-52 superheavy transport aircraft (this project is not to be confused with the 1960s strategic missile strike aircraft of the same name, which was based on the M-50 *Bounder* bomber!), the first direction of work envisaged the use of stock airframe components from the 3M strategic bomber. The designers came to the conclusion that this called for using a twin-fuselage layout in which the bulky space vehicle systems could be carried externally under an arched wing centre section. Only the port fuselage featured a flight deck.

Several versions embodying this concept were developed. They were referred to as 3M2 with suffix numbers. The versions

The model of the VM-T with the Demonstrator suborbital launch system displayed at one of the MAKS air shows

differed in the configuration of the arched wing centre section joining the fuselages, whereas the wings outboard of the fuselages were identical to those of the standard bomber. Each 'half' usually retained the original engine installation, producing an eight-engine power plant for the resulting machine. Since the aircraft had a quad landing gear with four-wheel bogies, the wingtip-mounted outrigger struts were moved inward or deleted altogether.

In the **3M2-1** version, the wing centre section consisted of two halves having the same leading-edge sweepback as the original wings of the bomber. In combination, this produced a V planform looking like an aft-facing arrowhead, each half having constant chord. This made it possible to retain a symmetrical arrangement of the engines on each fuselage. The centre section was fitted with two sets of two-section trailing-edge flaps and spoilers.

In the **3M2-2** version, the arched wing centre section had a rectangular planform, with a corresponding alteration in the position of the inboard engine air intakes, which were no longer staggered.

The **3M2-3** version departed still further from the original airframes. Only the fuselages with their tail units were retained; the rectangular wing centre section was stripped of the engines, which were moved to the outer wing panels. The latter were redesigned, with shorter span and winglets added, and the original eight 11,000 kgp (24,250 lbst) Dobrynin VD-7M turbojets were replaced by four 23,430 kgp (51,650 lbst) Lotarev D-18T high-bypass turbofans, two on each outer wing.

The **3M2-3A** version differed in having a wing centre section that was flat, not arched, in head-on view. The shuttle orbiter, riding a fuel/oxidiser tank (similar to the MAKS-OS), sat on top of this wing centre section; upon release it would go 'up, up and away.'

The **3M2-4** version reverted basically to the 3M2-1 configuration, but with substantial alteration to the wings and the inboard engine packages, which were housed in common nacelles.

In all cases the wing centre section served as a place for the attachment of the payload, which consisted of a small reusable

Above: A model of the VM-T carrying an M-17 'balloon killer' aircraft. Note the winglets and the underwing aerodynamic surfaces.

Below: Four views of the 3M2-1 superheavy transport carrying a space shuttle, riding a booster rocket between the fuselages

Below right: The 3M2-2 differs on y in the wing centre section planform. Note the position of the underwing supporting struts.

space plane with its huge propellant tank. In all versions except the 3M2-1, the wings were fitted with winglets.

3M2-5 Superheavy Transport Aircraft (Project)

The 3M2-5 version differed from the preceding ones in being a single-fuselage aircraft, which had virtually no structural commonality with the original 3M bomber, apart from the nose section of the fuselage with the flight deck and the arrangement of the engines in the wing roots. However, in this case, four, not two, engines flanked the fuselage on each side; they were set much farther apart from the fuselage. The wings were completely redesigned, with additional trailing-edge flaps between the engines and the fuselage, the inboard portions having no trailing-edge sweep. The fuselage was also new – it was much wider aft of the flight deck, having an almost rectangular greater cross-section. The tail unit, with twin end plates, was patterned on that of the 3M-T (VM-T) transport conversion of the bomber. Similarly, the space plane was perched on top of the fuselage. The aircraft had a quad landing gear with side-by-side pairs of 3M nose gear and main gear units having a narrow track – and, oddly enough, no outrigger struts to support the wings.

All these projects proceeded concurrently with the practical work on the 3M-T (VM-T). Eventually, all attention of the design bureau came to be concentrated on this latter work, and the 3M2 series of projects failed to proceed further than the drawing board.

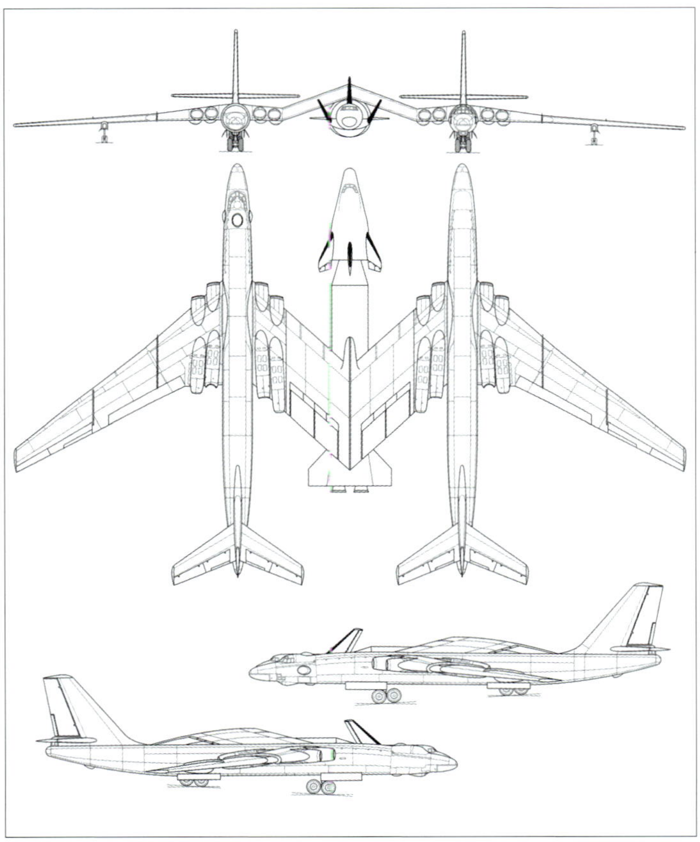

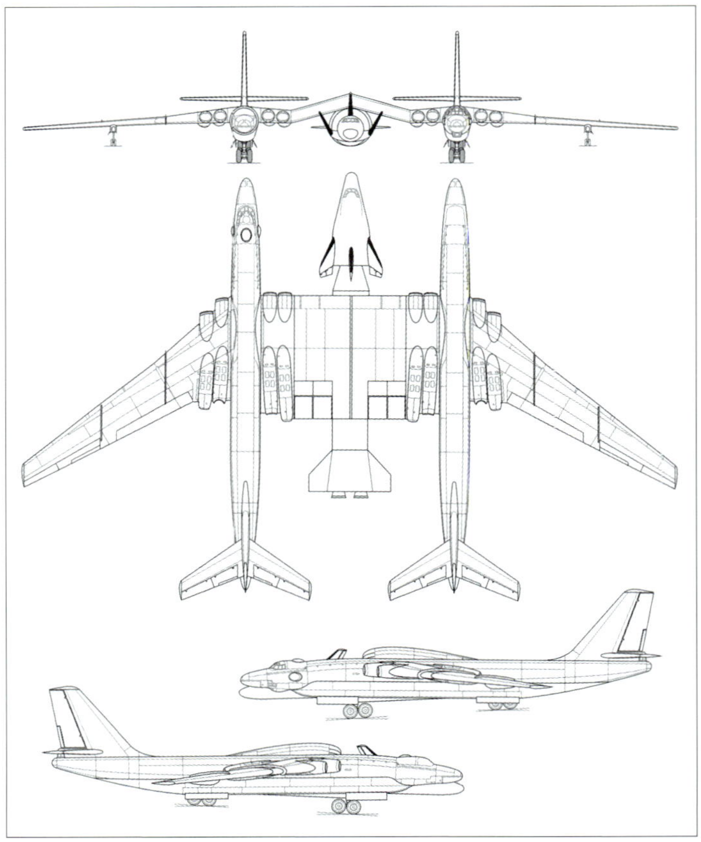

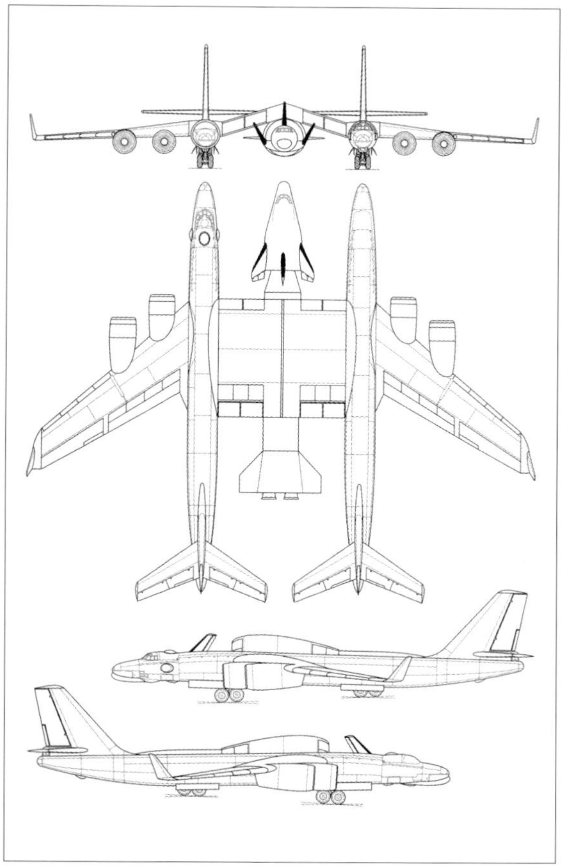

Left: The 3M2-3 version is similar to the 3M2-2 but has four D-18T turbofans instead of eight RD-3M-500A turbojets.

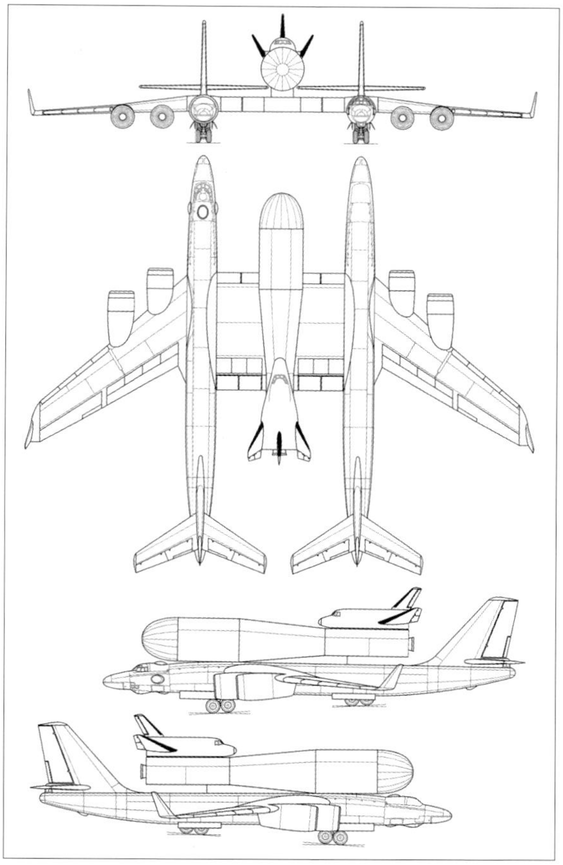

Right: The 3M2-3A version has the same power plant, but a new wing centre section that is not arched, the payload (a space shuttle riding a fuel tank) sitting on top rather than being carried underneath.

Below: The 3M2-4 version, with a wing arrangement similar to the 3M2-1 but redesigned inboard engine nacelles and a different payload

Below right: Four views of the completely different 3M2-5, with a single fat fuselage and four turbojets on each side.

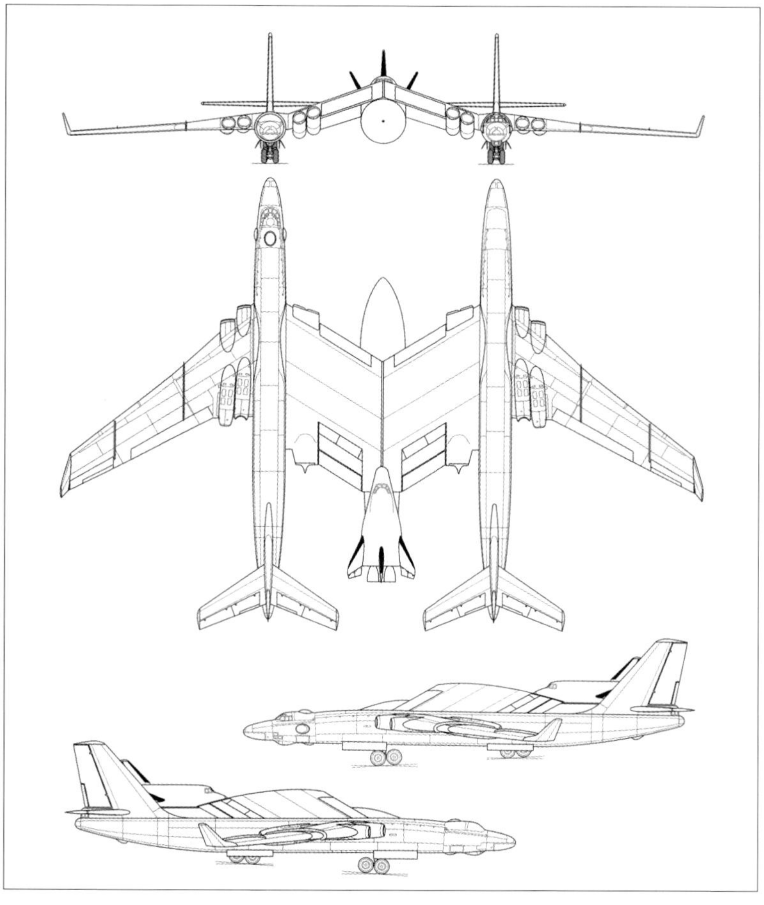

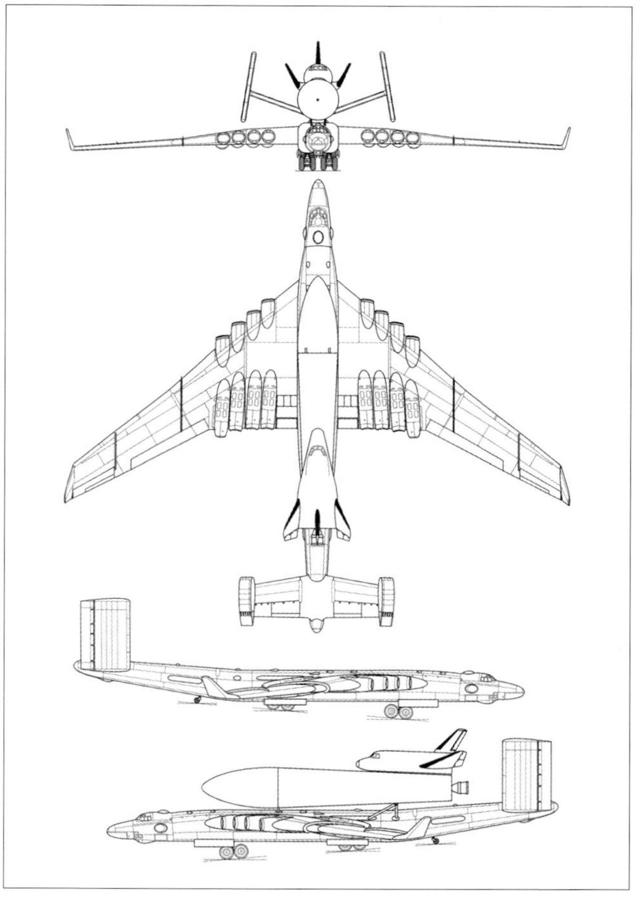

M-4 family specifications

	M-4 (prototype)	M-4 (early production)	M-4 (late production)	M-4-2
Year	1953	1954	1956	1956
Engine type	AM-3A	AM-3A	RD-3M-500A	RD-3M-500A
Takeoff thrust, kgp (lbst)	8,750 (19,290)	8,750 (19,290)	9,500/10,608 (20,940/23,386) [1]	9,500/10,608 (20,940/23,386) [1]
Length overall:				
less IFR probe	≈ 48.7 m (159 ft., $9^{21}/_{64}$ in.)	47.665 m (156 ft., $4^{37}/_{64}$ in.)	47.665 m (156 ft., $4^{37}/_{64}$ in.)	47.665 m (156 ft., $4^{37}/_{64}$ in.)
with IFR probe	–	–	–	–
Fuselage length	45.6 m (149 ft., $7^{7}/_{32}$ in.)	44.6 m (146 ft., $3^{29}/_{32}$ in.)	44.6 m (146 ft., $3^{29}/_{32}$ in.)	44.6 m (146 ft., $3^{29}/_{32}$ in.)
Height on ground	11.45 m (37 ft., $6^{25}/_{32}$ in.)	11.45 m (37 ft., $6^{25}/_{32}$ in.)	11.45 m (37 ft., $6^{25}/_{32}$ in.)	11.45 m (37 ft., $6^{25}/_{32}$ in.)
Wingspan	50.526 m (165 ft., $9^{7}/_{32}$ in.)	50.526 m (165 ft., $9^{7}/_{32}$ in.)	50.526 m (165 ft., $9^{7}/_{32}$ in.)	50.526 m (165 ft., $9^{7}/_{32}$ in.)
Stabiliser span	15.0 m (49 ft., $2^{35}/_{64}$ in.)	15.0 m (49 ft., $2^{35}/_{64}$ in.)	15.0 m (49 ft., $2^{35}/_{64}$ in.)	15.0 m (49 ft., $2^{35}/_{64}$ in.)
Landing-gear track	n.a.	n.a.	n.a.	n.a.
Landing-gear wheelbase	14.41 m (47 ft., $3^{21}/_{64}$ in.)	14.41 m (47 ft., $3^{21}/_{64}$ in.)	14.41 m (47 ft., $3^{21}/_{64}$ in.)	14.41 m (47 ft., $3^{21}/_{64}$ in.)
Wing area, m^2 (sq. ft.)	326.35 (3,512.8)	326.35 (3,512.8)	326.35 (3,512.8)	326.35 (3,512.8)
Flap area (total), m^2 (sq. ft.):				
hinged flaps	17.05 (183.52)	17.05 (183.52)	17.05 (183.52)	17.05 (183.52)
TsAGI flaps	n.a.	n.a.	n.a.	n.a.
Aileron area (total), m^2 (sq. ft.)	n.a.	n.a.	n.a.	n.a.
Vertical tail area, m^2 (sq. ft.)	45,725 (492.18)	45,725 (492.18)	45,725 (492.18)	45,725 (492.18)
Horizontal tail area, m^2 (sq. ft.)	n.a.	n.a.	n.a.	n.a.
Elevator area, m^2 (sq. ft.)	n.a.	n.a.	n.a.	n.a.
Empty weight, kg (lbs.)	n.a.	79,700 (175,705)	80,000 (176,370)	80,000 (176,370)
Takeoff weight, kg (lbs.):				
normal	131,800 (290,560)	138,500 (305,335)	140,000 (308,640)	140,000 (308,640)
maximum	181,500 (400,130)	181,500 (400,130)	184,500 (406,750)	184,500 (406,750)
Landing weight, kg (lbs.):				
normal	100,000 (220,450)	n.a.	n.a.	n.a.
maximum	n.a.	n.a.	n.a.	n.a.
Payload, kg (lbs.):				
normal	5,000 (11,020)	5,000 (11,020)	5,000 (11,020)	5,000 (11,020)
maximum	24,000 (52,910)	24,000 (52,910)	24,000 (52,910)	40,000 (88,180) [6]
Thrust/weight ratio with MTOW	0.19	0.19	0.21	0.21
Wing loading with MTOW, kg/m^2 (lbs. / sq. ft)	556 (114.0)	556 (114.0)	565 (115.8)	565 (115.8)
Top speed, km/h (mph):				
at sea level	n.a.	n.a.	n.a.	n.a.
at high altitude [2]	962/950 (597/590) [3]	947 (588)	947 (588)	950 (590)
Cruising speed, km/h (mph)	n.a.	n.a.	n.a.	n.a.
Unstick speed, km/h (mph)	n.a.	n.a.	n.a.	n.a.
Landing speed, km/h (mph)	n.a.	n.a.	n.a.	n.a.
Service ceiling, m (ft.)	12,500 (41,010)	11,000 (36,090)	11,000 (36,090)	11,000 (36,090)
Range with normal payload on internal fuel, km (miles)	9,800 (6,086)	8,100 (5,031)	>8,000 (4,968)	4,000 (2,484) [4]
Ditto with one fuel top-off, km (miles) [5]	–	–	–	–
Endurance	n.a.	n.a.	n.a.	n.a.
Takeoff run, m (ft.)	n.a.	2,610 (8,560)	2,610 (8,560)	2,610 (8,560)
Landing run, m (ft.)	n.a.	1,470 (4,820)	1,500 (4,920)	1,500 (4,920)

Notes:
1. The second figure is a two-m nute contingency rating; 2. Optimum cruise altitude;
3. Manufacturer's flight tests / state acceptance trials; 4. Combat radius; 5. Takeoff weight 190,000 kg (418,870 lbs.); 6. Fuel

3M (prototype)	3MN-1	3MS-1	3MN-2	3MD	VM-T
1956	1956	1958	1960	1960	1982
VD-7	VD-7B	RD-3M-500A	VD-7B	VD-7B	VD-7MD
11,000 (24,250)	9,580 (21,120)	9,500/10,608 (20,940/23,386) [1]	9,580 (21,120)	9,580 (21,120)	10,750 (23,700)
48.76 m (159 ft., 11 19/64 in.)	48.76 m (159 ft., 11 19/64 in.)	48.76 m (159 ft., 11 19/64 in.)	48.76 m (159 ft., 11 19/64 in.)	49.805 m (163 ft., 4 53/64 in.)	51.23 m (168 ft., 59/64 in.)
51.3 m (168 ft., 3 11/16 in.)	51.3 m (168 ft., 3 11/16 in.)	51.3 m (168 ft., 3 11/16 in.)	–	51.805 m (169 ft., 11 9/16 in.)	n.a.
46.7 m (153 ft., 2 19/32 in.)	46.7 m (153 ft., 2 19/32 in.)	46.7 m (153 ft., 2 19/32 in.)	46.7 m (153 ft., 2 19/32 in.)	47.752 m (156 ft., 8 in.)	n.a.
11.5 m (37 ft., 8 3/4 in.)	11.5 m (37 ft., 8 3/4 in.)	11.5 m (37 ft., 8 3/4 in.)	11.5 m (37 ft., 8 3/4 in.)	11.5 m (37 ft., 8 3/4 in.)	11.5 m (37 ft., 8 3/4 in.)
53.14 m (174 ft., 4 1/8 in.)	53.14 m (174 ft., 4 1/8 in.)	53.14 m (174 ft., 4 1/8 in.)	53.14 m (174 ft., 4 1/8 in.)	53.14 m (174 ft., 4 1/8 in.)	53.14 m (174 ft., 4 1/8 in.)
15.231 m (49 ft., 11 41/64 in.)	15.231 m (49 ft., 11 41/64 in.)	15.231 m (49 ft., 11 41/64 in.)	15.231 m (49 ft., 11 41/64 in.)	15.231 m (49 ft., 11 41/64 in.)	n.a.
52.342 m (171 ft., 8 43/64 in.)	52.342 m (171 ft., 8 43/64 in.)	52.342 m (171 ft., 8 43/64 in.)	52.342 m (171 ft., 8 43/64 in.)	52.342 m (171 ft., 8 43/64 in.)	52.342 m (171 ft., 8 43/64 in.)
14.41 m (47 ft., 3 21/64 in.)	14.41 m (47 ft., 3 21/64 in.)	14.41 m (47 ft., 3 21/64 in.)	14.41 m (47 ft., 3 21/64 in.)	14.41 m (47 ft., 3 21/64 in.)	14.41 m (47 ft., 3 21/64 in.)
351.78 (3,786.53)	351.78 (3,786.53)	351.78 (3,786.53)	351.78 (3,786.53)	359.82 (3,873.07)	351.78 (3,786.53)
17.05 (183.52)	17.05 (183.52)	17.05 (183.52)	17.05 (183.52)	17.05 (183.52)	17.05 (183.52)
43.954 (473.12)	43.954 (473.12)	43.954 (473.12)	43.954 (473.12)	43.954 (473.12)	43.954 (473.12)
16.8 (180.83)	16.8 (180.83)	16.8 (180.83)	16.8 (180.83)	16.8 (180.83)	16.8 (180.83)
45,725 (492.18)	45,725 (492.18)	45,725 (492.18)	45,725 (492.18)	45,725 (492.18)	n.a.
57.12 (614.83)	57.12 (614.83)	57.12 (614.83)	57.12 (614.83)	57.12 (614.83)	n.a.
15.58 (167.7)	15.58 (167.7)	15.58 (167.7)	15.58 (167.7)	15.58 (167.7)	n.a.
74,430 (164,090)	75,740 (166,975)	≈ 76,000 (167,550)	75,700 (166,890)	76,800 (169,310)	78,700 (173,500)
150,000 (330,690)	140,000 (308,640)	140,000 (308,640)	140,000 (308,640)	145,600 (320,990)	136,200 (300,260)
203,500 (448,630)	193,000 (425,485)	193,000 (425,485)	193,000 (425,485)	193,000 (425,485)	n.a.
n.a.	105,000 (231,480)	105,000 (231,480)	105,000 (231,480)	105,000 (231,480)	n.a.
n.a.	125,000 (275,570)	125,000 (275,570)	125,000 (275,570)	125,000 (275,570)	n.a.
5,000 (11,020)	5,000 (11,020)	5,000 (11,020)	5,000 (11,020)	5,000 (11,020)	n.a.
18,000 (39,680)	18,000 (39,680)	18,000 (39,680)	46,000 (101,410) [6]	18,000 (39,680)	32,000 (70,550)
0.22	0.20	0.20	0.20	0.20	0.32
578 (118.5)	549 (112.5)	549 (112.5)	549 (112.5)	≈ 520 (106.6)	387 (79.3)
n.a.	600 (372)	600 (372)	600 (372)	600 (372)	n.a.
950 (590)	925 (574)	925 (574)	910 (565)	970 (602)	n.a.
n.a.	780–800 (484–496)	780–800 (484–96)	780–800 (484–496)	780–800 (484–496)	580 (360)
n.a.	303 (188)	n.a.	n.a.	302 (187)	n.a.
n.a.	200–215 (124–133)	n.a.	n.a.	200–215 (124–133)	210–222 (130–136)
n.a.	12,750 (41,830)	12,250 (40,190)	12,250 (40,190)	12,150 (39,860)	10,200 (33,460)
≈ 12,000 (7,453)	11,800 (7,329)/ 11,050 (6,863) [5]	10,500 (6,521)	5,500 (3,416) [4]	10,950 (6,801)/ 10,800 (6,708) [5]	3,565 (2,214)
n.a.	13,800 (8,571)	n.a.	–	13,550 (8,416)	–
n.a.	18 hrs., 30 min. [7]	n.a.	n.a.	18 hrs., 10 min. [7]	≈ 17 hrs
2,800 (9,190)	2,900 (9,510)	≈ 2,800 (9,190)	≈ 2,800 (9,190)	2,850 (9,350)	1,200 (3,940)
1,650 (5,410)	1,350–1,520 (4,430–4,990)	1,350–1,520 (4,430–4,990)	1,350–1,520 (4,430–4,990)	1,350–1,520 (4,430–4,990)	1,800–1,950 (5,900–6,400)

7. With one fuel top-off

Appendix 2

M-4 Family Production List

The following is a production/fleet list of the M-4 and 3M variants. The split presentation of the c/ns is for the sake of convenience only.

C/n	Version	Tactical code / registration	Notes
M-4: c/n 5302023 = year of manufacture 1955, plant No. 23, meaning of the third digit unknown, 20th production aircraft and 23rd M-4 built			
00.01	M-4	none	First prototype, Myasishchev OKB; f/f on 20-1-1953. C/n sometimes reported as 4300001, which does not match the year of manufacture. Became: see next line.
		25 Red (?)	Also reported as 25 Black. Became: see next line.
	M-4-2		Demonstrator (nonfunctional HDU). Scrapped.
00.02	M-4	none	Static-test airframe, Myasishchev OKB
4.3.0.00.03	M-4	none	Second prototype, Myasishchev OKB; f/f in 1-1954. Became: see next line.
		101 Red	Myasishchev OKB, state acceptance trials in antishipping strike configuration. Scrapped.
4.3.0.01.04	M-4	none	First production aircraft, Myasishchev OKB; f/f in 8-1954. Used for ejection seat tests. Became: see next line.
		85 Red	Retrofitted with IFR probe ('M-4A') in late 1954 and used for trials, Myasishchev OKB. Used for record-setting flights as '103M.' Scrapped.
4.3.0.02.05	M-4	none	F/f in 11-1954. Myasishchev OKB. Became: see next line.
	M-4-2	71 Red	First prototype tanker, used for trials; Myasishchev OKB, last three digits of c/n removed during trials. Later delivered to 1096th or 1230th TBAP, Engels-2 AB. Crashed at Chagan AB on 16-10-1986, crew OK.
4.3.0.03.06	M-4	65 Red	F/f in 12-1954. Converted to: see next line.
	M-4-2		1230th APSZ
5.3.0.04.07	M-4	10 Red	40th or 79th TBAP, Ookraïnka AB. Became: see next line.
	M-4-2	64 Red (2)	Based at Engels. Crashed at Olen'ya AB on 1-6-1985, crew OK.
5.3.0.05.08	M-4	62 Red	Became: see next line.
	M-4-2		1230th APSZ. Stored at Engels (?), scrapped after 1996.
5.3.0.06.09	M-4	?	Crashed near Zhukovskiy on 21-3-1955 on delivery to first-line unit, crew killed
5.3.0.07.10	M-4	72 Red	Became: see next line.
	M-4-2		Stored at Engels (?), scrapped after 1996
5.3.0.08.11	M-4	?	Became: see next line.
	M-4-2		Stored at Engels (?), scrapped after 1996
5.3.0.09.12	M-4	?	1096th TBAP. Crashed at Engels-2 AB on 26-6-1956, crew killed.
5.3.0.10.13	M-4	?	Crashed at Zhukovskiy (?) on 16-6-1955 after in-flight fire, crew killed
5.3.0.11.14	M-4	24 Red	Used for state acceptance trials (1955). To 1096th TBAP; crashed on 22-8-1955, crew killed.
5.3.0.12.15	M-4	31 Red	Used for state acceptance trials (fuel dump system tests). Became: see next line.
	M-4-2		Converted in 2-1956? Used for trials of RSBN-6S Vstrecha SHORAN. Stored at Engels (?), scrapped after 1996.
5.3.0.13.16	M-4	64 Red (1)	Used for state acceptance trials? Based at Engels; damaged at Engels on 16-6-1976 but repaired. Became: see next line.
	M-4-2	?	Recoded 1980. Stored at Engels (?), scrapped after 1996.
5.3.0.14.17	M-4	?	Crashed on 17-8-1955, crew killed
5.3.0.15.18	M-4	70 Red	Retrofitted with IFR probe ('M-4A') in 2-1956 and used for state acceptance trials, Myasishchev OKB. To air force, based at Ookraïnka AB. Became: see next line.
	M-4-2	63 Red	Converted in 1961. Preserved in base museum at Ookraïnka AB.
5.3.0.16.19	M-4	97 Red?	Myasishchev OKB. Became: see next line.
	M-4-2		Second prototype tanker, converted in 2-1956 and used for state acceptance trials. Stored at Engels (?), scrapped after 1996.
5.3.0.17.20	M-4	?	Became: see next line.
	M-4-2		Stored at Engels (?), scrapped after 1996
5.3.0.18.21	M-4	?	Used for weapons tests (1956). Became: see next line.
	M-4-2		Stored at Engels (?), scrapped after 1996
5.3.0.19.22	M-4	?	Used for weapons tests (1956). Became: see next line.
	M-4-2		Stored at Engels (?), scrapped after 1996
5.3.0.20.23	M-4	?	Crashed near Omsk on 23-8-1957, crew killed

5.3.0.21.24	M-4	77 Red	Became: see next line.
	M-4-2		Stored at Engels (?), scrapped after 1996
5.3.0.22.25	M-4	75 Red	Crash-landed at Chagan AB on 10-2-1987, crew OK; repaired but not flown again and scrapped on site
5.3.0.23.26	M-4	?	Became: see next line.
	M-4-2		Stored at Engels (?), scrapped after 1996
5.3.0.24.27	M-4	08 Red	Became: see next line.
		63 Red	Became: see next line.
	M-4-2	74 Red	Based at Engels. Stored at Engels, scrapped after 1996.
5.3.0.25.28	M-4	03 Red	Became: see next line.
	M-4-2		Used for trials in 1958. Became: see next line.
		66 Red	1230th APSZ. Last flight on 21-1-1991, to Syzran' and preserved; later transported to Moscow for preservation at the Khroonichev Scientific & Production Centre but not reassembled, stored there in parts.
6.3.0.26.29	M-4	67 Red	Became: see next line.
	M-4-2		Stored at Engels (?), scrapped after 1996
6.3.0.27.30	M-4	?	Became: see next line.
	M-4-2		Stored at Engels (?), scrapped after 1996
6.3.0.28.31	M-4	55 Red	Became: see next line.
	M-4-2	60 Red	1230th APSZ. Preserved at DA Museum, Dyagilevo AB.
6.3.0.29.32	M-4	?	Became: see next line.
	M-4-2		Stored at Engels (?), scrapped after 1996
6.3.0.30.33	M-4	?	Became: see next line.
	M-4-2		Stored at Engels (?), scrapped after 1996
6.3.0.31.34	M-4	15 Red	Became: see next line.
	M-4-2	61 Red	Stored at Zhukovskiy, scrapped in 1994–95
6.3.0.32.35	M-4	?	Crashed at Zhukovskiy on 18-3-1958, crew killed
3M et seq. **System 1 (batch 2 only): c/n 6320301 = year of manufacture 1956, plant No. 23, batch 2, 3rd aircraft in the batch, meaning of the last two digits unknown**			
5.3.2.01.01	3M	none	First prototype, Myasishchev OKB; f/f on 27-3-1956. Used for state acceptance trials from 1-1958 to 12-1958, also for record-setting flights as the '201M.' Became: see next line.
		60 Red?	
5.3.2.02.01	3M	?	Static-test airframe, Myasishchev OKB
6.3.2.03.01	3MN	51 Red	Myasishchev OKB; f/f in 10-1956. First nuclear-capable (Mod. 2), used for state acceptance trials from 6-1958 to 5-1959. Became: see next line.
	3MN-2		Prototype of tanker version, used for state acceptance trials (1963). Later to the air force; stored at Engels (?), scrapped after 1996.
6.3.2.04.01	3M	none	Myasishchev OKB, f/f in 9-1956; used for refuelling tests and drop tank tests. Became: see next line.
		60 Red	Became: see next line.
		80 Red	
6.3.2.05.01	3M	80 Red	Used for tests with AM-3A engines. Became: see next line.
		23 Red	Became: see next line.
		83 Red	Stored at Engels (?), scrapped after 1996
6.3.2.06.01	3M	26 Red?	Used for ejection system tests in 2957
6.3.2.07.01	3M	?	
6.3.2.08.01	3M	86 Red	Used for tests with AM-3A engines. Became: see next line.
	3MS-2	24 Red	1230th APSZ. Stored at Engels, scrapped after 1997.
6.3.2.09.01	3MS	34 Red	First 3MS, RD-3M-500A engines. DBR on 20-9-1958 after uncontained engine failure, 1 crewman killed.
6.3.2.10.01	3MS	33 Red	
System 2: c/n 6300304 = year of manufacture 1956, plant No. 23, Batch 3, 4th aircraft in the batch			
6.3.003.01	3MN	58 Red	Pattern aircraft for series production. Also reported as a 3MS! Became: see next line.
	3MN-2		1230th APSZ. Stored at Engels, scrapped after 1996.
6.3.003.02	3M	?	Stored at Engels (?), scrapped after 1996
6.3.003.03	3M	87 Red	Crashed at Engels-2 AB on 22-11-1957, crew killed
6.3.003.04	3M	23 Red	Stored at Engels (?), scrapped after 1996

6.3.003.05	3M	?	Stored at Engels (?), scrapped after 1996
6.3.004.01	3M	80 Red?	Stored at Engels (?), scrapped after 1996
?.3.004.02	3M	?	Stored at Engels (?), scrapped after 1996
?.3.004.03	3M	?	Stored at Engels (?), scrapped after 1996
7.3.004.04	3MS	46 Red	Became: see next line.
	3MS-2?	25 Red	Stored at Engels, scrapped after 1997
7.3.004.05	3M	26 Red	Also reported as 80 Red! Stored at Engels (?), scrapped after 1996.
7.3.005.01	3MS	83 Red	Became: see next line.
	3MS-2		
7.3.005.02	3M	17 Red	Myasishchev OKB, used for test work; to air force by 1987. Stored at Engels, scrapped after 1996.
7.3.005.03	3MN	?	Became: see next line.
	3M-5	no code	Myasishchev OKB, development aircraft, converted in 1973; scrapped at Zhukovskiy in 10-1989
7.3.005.04	3M	85 Red	Became: see next line.
	3MS-2		1230th APSZ. Stored at Engels, scrapped after 1996.
7.3.005.05	3M	23 Red	Became: see next line.
	3MS-2		1230th APSZ. Placed in storage in 1985. Became: see next line.
		90 Red	Reactivated in 1987 with new code, 1230th APSZ. Became: see next line.
		10 Red	Recoded in 1991, 1096th TBAP. Collided with another 3M and crashed on 16-5-1992, crew killed.
7.3.006.01	3MS	09 Black	Became: see next line.
		24 Red	Became: see next line.
	3MS-2	21 Red	
7.3.006.02	3MS	22 Red	Became: see next line.
		82 Red	Stored at Engels, scrapped after 1996
7.3.006.03	3M	58 Red	Stored at Engels, scrapped 1998
7.3.006.04	3M	27 Red	Became: see next line.
	3MS-2?		Stored at Engels, scrapped after 1996
7.3.006.05	3MS	?	Became: see next line.
	3MS-2	19 Red	Crashed on 16-5-1992, part of the crew killed
7.3.007.01	3M	none	First nuclear-capable (Mod. 2), used for state acceptance trials (nuclear protection tests), 6-1959 to 12-1959. To SibNIA, used for repeat static tests.
7.3.007.02	3MS	?	Became: see next line.
	3MS-2	29 Red	1096th TBAP. Crashed on 15-4-1972, crew killed.
7.3.007.03	3M	?	Stored at Engels, scrapped after 1996. To SibNIA for repeat static tests.
7.3.007.04	3M	?	Damaged by fire at Akhtoobinsk (Vladimirovka AB) on 14-2-1958 during state acceptance trials
7.3.007.05	3M	none	Used for tests with RSBN-6S SHORAN. Stored at Engels (?), scrapped after 1996.
7.3.008.01	3MS	63 Red	Also reported as 3MN! Became: see next line.
	3MS-2	24 Red	Also reported as 3MN-2! 1230th APSZ. Stored at Engels, scrapped after 1996.
7.3.008.02	3M	12 Red	Became: see next line.
		25 Red	Stored at Engels, scrapped after 1996
7.3.008.03	3MS	?	Became: see next line.
	3MS-2	28 Red	1230th APSZ. Stored at Engels, scrapped after 1996.
7.3.008.04	3M	?	Stored at Engels, scrapped after 1996
7.3.008.05	3MS	14 Red	Became: see next line.
	3MS-2		Converted in 1967. Last flight on 21-3-1994; preserved at DA Museum, Engels-2 AB.
7.3.009.01	3MS	?	
7.3.009.02	3M	?	Actual mfd 1958
7.3.009.03	3MN	31 Red	1230th TBAP, unit CO's aircraft. Scrapped at Engels.
7.3.009.04	3M	84 Red	Stored at Engels, scrapped after 1996
7.3.009.05	3M	54 Red	Stored at Engels, scrapped after 1996
8.3.010.01	3MS	33 Red	Became: see next line.
		52 Red	Stored at Engels, scrapped after 1996
8.3.010.02	3MS	?	Stored at Engels, scrapped after 1996
8.3.010.03	3MS	?	Stored at Engels, scrapped after 1996
8.3.010.04	3MN	56 Red	Stored at Engels, scrapped after 1996
8.3.010.05	3MS	?	Stored at Engels, scrapped after 1996

8.3.011.01	3MN-1	none	Became: see next line.
	3ME	none	Myasishchev OKB, converted in 3-1959. DBR at Zhukovskiy on 12-5-1960.
8.3.011.02	3MS	?	Stored at Engels, scrapped after 1996
8.3.011.03	3MS	?	Stored at Engels, scrapped after 1996
8.3.011.04	3MS	?	Stored at Engels, scrapped after 1996
8.3.011.05	3MS	?	Stored at Engels, scrapped after 1996
8.3.012.01	3MS	?	Stored at Engels, scrapped after 1996
8.3.012.02	3MS	?	Stored at Engels, scrapped after 1996
8.3.012.03	3MS	?	Stored at Engels, scrapped after 1996
8.3.012.04	3MS	?	Stored at Engels, scrapped after 1996
8.3.012.05	3MS	?	Stored at Engels, scrapped after 1996
9.3.013.01	3MS		Converted to: see next line.
	3MD	?	Myasishchev OKB. First prototype, mfd on 15-11-1959, f/f on 25-11-1959. Used for tests of VD-7P engine. To air force; crashed on 5-5-1965, crew killed.
9.3.013.02	3MS		Converted to: see next line.
	3MD	?	Myasishchev OKB. Second prototype, mfd in 1960.
9.3.013.03	3MS	?	Stored at Engels, scrapped after 1996
9.3.013.04	3MS	?	Stored at Engels, scrapped after 1996
9.3.013.05	3MS	?	Stored at Engels, scrapped after 1996
9.3.014.01	3MS	28 Red	Used for development work by the Myasishchev OKB
9.3.014.02	3MN	31 Red	Became: see next line.
	3MN-2	?	1230th APSZ. Became: see next line.
	VM-T	CCCP-01402	Myasishchev OKB, f/f on 29-4-1981. Became: see next line.
		RA-01402	Stored at Dyagilevo AB
9.3.014.03	3MS	?	Stored at Engels, scrapped after 1996
9.3.014.04	3MN	50 Red	Stored at Engels, scrapped after 1996
9.3.014.05	3MS	?	Stored at Engels, scrapped after 1996
9.3.015.01	3MS-1	?	Stored at Engels, scrapped after 1996
9.3.015.02	3MN	?	Became: see next line.
	3MN-2	?	1230th APSZ. Became: see next line.
	VM-T	CCCP-01502	Myasishchev OKB; D/D on 25-1-1982, f/f in 3-1982. Became: see next line.
		êî-01502	(RF-01502) Reregistered by 8-1993. Stored at Zhukovskiy.
9.3.015.03	3MS	?	
9.3.015.04	3MN	?	Became: see next line.
	3MN-2	?	1230th APSZ. Became: see next line.
	VM-T	none	Myasishchev OKB, static-test airframe; to TsAGI and SibNIA
9.3.015.05	3MS	?	Stored at Engels, scrapped after 1996
9.3.016.01	3MN	?	Became: see next line.
	3MN-2	55 Red	1230th APSZ. Stored at Engels, scrapped after 1996.
?.3.016.02	3MS	?	Stored at Engels, scrapped after 1996
?.3.016.03	3MS	?	Stored at Engels, scrapped after 1996
?.3.016.04	3MS	?	Stored at Engels, scrapped after 1996
?.3.016.05	3MS	?	Stored at Engels, scrapped after 1996
0.3.017.01	3MS	55 Red	Last production 3MS. Stored at Engels, scrapped after 1996.
0.3.017.02	3MD	?	First production 3MD. Stored at Engels, scrapped after 1996.
0.3.017.03	3MD	?	Stored at Engels, scrapped after 1996
0.3.017.04	3MD	10 Red	Stored at Engels, scrapped after 1996
0.3.017.05	3MD	?	Stored at Engels, scrapped after 1996
0.3.018.01	3MD	64 Red	Crashed on 12-7-1961 (also reported as 12-6-1961!) after engine failure, crew killed
0.3.018.02	3MD	?	Stored at Engels, scrapped after 1996
0.3.018.03	3MD	?	Stored at Engels, scrapped after 1996
0.3.018.04	3MD	30 Red	Based at Ookraïnka AB. Stored at Engels from 1-1985 to 7-3-1986, redeployed to Mozdok. Last flight on 8-7-1986 and preserved Soviet air force Museum, Monino.
0.3.018.05	3MD	?	Stored at Engels, scrapped after 1996

Appendix 3

Accident Attrition

Unfortunately the M-4 family has suffered heavy accident attrition. According to available information, 31 of the 125 aircraft built – 17 of the 35 M-4s, 12 of the 69 *Bison-Bs*, and two of the 11 3MDs – were lost in fatal and nonfatal accidents between 1955 and 1992. The causes varied – critical hardware failures, pilot error (especially in the early days, when the M-4's handling had not yet been studied in full), or the negligence of the ground staff. Known accidents are listed below.

• The first total hull loss occurred as early as 21 March 1955, when a factory test crew was ferrying the sixth production M-4 (tactical code unknown, c/n 5300609) from Zhukovskiy to Engels. En route the aircraft encountered a storm front; plant No. 23 test pilots Lev P. Vinogradov and Leonid V. Soomtsov attempted a forced landing, whereupon the bomber broke up, killing navigator / radar operator S. I. Leonov. A different account of the accident says the forced landing was occasioned by decompression of the forward crew cabin at 5,000 m (16,400 ft.) during a test flight from Zhukovskiy and that the pilots brought the bomber home but were foiled by a sudden blizzard, which obscured the runway. After three unsuccessful attempts the aircraft made an off-field landing, touching down too fast and breaking up.

• On 16 June 1955, M-4 c/n 5301013 (tactical code unknown) developed a fuel leak in one of the engine bays during a predelivery test flight from Zhukovskiy. A fire ensued; the bomber turned back for a forced landing but crash-landed 500 m (1,640 ft.) short of the runway threshold, killing one of the crew. The blame rested with the OKB; in a misguided attempt to cut structural weight, the designers had deleted some of the fixtures holding the fuel line, which started flexing and eventually broke.

• On 22 August 1955, a 1096th TBAP M-4 ('24 Red,' c/n 5301114) was lost in similar circumstances. The bomber became airborne with slight right bank after a takeoff run of some 1,500 m (4,920 ft.). The bank angle grew steadily, reaching 60°–80° by the time the aircraft climbed to 25–40 m (80–130 ft.; eyewitness accounts varied), whereupon the bomber sideslipped and crashed in a fireball, killing the entire crew (captain Il'ya N. Pronin, copilot Vladimir K. Kokkinaki, navigator V. V. Vasil'yev, flight engineer D. F. Koval'skiy, radio operator K. I. Petukhov, and gunners A. I. Zaitsev and S. I. Leont'yev). Pilot error was cited as the cause, but the underlying reason was the M-4's insufficient roll stability at critical AoAs reached when the nose gear bogie tilted during rotation.

• On 26 June 1956, an M-4 operated by the 1096th TBAP (tactical code unknown, c/n 5300912) crashed while taking off from Engels-2 AB. Immediately after takeoff, the aircraft climbed too sharply, reaching an altitude of 20 m (65 ft.), instead of the 3–5 m (10–16 ft.) that was appropriate at that point, before it stalled and crashed, killing all on board – Col. A. V. Verbitskiy, Maj. N. P. Gorchakov, Lt.-Col. I. L. Spitsyn, Maj. V. K. Yelpatov, Capt. L. I. Lezhnyov, and Lt. (SG) I. N. Glebov. Verbitskiy was the regiment's CO, and his death was a sore loss for the unit. The cause of the crash could not be established because the M-4 had no flight data recorder and cockpit voice recorder. The members of the accident investigation panel voiced conflicting views, with air force representatives claiming that the prescribed takeoff procedure had been observed and that the tilting bogie nose gear unit was the culprit; conversely, the aircraft industry spokesmen blamed the crew.

• On 17 August 1957, an M-4 (tactical code unknown, c/n 5301417) stalled and crashed in similar circumstances to the 22 August 1955 accident.

• On 23 August 1957, M-4 c/n 5302023 (tactical code unknown) crashed near Omsk in western Siberia, killing the crew of eight. Again, debates raged as the accident was being investigated; the VVS and the aircraft industry hurled accusations at each other, trying to pass the buck rather than to establish the truth. One side claimed the hydraulic control actuators had jammed, and the other side cited pilot error as the reason, but the actual cause was never determined for want of documentary evidence from 'black boxes.'

• The first loss of a *Bison-B* occurred on 22 November 1957, when the crew of 3M '87 Red' (c/n 6300303) attempted to take off at Engels-2 AB with the control gust locks still engaged. The bomber stalled and crashed immediately after liftoff, killing the crew – Maj. V. I. Shilik, Capt. L. Z. Kozodayev, Capt. Yu. M. Soltynskiy, Lt. (SG) Ye. A. Golovin, Lt. (SG) V. M. Zemlyanukhin, Lt. V. I. Glazko, and Sgt (SG) P. A. Gagin.

• On 14 February 1958 the wheel brakes of a 3M undergoing state acceptance trials (tactical code unknown, c/n 7300704) overheated during taxiing at the GK NII VVS facility in Akhtoobinsk (Vladimirovka AB). A hydraulic line burst, causing the hydraulic fluid to ignite upon coming into contact with the brakes; the fire quickly spread to the rear fuselage and tail unit. Despite the efforts of the fire brigade, the entire airframe structure aft of the wings was destroyed before the blaze could be extinguished. Still, the trials continued with another 3M coded '25 Red' (c/n 7300802) replacing the lost aircraft.

• On 18 March 1958 the final production M-4 (tactical code unknown, c/n 6303235) was lost during a postmodification test flight from Zhukovskiy. The aircraft was making a maximum-TOW takeoff in manual-control mode (with the hydraulic actuators switched off), the mission being to check its flight safety in this mode. During the takeoff run, the wind suddenly changed from 70° to 90° to the direction of flight and grew stronger. The rudder trim tab deflected almost fully to starboard, causing an uncommanded rudder hard-over to port with a force of 300 kgf (660 lbf), which the pilots were unable to counter. As it became airborne 48 seconds after brake release, the aircraft yawed and rolled uncontrollably to port; 10 seconds later, the port wing struck the ground and the bomber crashed. Crew captain Sergey F. Mashkovskiy, copilot Erik N. Ponomaryov, navigator Ivan F. Ovcharenko, flight engineer Aleksey N. Koterev, test engineer Yuriy S. Kazanskiy, and radio operator Chizhikov died in the blaze; tail gunner Mikhail D. Mikhailov survived with injuries, being thrown out of the aircraft.

• On 20 September 1958 the first 3MS ('34 Red,' c/n 6320901) was ground-running its RD-3M-500A engines when one of the engines suffered an uncontained failure. A departing first-stage compressor blade punctured a fuel tank, causing a massive fire that destroyed the aircraft completely; at least one person died.

• As already mentioned in chapter 3, on 12 May 1960 the sole prototype of the 3ME missile strike aircraft (no code, c/n 8301101) was damaged beyond repair at the OKB's flight test facility in Zhukovskiy in a ground collision with the prototype of the M-50 bomber ('012 Blue'), which had jumped chocks during a full-power engine run. Unfortunately, the 3ME's radio operator, Kruchinin, was killed and two more persons suffered minor injuries. Since the plans to launch production of the 3ME had been put on hold anyway, the aircraft was written off.

• On 12 July 1960, 3MD '64 Red' (c/n 9301801) suffered an uncontained engine failure and a massive fire on takeoff from Engels-2 AB. None of the crew attempted to eject, and the aircraft crashed near Gheneral'skoye village, northeast of the base, killing all on board.

• In October 1960, a 3M made a successful emergency landing at Engels-2 AB after a midair collision, losing two-thirds of the port tailplane and the rudder.

• On 7 December 1961, a 3M of the 201st TBAD captained by Maj. P. N. Morozov crashed into a river shortly after taking off from Siauliai AB, killing all seven crew members. A different account of the tragedy says the crash occurred on 13 July 1961 at 1300 hrs (!).

• On 25 December 1961, a 40th TBAP M-4 captained by Maj. D. M. Chuvashov started rolling to port during a high-gross-weight takeoff. About 1,400 m (4,590 ft.) from the runway threshold, the port wing scraped the ground, causing the aircraft to veer off the runway onto the rough shoulder. About 200 m (660 ft.) later, the port outer wing broke off; yawing even sharper to port, the aircraft slammed into a parked 3M bomber, and both bombers erupted in flames. Chuvashov and six other crewmen (Capt. V. M. Barinov, Capt. M. Kh. Abdulkarimov, Lt. (SG) A. V. Khrooshch, Lt. (SG) P. S. Petrov, Ens. A. Tkachuk, and Sgt (SG) V. A. Sveshnikov) died in the postcrash fire; only the tail gunner escaped. The 3M's tail gunner, however, lost his life. Insufficient structural strength of the wings was cited as the cause.

• On 8 March (some sources say 8 April) 1963, two 40th TBAP M-4s captained by Lt.-Col. I. N. Sereda (lead aircraft) and Maj. N. G. Gil'mitdinov (wingman) collided during a maximum-range sortie at night; the wingman was flying too close and clipped the leader's tail while switching from echelon port to echelon starboard formation. Ten crewmen aboard the two bombers (both crew captains plus Capt. M. Svistookhin, Capt. V. K. Potapkin, Lt. (SG) V. Ye. Strepakov, and Lt. (SG) V. V. Anokhin in the lead aircraft and Capt. A. V. Sobolev, Capt. A. S. Makov, Lt. V. A. Gaidai, and Lt. (SG) N. A. Raketskiy) died; six others ejected successfully. The wingman's poor airmanship and lack of discipline were cited as the cause; Maj. Gil'mitdinov already had a history of 'hot-dogging' at the time of the accident, but no action had been taken by his superiors until it was too late.

• On 4 May (some sources say 5 May) 1965, a 3MD (tactical code unknown, c/n 9301301) captained by the 79th TBAP CO, Col. V. I. Boldinskiy, strayed from the desired track and collided with a mountain while flying a low-level mission over the Sea of Okhotsk at 600 m (1,970 ft.). There were no survivors.

• On 27 May 1966, another case of poor discipline ended in a collision – luckily without fatal consequences. A 1006th (?) TBAP Tu-95KM captained by Lt.-Col. Gershunenko and an M-4-2 tanker (identity unknown) captained by Maj. Vasil'yev were flying a mission involving two fuel transfers and four 'dry' contacts. After this, the *Bear* was to pass the tanker on the right with a horizontal and vertical separation of 300 m (990 ft.), for the purpose of photographing it. Yet, considering himself to be a top-class pilot, Gershunenko deliberately ignored instructions and overtook the tanker with a much-smaller separation. After assuming line abreast formation, he stopped observing the tanker and concentrated on the flight deck equipment (which was actually the copilot's responsibility). Neither did the other crew members keep an eye on the M-4, believing the captain was doing this.

Losing his concentration briefly, Gershunenko inadvertently banked the aircraft slightly to port with a gentle climb. As a result, the Tu-95KM drifted toward the tanker, striking the latter's forward fuselage underside with its tail unit. The M-4's crew, too, had not been watching the Tu-95 closely enough and was late in reacting – the collision came as a total surprise for everyone. As a result, the *Bear* lost part of the fin, half the rudder, and part of the starboard tailplane; the tanker suffered damage to two of the four engines, the nose gear doors, and the fuel system. Both aircraft landed safely at the nearest bases.

• On 25 January 1971, an M-4 of the 1230th TBAP captained by Maj. V. Ya. Bondarenko was making a positioning flight from Vorkuta to Engels. Immediately after takeoff, the aircraft started rolling to port at an altitude of 5–8 m (16–26 ft.). When the bank angle reached 60°–65°, the aircraft clipped a pile of snow with the port wingtip and crashed inverted 720 m (2,362 ft.) from the unstick point. Seven of the eight crew perished. The aircraft's inherent propensity to reach excessive AoAs on takeoff was cited as the cause.

• In 1971 an M-4-2 tanker of the 201st TBAD was left with low fuel after refuelling a Tu-95KM of the 106th TBAD, operating from Olen'ya AB, beyond the Arctic Circle. Heading back to the shore, the crew requested an emergency landing at Olen'ya. As luck would have it, an unserviceable Tu-95 was stuck on the runway there and could not be moved; yet, the runway was wide enough to pass it, and, with no other landing options, the captain of the M-4-2 decided to take a chance. Unfortunately, the profile of the runway was such that the middle point was at much-higher elevation than the ends, obscuring the stuck Tu-95 from the tanker's crew until it was too late. Seeing the obstacle ahead, the captain ground-looped the aircraft to avoid a collision; rolling onto rough ground beside the runway, the M-4-2 broke up and exploded, killing the crew.

• On 15 April 1972 (one source says 1971!), a 1096th TBAP 3MS-2 ('29 Red,' c/n 7300702) overshot while landing at Engels-2 AB in a stiff crosswind. Touching down nose gear first 930 m (3,050 ft.) from the runway threshold, the aircraft started bouncing; on the third rebound it stalled and crashed, killing the entire crew – Maj. G. A. Lybin, Capt. Yu. S. Davydenko, Capt. Ye. M. Kashkaryov, Lt. (SG) G. I. Khmelinin, Capt. I. S. Korepanov, Sgt. (JG) P. P. Burykin, and Capt. A. A. Pleshachkov.

• On 17 September 1974, a mainwheel brake cylinder of a 40th TBAP M-4-2 captained by Maj. L. N. Stanislavskiy burst at an altitude of 4,500 m (14,760 ft.) soon after gear retraction, with the leaking hydraulic fluid catching fire (the brake cylinder had overheated because the aircraft had been taxiing too fast, requiring vigorous braking). The fire extinguishers were used, but to no avail; the fire spread to the adjacent fuel tanks, which exploded, blowing out the mainwheel well doors. As the burning tanker flew on, 2 minutes, 15 seconds later, Stanislavskiy ordered the crew to eject, which they did. Moments later, a second explosion destroyed the aircraft. The captain (who was caught by the blast) and tail gunner Pvt. A. F. Ooglanov (who was unable to eject) perished.

• On 11 March 1975, a 40th TBAP M-4 captained by Maj. V. L. Klemeshev suffered a landing-gear failure on takeoff for a night practice sortie, the main strut jamming halfway through retraction and refusing to extend – both normally and in emergency mode. After circling to burn off fuel, the aircraft made an emergency landing on the auxiliary dirt airstrip at Ookraïnka AB; the nose gear collapsed but the crew was unhurt. The cause was traced to a short circuit in the gear control wiring.

• On 13 May 1975, a 1230th TBAP 3M captained by Maj. S. A. Krylov suffered an in-flight fire in the bomb bay, which triggered an explosion, the aircraft breaking up in midair. Six of the seven crewmen died in the blast; only the tail gunner managed to eject. The fire had been caused by a broken wing-deicing air duct, the escaping hot air igniting the insulation of the wiring bundles in the bomb bay.

• On 16 June 1976, M-4 '64 Red' (c/n 5301316) stalled and crash-landed at Engels at night, suffering damage to the starboard outer wing, which struck the ground first. The aircraft was captained by Lt. (SG) Ivan I. Kozlovskiy; the crew also included copilot Vikentiy I. Kurtsevich, navigator Vladimir Avakoomov, flight engineer Anatoliy Veselovskiy, and tail gunner Nikolay Tiatoorov. A replacement wing was flown in by an An-22 transport from Ookraïnka AB, where another M-4 had wrecked its port wing and become a spares source. The tanker was repaired and returned to service.

• On 17 April 1978, a 40th TBAP M-4 captained by Capt. N. I. Bogomolov was coming in to land at Ookraïnka AB when the captain lost concentration and let the speed decay past stalling speed. Rolling left, the aircraft struck the ground with the port outrigger strut first at 210 km/h (130 mph), then came down on the nose gear unit, bounced, veered off the runway onto rough ground to the left, and broke up. The aircraft was consumed by the postcrash fire, but the crew walked away.

• On 8 June (some sources say 8 July) 1978, the No. 2 engine of a 40th TBAP 3M captained by Maj. V. V. Oosov caught fire during a night practice flight around the circuit of Ookraïnka AB. The nervous copilot, Maj. V. M. Smirnov, shut down the wrong engine by mistake and then triggered the fire extinguishers for the same (good but inoperative) engine. The pilots then panicked and ejected without taking time to deflate some of the escape hatch seals; radio operator Kharchenko and the flight engineer were thus unable to eject and perished when the bomber crashed.

• On 8 August 1984, a 1096th TBAP 3M lost control a few minutes after takeoff from Engels-2 AB when a broken engine bleed air duct directed hot air at the engine turbostarter petrol tank, which exploded violently, blowing off a big chunk of wing skin and part of the flap – thereby causing the port wing to lose lift. The pilots requested an emergency landing, but the aircraft rolled and impacted with 60°–70° left bank, narrowly missing the aforementioned Gheneral'skoye village. Aircraft captain Col. V. R. Tukhvatoollin, Ens. A. V. Vorotnikov, Maj. V. A. Kozionov, Capt. K. K. Mironov, and Capt. N. V. Shavrin did not eject and died in the crash; radar operator A. I. Yefishin and the tail gunner ejected safely.

• On 1 July (some sources say 1 June) 1985, a 1096th TBAP M-4-2 ('64 Red,' c/n 5300407) came in too low while landing at Olen'ya AB. The aircraft was captained by Capt. Vladimir S. Timchenko, with copilot Lt. Sergey V. Lyul'chenko, navigator Sergey S. Mit'ko, radar operator Lt. (SG) Matveyev, flight engineer Capt. Konovalov, radio operator Ens. Zinoviy N. Laskovyy, and tail gunner Ens. Mikhailov. Seeing that the aircraft was flaring out 150 m (490 ft.) short of the runway, the ATC officer called to the pilots, *'Hold her level.'* Responding, the captain hauled back on the control yoke but overdid it; the M-4 pitched up into a stalled position before dropping the port wing and crashing 23 m (75 ft.) short of the runway. The port wing broke at rib 22 and the escaping fuel ignited; the burning aircraft ground-looped, collapsing the nose gear before it came to a halt 740 m (2,430 ft.) from the runway threshold and 50 m (164 ft.) off centre. The crew vacated the aircraft without injury.

• On 16 October 1986, M-4-2 '71 Red' (c/n 4300205) was lost in near-identical circumstances. Captained by Maj. V. D. Arkhipov, the tanker was landing at Chagan AB, near Semipalatinsk; everything appeared normal until the aircraft passed the inner marker beacon. Then the crew lost awareness of the speed, and the aircraft decelerated until it stalled 170 m (560 ft.) short of the runway. Coming down with 26° right bank, the aircraft impacted with the port wing first and then landed hard on the nose gear unit. Ground-looping to the right, the aircraft came to rest on the runway shoulder 600 m (1,970 ft.) from the runway threshold and 250 to the right of the runway axis. Unfortunately, before it did so it struck an APM-90 ground support vehicle (a high-powered searchlight on a ZiL-130 4x2 general-purpose lorry used for runway illumination) with the starboard wing; the driver was fatally injured, dying before reaching the hospital. The aircraft was destroyed by the postcrash fire, with only the port wing remaining relatively intact; the crew escaped unhurt.

• On 10 February 1987, M-4-2 '75 Red' (c/n 5302225), captained by Maj. V. N. Kovalenko, had its main gear unit collapse while landing at Chagan AB. The rear fuselage scraped hard against the runway, a fire breaking out and the tail gunner's station snapping off at the manufacturing break. The crew was unhurt; tail gunner Ens. M. M. Il'yin was trapped in his pressure cabin until flight engineer Oleg Gribalyov came to the rescue, extracting him through the escape hatch. The cause was poor electric contact in the gear control circuitry and failure of one of the two electric motors on the MPSh-16M gear actuator, which doubled the gear extension time.

As it often happens, to avoid having this recorded as a total hull loss (with repercussions for everyone involved), the tanker was repaired by replacing the wrecked tail gunner's station with an intact one from a cannibalised 3M; the aircraft was photographed and a report was filed to the VVS top command. Yet, no one was willing to run the risk of flying this tanker with an 'ass transplant,' and the aircraft was not flown again; eventually it was quietly struck off charge and scrapped on site.

• The last accident involving the *Bison* occurred on 16 May 1992, when two 3MS-2 tankers from the 1096th TBAP – '10 Red' (c/n 7300505), captained by A. M. Sayuk, and '16 Red' (c/n 7300605) captained by S. Chelyapin – collided in midair at 6,000 m (19,685 ft.) near Saratov when flying in adverse weather. Seconds later, '10 Red' exploded in midair; its rear fuselage floated earthward on the brake parachutes, landing near Krasnorechenskoye village (Saratov Region). The tail gunner, Ens. V. A. Shcherbich, was still alive but succumbed to his injuries soon afterward. Only five of the two tankers' 14 crew members managed to eject, two of them perishing in the process or in hospital. Thus, the only survivors were Chelyapin, his copilot, and his radar operator, A. F. Mostovoy.

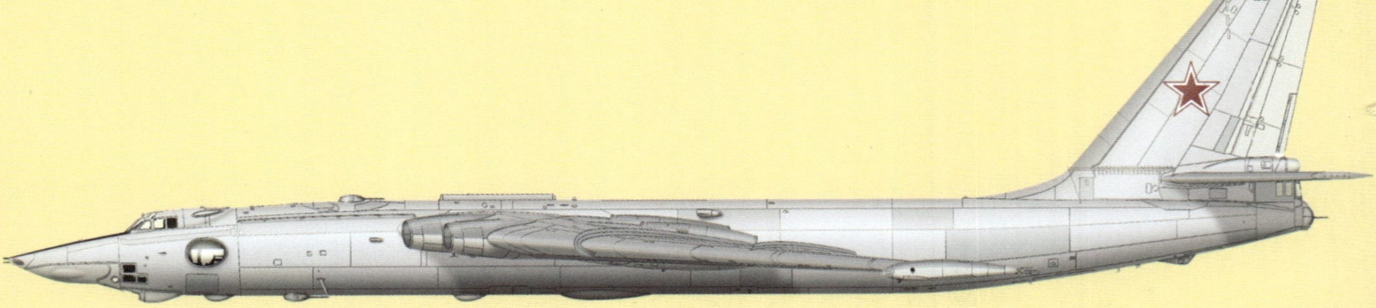
25

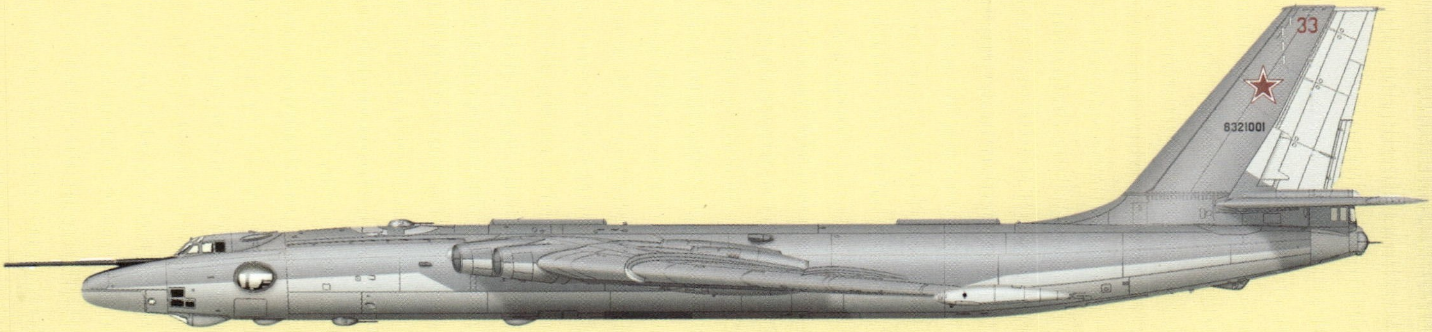
33
6321001

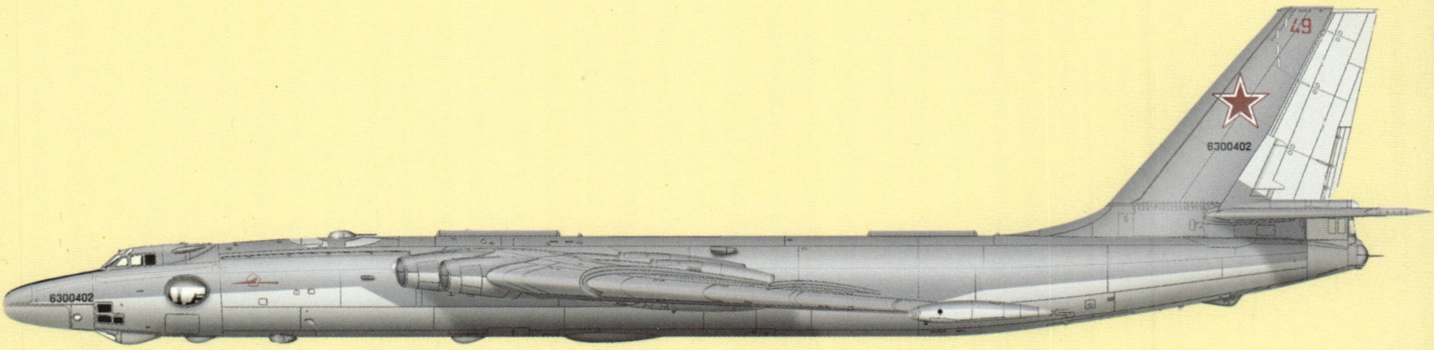
49
6300402
6300402

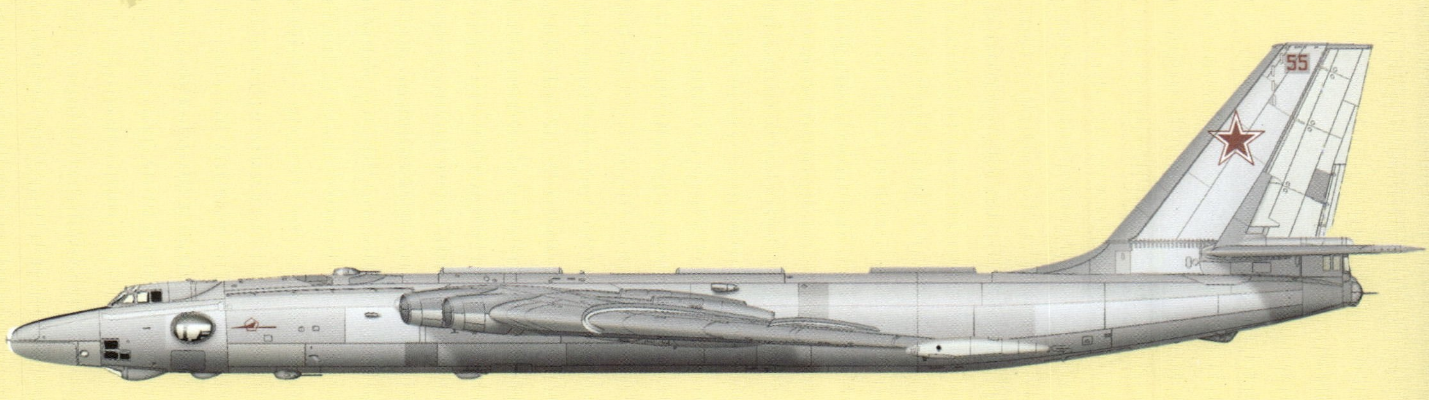
55